7 Springer Series in Solid-State Sciences

Edited by Peter Fulde

Springer Series in Solid-State Sciences

Editors: M. Cardona P. Fulde H.-J. Queisser

Volume 40 **Semiconductor Physics** - An Introduction By K. Seeger

Volume 41 **The LMTO Method** By H.L. Skriver

Volume 42 **Crystal Optics with Spatial Dispersion and the Theory of Excitations**
By V.M. Agranovich and V.L. Ginzburg

Volume 43 **Resonant Nonlinear Interactions of Light with Matter**
By V.S. Butylkin, A.E. Kaplan, Yu.G. Khronopulo, and E.I. Yakubovich

Volume 44 **Elastic Media with Microstructure II** Three-Dimensional Models
By I.A. Kunin

Volume 45 **Electronic Properties of Doped Semiconductors**
By B.I. Shklovsky and A.L. Efros

Volume 46 **Topological Disorder in Condensed Matter**
Editors: F. Yonezawa and T. Ninomiya

Volumes 1 - 39 are listed on the back inside cover

E. N. Economou

Green's Functions in Quantum Physics

Second Corrected and Updated Edition

With 52 Figures

Springer-Verlag
Berlin Heidelberg New York Tokyo 1983

Professor *Eleftherios N. Economou,* PhD
Department of Physics, University of Crete,
Heraklion, Crete, Greece

Series Editors:
Professor Dr. Manuel Cardona
Professor Dr. Peter Fulde
Professor Dr. Hans-Joachim Queisser
Max-Planck-Institut für Festkörperforschung, Heisenbergstraße 1
D-7000 Stuttgart 80, Fed. Rep. of Germany

Science
QC
174.17
G68
E25
1983

ISBN 3-540-12266-4 2. Aufl. Springer-Verlag Berlin Heidelberg New York Tokyo
ISBN 0-387-12266-4 2nd ed. Springer-Verlag New York Heidelberg Berlin Tokyo

ISBN 3-540-09154-8 1. Auflage Springer-Verlag Berlin Heidelberg New York
ISBN 0-387-09154-8 1st edition Springer-Verlag New York Heidelberg Berlin

Library of Congress Cataloging in Publication Data. Economou, E. N., 1940-. Green's functions in quantum physics. (Springer series in solid-state sciences ; v. 7). Bibliography: p. Includes index. 1. Green's functions. 2. Quantum theory. I. Title. II. Series. QC174.17.G68E25 1983 530.1'43 83-4739

Offset printing and bookbinding: Beltz Offsetdruck, 6944 Hemsbach/Bergstr.
2153/3130-543210

to Sophia

Preface to the Second Edition

In this edition the second and main part of the book has been considerably expanded as to cover important applications of the formalism.

In Chap.5 a section was added outlining the extensive role of the tight-binding (or equivalently the linear combination of atomic-like orbitals) approach to many branches of solid-state physics. Some additional information (including a table of numerical values) regarding square and cubic lattice Green's functions were incorporated.

In Chap.6 the difficult subjects of superconductivity and the Kondo effect are examined by employing an appealingly simple connection to the question of the existence of a bound state in a very shallow potential well. The existence of such a bound state depends entirely on the form of the unperturbed density of states near the end of the spectrum: if the density of states blows up there is always at least one bound state. If the density of states approaches zero continuously, a critical depth (and/or width) of the well must be reached in order to have a bound state. The borderline case of a finite discontinuity (which is very important to superconductivity and the Kondo effect) always produces a bound state with an exponentially small binding energy.

Chapter 7 has been expanded to cover details of the new and fast developing field of wave propagation in disordered media. The coherent potential approximation (a simple but powerful method) is presented with an extensive list of references to the current literature. Then the electrical conductivity is examined both because it is an interesting quantity in its own right and because it plays a central role in demonstrating how disorder can create a qualitatively different behavior. Since the first edition of this book, very significant advances in the field of random media have taken place. An effort has been made to present in a simple way the essential points (for the reader with a casual interest in this subject) and to review the current literature (for the sake of the reader whose research activities are or will be related to the field of disordered systems).

In this edition each chapter is preceded by a short outline of the material to be covered and it is concluded by a summary containing the most important equations numbered as in the main text.

I would like to thank A. Andriotis and A. Fertis for pointing out to me several misprints in the first edition. I would also like to express my gratitude to Exxon Research and Engineering Company for its hospitality during the conclusion of this work.

Heraklion, Crete
January, 1983

E.N. Economou

Preface to the First Edition

This text grew out of a series of lectures addressed to solid-state experimentalists and graduate students beginning their research career in solid-state physics.

The first part consisting of Chaps.1 and 2 is a rather extensive mathematical introduction which covers the material about Green's functions usually included in a graduate course on mathematical physics. Emphasis is given to those topics which are of significance in quantum physics. On the other hand, little attention is given to the important question of determining the Green's functions associated with boundary conditions on surfaces at finite distance from the source.

The second and main part of the book is, in my opinion, a first effort to collect in a systematic but concise way various topics of quantum physics, where the Green's functions (as defined in Part I) can be successfully applied. Chapter 3 is a direct application of the formalism developed in Part I. In Chap.4 the perturbation theory for Green's functions is presented and applied to scattering and to the question of formation of bound states. Next, the Green's functions for the so-called tight-binding Hamiltonian (TBH) are calculated. The TBH is of central importance for solid-state physics because it is the simplest example of wave propagation in periodic structures. It is also important for quantum physics in general, because it is rich in physical phenomena (e.g., negative effective mass, creation of a bound state by a repulsive perturbation) and, at the same time, simple in its mathematical treatment. Thus one can derive *simple*, *exact* expressions for scattering cross sections and for bound and resonance levels. The multiple scattering formalism is presented within the framework of the TBH, and it is applied to questions related with the behavior of disordered systems (such as amorphous semiconductors). The material of Part II is of interest not only to solid-state physicists but to students of a graduate course in quantum mechanics (or scattering theory) as well.

In Part III, with the help of the second quantization formalism, the many-body Green's functions are introduced and utilized in extracting physical information about interacting many particle systems. Many excellent books have been devoted to the material of Part III (e.g., FETTER and WALECKA: *Quantum Theory of Many-Particle Systems* [4.1]). Thus the present treatment must be viewed as a brief introduction to the subject; this introduction may help the solid-state theorist approach the existing thorough treatments of the subject and the solid-state experimentalist to become acquainted with the formalism.

I would like to thank Nuclear Research Center "Demokritos" and the Greek Atomic Energy Commission for their hospitality during the writing of the second half of this book.

Athens, Greece, November 1978 E.N. Economou

Contents

Part III: **Green's Functions in Many-Body Systems**

Part I

Green's Functions in Mathematical Physics

1. Time-Independent Green's Functions

In this chapter, the time-independent Green's functions are defined, their main properties are presented, methods for their calculation are briefly discussed, and their use in problems of physical interest is summarized.

1.1 Formalism

Green's functions can be defined as solutions of inhomogeneous differential equations of the type[1]

$$[z - L(\underline{r})]\, G(\underline{r},\underline{r}';z) = \delta(\underline{r} - \underline{r}') \qquad (1.1)$$

subject to certain boundary conditions for $\underline{r}$ or $\underline{r}'$ lying on the surface S of the domain Ω of $\underline{r},\underline{r}'$. Here we assume that z is a complex variable with $\lambda \equiv \mathrm{Re}\{z\}$ and $s \equiv \mathrm{Im}\{z\}$; and that $L(\underline{r})$ is a time-independent, linear, Hermitian differential operator which possesses a complete set of eigenfunctions $\{\phi_n(\underline{r})\}$, i.e.,

$$L(\underline{r})\phi_n(\underline{r}) = \lambda_n\phi_n(\underline{r}) \quad , \qquad (1.2)$$

where $\{\phi_n(\underline{r})\}$ satisfy the same boundary conditions as $G(\underline{r},\underline{r}';z)$. The set $\{\phi_n\}$ can be considered as orthonormal without loss of generality, i.e.,

$$\int_\Omega \phi_n^*(\underline{r})\phi_m(\underline{r})d\underline{r} = \delta_{nm} \quad . \qquad (1.3)$$

The completeness of the set $\{\phi_n(\underline{r})\}$ means that

---

[1] Several authors write the right-hand side of (1.1) as $4\pi\delta(\underline{r}-\underline{r}')$ or $-4\pi\delta(r-r')$.

$$\sum_n \phi_n(\underline{r})\phi_n^*(\underline{r}') = \delta(\underline{r} - \underline{r}') \quad . \tag{1.4}$$

Note that n may stand for a set of indices which can take either discrete values (for the discrete part of the spectrum of L, if any) and/or continuous values (for the continuous part of the spectrum of L, if any). Similarly, the symbol $\sum_n$ should be interpreted as $\sum_n' + \int dn$, where $\sum_n'$ indicates a genuine summation over the eigenfunctions belonging to the discrete spectrum (if any) and $\int dn$ is a (multiple) integral over the continuous spectrum (if any).

Working with Green's functions is greatly facilitated by introducing an abstract vector space, a particular representation of which are the various functions we are dealing with. The most convenient way of achieving this is by using Dirac's bra and ket notation, according to which one can write

$$\phi_n(\underline{r}) \equiv \langle\underline{r}|\phi_n\rangle \quad \text{, etc.} \tag{1.5}$$

$$\delta(\underline{r} - \underline{r}')L(\underline{r}) \equiv \langle\underline{r}|L|\underline{r}'\rangle \tag{1.6}$$

$$G(\underline{r},\underline{r}';z) \equiv \langle\underline{r}|G(z)|\underline{r}'\rangle \tag{1.7}$$

$$\langle\underline{r}|\underline{r}'\rangle = \delta(\underline{r} - \underline{r}') \tag{1.8}$$

$$\int d\underline{r}|\underline{r}\rangle\langle\underline{r}| = 1 \quad . \tag{1.9}$$

n the new notation we can write

$$(z - L)G(z) = 1 \tag{1.1'}$$

$$L|\phi_n\rangle = \lambda_n|\phi_n\rangle \tag{1.2'}$$

$$\langle\phi_n|\phi_m\rangle = \delta_{nm} \tag{1.3'}$$

$$\sum_n |\phi_n\rangle\langle\phi_n| = 1 \quad . \tag{1.4'}$$

The ordinary $\underline{r}$-representation is recaptured by using (1.5-9); e.g., taking the $\langle\underline{r}|$, $|\underline{r}'\rangle$ matrix element of (1.1') we have

$$\langle\underline{r}|(z - L)G(z)|\underline{r}'\rangle = \langle r|1|\underline{r}'\rangle = \langle\underline{r}|\underline{r}'\rangle = \delta(\underline{r} - \underline{r}') \quad ,$$

and thus

$$zG(\underline{r},\underline{r}';z) - \langle \underline{r}|LG(z)|\underline{r}'\rangle$$

$$= zG(\underline{r},\underline{r}';z) - \int d\underline{r}'' \langle \underline{r}|L|\underline{r}''\rangle\langle \underline{r}''|G(z)|\underline{r}'\rangle$$

$$= zG(\underline{r},\underline{r}';z) - L(\underline{r})G(\underline{r},\underline{r}';z) = \delta(\underline{r} - \underline{r}')$$

which is identical to (1.1). The usefulness of the bra and ket notation is that (i) the intermediate algebraic manipulations are facilitated and (ii) one is not restricted to the $\underline{r}$-representation (e.g., one can express all equations in the $\underline{k}$-representation which is equivalent to taking the Fourier transform with respect to $\underline{r}$ and $\underline{r}'$ of the original equations).

If all eigenvalues of $z - L$ are nonzero, i.e., if $z \neq \{\lambda_n\}$, then one can solve (1.1') formally as

$$G(z) = \frac{1}{z - L} \quad . \tag{1.10}$$

Multiplying (1.10) by (1.4') we obtain

$$G(z) = \frac{1}{z - L} \sum_n |\phi_n\rangle\langle\phi_n| = \sum_n \frac{1}{z - L} |\phi_n\rangle\langle\phi_n| = \sum_n \frac{|\phi_n\rangle\langle\phi_n|}{z - \lambda_n} \quad . \tag{1.11}$$

The last step follows from the general relation $F(L)|\phi_n\rangle = F(\lambda_n)|\phi_n\rangle$ valid by definition for any function F. Equation (1.11) can be written more explicitly as

$$G(z) = \sum_n{}' \frac{|\phi_n\rangle\langle\phi_n|}{z - \lambda_n} + \int dn \, \frac{|\phi_n\rangle\langle\phi_n|}{z - \lambda_n} \tag{1.12}$$

or in the $\underline{r}$-representation

$$G(\underline{r},\underline{r}';z) = \sum_n{}' \frac{\phi_n(\underline{r})\phi_n^*(\underline{r}')}{z - \lambda_n} + \int dn \, \frac{\phi_n(\underline{r})\phi_n^*(\underline{r}')}{z - \lambda_n} \quad . \tag{1.13}$$

Since L is a Hermitian operator, all of its eigenvalues $\{\lambda_n\}$ are real. Hence, if $\mathrm{Im}\{z\} \neq 0$, $z \neq \{\lambda_n\}$ which means that $G(z)$ is an analytic function in the complex z-plane except at those points or portions of the real z-axis which correspond to the eigenvalues of L. As can be seen from (1.12) or (1.13), $G(z)$ exhibits simple poles at the position of the discrete eigenvalues of L; the inverse is also true: *the poles of* $G(z)$ *give the discrete eigenvalues of* L. If $z = \lambda$, where λ belongs to the continuous spectrum of L, $G(\underline{r},\underline{r}';\lambda)$ is not well defined since the integrand in (1.13) has a pole. However, one can attempt to define $G(\underline{r},\underline{r}';\lambda)$ by a limiting procedure. In the usual case, where the eigenstates associated with the continuous spectrum are propagating or extended (i.e., not decaying as $r \to \infty$), the side limits of $G(\underline{r},\underline{r}';\lambda \pm is)$ as $s \to 0^+$ exist but are different from each other. Thus, this type of continuous spectrum produces a branch cut in $G(z)$ along the real z-axis. We will see later that in disordered systems there is the possibility of a continuous spectrum associated with localized eigenstates (i.e., states decaying fast enough as $r \to \infty$ so that the normalized $\{\phi_n(\underline{r})\}$ approach a nonzero limit as $\Omega \to \infty$). For such an unusual spectrum even the side limits $\lim_{s\to 0^+} G(\underline{r},\underline{r}';\lambda \pm is)$, do not exist; the line of singularity corresponding to such a spectrum is not a branch cut but what is called a natural boundary. In what follows we restrict ourselves to the normal case of a continuous spectrum consisting of extended eigenstates. For λ belonging to such a spectrum we define two Green's functions as follows

$$G^+(\underline{r},\underline{r}';\lambda) \equiv \lim_{s\to 0^+} G(\underline{r},\underline{r}';\lambda+is) \tag{1.14}$$

$$G^-(\underline{r},\underline{r}';\lambda) \equiv \lim_{s\to 0^+} G(\underline{r},\underline{r}';\lambda-is) \tag{1.15}$$

with similar definitions for the corresponding operators $G^+(\lambda)$, $G^-(\lambda)$. From (1.13) one can easily see that

$$G^*(\underline{r},\underline{r}';z) = G(\underline{r}',\underline{r};z^*) \quad . \tag{1.16}$$

If z is real, $z = \lambda$, and $\lambda \neq \{\lambda_n\}$, it follows from (1.16) that $G(\underline{r},\underline{r}';\lambda)$ is Hermitian; in particular, $G(\underline{r},\underline{r};\lambda)$ is real. On the other hand, for λ belonging to the continuous spectrum, we have from (1.16) that

$$G^-(\underline{r},\underline{r}';\lambda) = [G^+(\underline{r}',\underline{r};\lambda)]^* , \tag{1.17}$$

which shows that

$$\mathrm{Re}\{G^-(\underline{r},\underline{r};\lambda)\} = \mathrm{Re}\{G^+(\underline{r},\underline{r};\lambda)\} \tag{1.18}$$

and

$$\mathrm{Im}\{G^-(\underline{r},\underline{r};\lambda)\} = -\,\mathrm{Im}\{G^+(\underline{r},\underline{r};\lambda)\} \quad . \tag{1.19}$$

Using the identity

$$\lim_{y\to 0^+} \frac{1}{x \pm iy} = P\,\frac{1}{x} \mp i\pi\delta(x) \tag{1.20}$$

and (1.13) we can express the discontinuity

$$\tilde{G}(\lambda) \equiv G^+(\lambda) - G^-(\lambda) \tag{1.21}$$

as

$$\begin{aligned}\tilde{G}(\underline{r},\underline{r}';\lambda) &= -\,2\pi i \sum_n \delta(\lambda - \lambda_n)\phi_n(\underline{r})\phi_n^*(\underline{r}') \\ &= -\,2\pi i \sum_n{}' \delta(\lambda - \lambda_n)\phi_n(\underline{r})\phi_n^*(\underline{r}') - 2\pi i \int \delta(\lambda - \lambda_n)\phi_n(\underline{r})\phi_n^*(\underline{r}')dn \ .\end{aligned} \tag{1.22}$$

For the diagonal matrix element we obtain from (1.13) and (1.20)

$$G^{\pm}(\underline{r},\underline{r};\lambda) = P\sum_n \frac{\phi_n(\underline{r})\phi_n^*(\underline{r})}{\lambda - \lambda_n} \mp i\pi \sum_n \delta(\lambda - \lambda_n)\phi_n(\underline{r})\phi_n^*(\underline{r}) \quad . \tag{1.23}$$

Integrating (1.23) over $\underline{r}$ we have

$$\mathrm{Tr}G^{\pm}(\lambda) = P\sum_n \frac{1}{\lambda - \lambda_n} \mp i\pi \sum_n \delta(\lambda - \lambda_n) \quad . \tag{1.24}$$

The quantity $\sum_n \delta(\lambda - \lambda_n)$ is the density of states (DOS) at λ , $N(\lambda)$; $N(\lambda)d\lambda$ gives the number of states in the interval $[\lambda,\lambda + d\lambda]$. The quantity

$$\begin{aligned}\rho(\underline{r};\lambda) \equiv \sum_n \delta(\lambda - \lambda_n)\phi_n(\underline{r})\phi_n^*(\underline{r}) &= \sum_n{}' \delta(\lambda - \lambda_n)\phi_n(\underline{r})\phi_n^*(\underline{r}) \\ &+ \int \delta(\lambda - \lambda_n)\phi_n(\underline{r})\phi_n^*(\underline{r})dn\end{aligned} \tag{1.25}$$

is the DOS per unit volume. Obviously,

$$N(\lambda) = \int\rho(\underline{r};\lambda)d\underline{r} \quad . \tag{1.26}$$

Using (1.22-25) we obtain

$$\rho(\underline{r};\lambda) = \mp \frac{1}{\pi} \, \mathrm{Im}\{G^{\pm}(\underline{r},\underline{r};\lambda)\} = \frac{-1}{2\pi i} \tilde{G}(\underline{r},\underline{r};\lambda) \tag{1.27}$$

and

$$N(\lambda) = \mp \frac{1}{\pi} \, \mathrm{Im}\{\mathrm{Tr}G^{\pm}(\lambda)\} \quad . \tag{1.28}$$

$G(z)$ can be expressed in terms of the discontinuity $\tilde{G}(\lambda) \equiv G^{+}(\lambda) - G^{-}(\lambda)$:

$$G(\underline{r},\underline{r}';z) = \sum_n \phi_n(\underline{r})\phi_n^*(\underline{r}')/(z - \lambda_n)$$

$$= \int_{-\infty}^{\infty} d\lambda \sum_n \delta(\lambda - \lambda_n)\phi_n(\underline{r})\phi_n^*(\underline{r}')/(z - \lambda) = \frac{i}{2\pi} \int_{-\infty}^{\infty} d\lambda \, \frac{\tilde{G}(\underline{r},\underline{r}';\lambda)}{z - \lambda} \tag{1.29}$$

where (1.22) was taken into account. In particular, for the diagonal matrix elements of G we have

$$G(\underline{r},\underline{r};z) = \int_{-\infty}^{\infty} d\lambda \, \frac{\rho(\underline{r};\lambda)}{z - \lambda} \quad . \tag{1.30}$$

Note that $\rho(\underline{r};\lambda)$ versus λ may consist of a sum of δ functions (corresponding to the discrete spectrum of L) and a continuous function (corresponding to the continuous spectrum of L) as shown in (1.25). Equation (1.30) shows that the DOS per unit volume (i.e., the imaginary part of $\mp G^{\pm}(\underline{r},\underline{r};\lambda)/\pi$) enables us to calculate $G(\underline{r},\underline{r};z)$ (both Re{G} and Im{G}) for all values of $z = \lambda + is$.

Consider the expression

$$\underline{E}(z) \equiv E_x(x,y) - iE_y(x,y) = \int_C dz' \, \frac{2\rho(z')}{z - z'}$$

giving the x and y component of the electric field $\underline{E}(x,y)$ in two-dimensional (2-D) space in terms of the charge density $\rho(z') \equiv \rho(x',y')$ along the line C. Comparing with (1.30), we see that $G(\underline{r},\underline{r};z)$ can be thought of as the electric field generated by a positive charge distribution on the x-axis given by one half the DOS per unit volume, $\rho/2$. More explicitly the correspondence is: $\mathrm{Re}\{G(\underline{r},\underline{r};z)\} \leftrightarrow E_x(z)$, $\mathrm{Im}\{G(\underline{r},r;z)\} \leftrightarrow -E_y(z)$, $\rho(\underline{r};\lambda) \leftrightarrow 2\rho(z')$, $z \equiv x + iy \leftrightarrow z \equiv \lambda + is$, $z' \equiv x' + iy' \leftrightarrow \lambda'$. This analogy is often helpful in visualizing the z dependence of $G(\underline{r},\underline{r};z)$ for complex values of z. For example, we see immediately that $\mathrm{Re}\{G^{+}(\underline{r},\underline{r};\lambda)\} = \mathrm{Re}\{G^{-}(\underline{r},\underline{r};\lambda)\}$ while $\mathrm{Im}\{G^{+}(\underline{r},\underline{r};\lambda)\} = -\mathrm{Im}\{G^{-}(\underline{r},\underline{r};\lambda)\}$ with $\mathrm{Im}\, G^{+}(\underline{r},\underline{r};\lambda)$ being always negative or zero. Of course, when λ is not an eigenvalue of L, $G(\underline{r},\underline{r};\lambda)$ is real; it satisfies the relation

$$dG(\underline{r},\underline{r};\lambda)/d\lambda = -\langle\underline{r}|(\lambda - L)^{-2}|\underline{r}\rangle < 0 \quad . \tag{1.31}$$

To prove (1.31) we write $dG(\underline{r},\underline{r};\lambda)/d\lambda = d[\langle\underline{r}|(\lambda - L)^{-1}|\underline{r}\rangle]/d\lambda = -\langle\underline{r}|(\lambda - L)^{-2}|\underline{r}\rangle$, which is negative since, for λ real and not coinciding with any eigenvalue of L, $(\lambda - L)^{-2}$ is a positive definite operator.

To summarize our findings: $G(\underline{r},\underline{r}';z)$ is analytic in the complex z-plane except on portions or points of the real axis. The positions of the poles of $G(\underline{r},\underline{r}';z)$ on the real axis give the eigenvalues of L. The residue at each pole gives the product $\phi_n(\underline{r})\phi_n(\underline{r}')$ where $\phi_n(\underline{r})$ is the corresponding (non-degenerate) eigenfunction. The branch cuts of $G(\underline{r},\underline{r}';\lambda)$ along the real λ-axis correspond to the continuous spectrum of L, and the discontinuity of the diagonal matrix element $G(\underline{r},\underline{r};\lambda)$ across the branch cut gives the DOS per unit volume times $-2\pi i$. Note that the analytic continuation of $G(\underline{r},\underline{r}';z)$ across the branch cut does not coincide with $G(\underline{r},\underline{r}';z)$, and it may develop singularities in the complex z-plane.

Knowledge of the Green's function $G(\underline{r},\underline{r}';z)$ permits us to obtain immediately the solution of the general inhomogeneous equation

$$[z - L(\underline{r})]u(\underline{r}) = f(\underline{r}) \quad , \tag{1.32}$$

where the unknown function $u(\underline{r})$ satisfies on S the same boundary conditions as $G(\underline{r},\underline{r}';z)$; $f(\underline{r})$ is a given function. By taking into account (1.1), it is easy to show that the solution of (1.32) is

$$u(\underline{r}) = \int G(\underline{r},\underline{r}';z)f(\underline{r}')d\underline{r}' \quad ; \quad z \neq \{\lambda_n\} \tag{1.33a}$$

$$= \int G^{\pm}(\underline{r},\underline{r}';\lambda)f(\underline{r}')d\underline{r}' + \phi(\underline{r}) \quad ; \quad z = \lambda \quad , \tag{1.33b}$$

where λ belongs to the branch cut of $G(z)$ (i.e., λ belongs to the continuous spectrum of L) and $\phi(\underline{r})$ is the general solution of the corresponding homogeneous equation. If z coincides with a discrete eigenvalue of L, let us say λ_n, there is no solution of (1.32) unless $f(\underline{r})$ is orthogonal to all eigenfunctions associated with λ_n. If $u(\underline{r})$ describes physically the response of a system to a source $f(\underline{r})$, then $G(\underline{r},\underline{r}')$ describes the response of the same system to a unit point source located at $\underline{r}'$. Note that the symmetry relation (1.16) is a generalized reciprocity relation: the response at $\underline{r}$ from a source at $\underline{r}'$ is essentially the same as the response at $\underline{r}'$ from a source at $\underline{r}$. Equation (1.33a) means that the response to the general source $f(\underline{r})$ can be expressed as the sum of the responses to point sources distributed according to $f(\underline{r})$.

1.2 Examples

In this section we consider the case where $L(\underline{r}) = -\nabla^2$ and the domain Ω extends over the whole real space. The boundary condition is that the eigenfunctions of L must be finite at infinity. Then the eigenfunctions are

$$\langle \underline{r}|\underline{k}\rangle = \frac{1}{\sqrt{\Omega}} e^{i\underline{k}\underline{r}} \tag{1.34}$$

and the eigenvalues are

$$\lambda_n = \underline{k}^2 \tag{1.35}$$

where the components of the vector $\underline{k}$ are real. Thus, the spectrum is continuous, extending from 0 to $+\infty$. The Green's function can be obtained by either solving the defining equation, which in the present case is

$$(z + \nabla_r^2)G(\underline{r},\underline{r}';z) = \delta(\underline{r} - \underline{r}') \quad , \tag{1.36}$$

or from (1.13), which in the present case can be written as

$$G(\underline{r},\underline{r}';z) = \sum_{\underline{k}} \frac{\langle \underline{r}|\underline{k}\rangle\langle \underline{k}|\underline{r}'\rangle}{z - k^2} = \int \frac{d\underline{k}}{(2\pi)^d} \frac{e^{i\underline{k}(\underline{r} - \underline{r}')}}{z - k^2} \quad , \tag{1.37}$$

where d is the dimensionality[2]. For d = 3, we will use (1.37) to evaluate G, while for d = 2 or 1 we will compute G from (1.36).

1.2.1 Three-Dimensional Case (d = 3)

If $\underline{\rho}$ is the difference $\underline{r} - \underline{r}'$ and θ the angle between $\underline{k}$ and $\underline{\rho}$, we can write (1.37) as

$$\begin{aligned} G(\underline{r},\underline{r}';z) &= \frac{1}{4\pi^2} \int_0^\infty \frac{k^2 dk}{z - k^2} \int_0^\pi d\theta \sin\theta \, e^{ik\rho\cos\theta} \\ &= \frac{1}{4\pi^2} \int_0^\infty \frac{k^2 dk}{z - k^2} \frac{e^{ik\rho} - e^{-ik\rho}}{ik\rho} \\ &= \frac{1}{4i\pi^2\rho} \int_{-\infty}^{\infty} \frac{ke^{ik\rho}}{z - k^2} dk \quad . \end{aligned} \tag{1.38}$$

[2] One can prove that $\sum_{\underline{k}} \rightarrow [\Omega/(2\pi)^d] \int d\underline{k}$ as $\Omega \rightarrow \infty$

The integration path can be closed by an infinite semicircle in the upper half plane. Unless z is real and nonnegative, one of the poles (denoted by $\sqrt{z}$) of the integrand in (1.38) has a positive imaginary part and hence lies within the integration contour, and the other (denoted by $-\sqrt{z}$) has a negative imaginary part and lies outside the integration contour. By employing the residue theorem we obtain from (1.38)

$$G(\underline{r},\underline{r}';z) = -\frac{e^{i\sqrt{z}|\underline{r}-\underline{r}'|}}{4\pi|\underline{r}-\underline{r}'|} \quad ; \quad \mathrm{Im}\{\sqrt{z}\}>0 \quad . \tag{1.39}$$

If $z = \lambda$, where $\lambda \geq 0$ (i.e., if z coincides with the eigenvalues of $-\nabla^2$), the two poles lie on the integration contour and G is not well defined. The side limits $G^{\pm}(\underline{r},\underline{r}';\lambda)$ are well defined and are given by

$$G^{\pm}(\underline{r},\underline{r}';\lambda) = -\frac{e^{\pm i\sqrt{\lambda}|\underline{r}-\underline{r}'|}}{4\pi|\underline{r}-\underline{r}'|} \quad ; \quad \sqrt{\lambda}, \quad \lambda \geq 0 \quad . \tag{1.40}$$

For $z = \lambda$ where $\lambda < 0$ we obtain from (1.39)

$$G(\underline{r},\underline{r}';\lambda) = -\frac{e^{-\sqrt{|\lambda|}\,|\underline{r}-\underline{r}'|}}{4\pi|\underline{r}-\underline{r}'|} \quad ; \quad \lambda < 0, \quad \sqrt{|\lambda|} > 0 \quad . \tag{1.41}$$

For the particular case $z = 0$ we have

$$G(\underline{r},\underline{r}';0) = -\frac{1}{4\pi|\underline{r}-\underline{r}'|} \cdot \tag{1.42}$$

As can be seen from (1.36), $G(\underline{r},\underline{r}';0)$ is the Green's function corresponding to Laplace's equation, i.e.,

$$\nabla_r^2\, G(\underline{r},\underline{r}';0) = \delta(\underline{r}-\underline{r}') \quad . \tag{1.43}$$

By applying (1.33b), we can write the general solution of Poisson's equation

$$\nabla^2 V(\underline{r}) = -4\pi\rho(\underline{r})$$

as

$$V(\underline{r}) = \int G(\underline{r},\underline{r}';0)(-4\pi)\rho(\underline{r}')\,d\underline{r}' + C$$

$$= \int \frac{\rho(\underline{r}')\,d\underline{r}'}{|\underline{r} - \underline{r}'|} + C \quad . \tag{1.44}$$

The constant has been added since the most general eigenfunction of $-\nabla^2$ corresponding to eigenvalue 0 is a constant, as can be seen from (1.34,35). Equation (1.44) is the basic result in electrostatics.

1.2.2 Two-Dimensional Case (d = 2)

Because of symmetry considerations, $G(\underline{r},\underline{r}';z)$ is a function of the magnitude of the 2-D vector $\underline{\rho} = \underline{r} - \underline{r}'$ and z. Furthermore, it satisfies the homogeneous equation

$$(z + \nabla^2)G(\rho;z) = 0 \text{ for } \rho \neq 0 \quad . \tag{1.45}$$

The δ-function source can be transformed to an equivalent boundary condition as $\rho \to 0$; by applying Gauss' theorem

$$\int_0^\rho \nabla^2 G\, 2\pi\rho' d\rho' = 2\pi\rho \frac{\partial G}{\partial \rho}$$

we obtain from (1.36)

$$2\pi\rho \frac{\partial G}{\partial \rho} + 2\pi z \int_0^\rho G\rho' d\rho' = 1$$

which as $\rho \to 0$ leads to

$$G(\rho) \xrightarrow[\rho \to 0]{} \frac{1}{2\pi} \ln(\rho) + \text{const.} \tag{1.46}$$

Furthermore, $G(\rho)$ must satisfy the condition

$$G(\rho) \xrightarrow[\rho \to \infty]{} 0 \quad . \tag{1.47}$$

The only solution of (1.45) which is symmetric and satisfies the boundary conditions (1.46,47) is

$$G(\underline{r},\underline{r}';z) = \frac{-i}{4} H_0^{(1)}(\sqrt{z}|\underline{r} - \underline{r}'|) \quad ; \quad Im\{\sqrt{z}\} > 0 \tag{1.48}$$

where $H_0^{(1)}$ is the Hankel function of zero order of the first kind[3]. This can be seen from the fact that the general solution of (1.45) is a superposition of terms like $[A_n H_n^{(1)}(\sqrt{z}\rho) + B_n H_n^{(2)}(\sqrt{z}\rho)] \exp(\pm in\vartheta)$. Since we are looking for a ϑ-independent solution, $n = 0$; furthermore, the Hankel function $H_0^{(2)}(\sqrt{z}\rho)$, for $Im\{\sqrt{z}\}>0$, blows up as $\rho \to \infty$ and must be excluded. Finally, (1.46) fixes the coefficient A_0. For $z = \lambda$, where $\lambda \geq 0$, (i.e., for z coinciding with the spectrum of $-\nabla^2$) $Im\{\sqrt{z}\} = 0$, and only the side limits are well defined as

$$G^{\pm}(\underline{r},\underline{r}';\lambda) = \frac{-i}{4} H_0^{(1)}(\pm\sqrt{\lambda}\rho) \; ; \; \lambda > 0 \; , \; \sqrt{|\lambda|} > 0 \quad , \tag{1.49}$$

where

$$H_0^{(1)}(-\sqrt{\lambda}\rho) = - H_0^{(2)}(\sqrt{\lambda}\rho) \quad . \tag{1.50}$$

Equation (1.48) for the particular case when $z = -|\lambda|$ can be recast as

$$G(\underline{r},\underline{r}';-|\lambda|) = -\frac{1}{2\pi} K_0(\sqrt{|\lambda|}\,|\underline{r} - \underline{r}'|) ; \quad \sqrt{|\lambda|} > 0 \quad , \tag{1.51}$$

where K_0 is the modified Bessel function of zero order[3].

The Green's function corresponding to the 2-D Laplace equation can be obtained from (1.48) by letting $z \to 0$ and keeping the leading $|\underline{r} - \underline{r}'|$ -dependent term. We find that

$$G(\underline{r},\underline{r}';0) = \frac{1}{2\pi} \ln|\underline{r} - \underline{r}'| + \text{const.} \tag{1.52}$$

The solution of Poisson's equation in 2-D is then

$$V(\underline{r}) = -2 \int \rho(\underline{r}') \ln|\underline{r} - \underline{r}'| \, d\underline{r}' + \text{const.} \tag{1.53}$$

Taking the $-\nabla$ of (1.53) we obtain the expression for the 2-D electric field given before.

---

[3] For definition and properties of Bessel and Hankel functions see [3.1].

1.2.3 One-Dimensional Case (d = 1)

The basic equation (1.36) becomes

$$\left(z + \frac{d^2}{dx^2}\right) G(x,x';z) = \delta(x - x') \quad . \tag{1.54}$$

For x<x' we have $G = A\exp(-i\sqrt{z}x)$ with $\mathrm{Im}\{\sqrt{z}\}>0$, while for x>x' we obtain $G = B\exp(i\sqrt{z}x)$; the choice of signs in the exponents makes sure that $G \to 0$ as $|x| \to \infty$. By integrating (1.54) we find that $G(x'^-,x';z) = G(x'^+,x';z)$ and $(dG/dx)_{x=x'^+} - (dG/dx)_{x=x'^-} = 1$. We thus determine the constants A and B. We obtain finally

$$G(x,x';z) = \frac{e^{i\sqrt{z}|x - x'|}}{2i\sqrt{z}} \quad ; \quad \mathrm{Im}\{\sqrt{z}\} > 0 \quad . \tag{1.55}$$

For $z = \lambda \geq 0$ (i.e., within the continuous spectrum of $-d^2/dx^2$) we have for the side limits

$$G^{\pm}(x,x';\lambda) = \mp \frac{i}{2\sqrt{\lambda}} e^{\pm i\sqrt{\lambda}|x - x'|} \quad ; \quad \lambda \geq 0 \quad , \quad \sqrt{\lambda} > 0 \quad . \tag{1.56}$$

For $z = -|\lambda|$ we obtain from (1.55)

$$G(x,x'; -|\lambda|) = -\frac{1}{2\sqrt{|\lambda|}} e^{-\sqrt{|\lambda|}|x - x'|} \quad ; \quad \sqrt{|\lambda|} > 0 \quad . \tag{1.57}$$

The Green's function for the 1-D Laplace equation can be found either by solving (1.54) for z = o directly or by taking the limit of $(G^+ + G^-)/2$ as $\lambda \to 0$. We find

$$G(x,x';0) = \frac{1}{2}|x - x'| + \text{const.} \tag{1.58}$$

1.2.4 Finite Domain Ω

The problem of determination of G becomes more tedious when the surface S bounding our domain Ω consists in part (or in total) by pieces at finite distance from the point $\underline{r}'$ of the source. One can follow any one of the following methods to determine G:

1) Use the general equation (1.13) where the eigenvalues and eigenfunctions are the ones associated with the boundary conditions on S.

2) Write G as $G = G^{\infty} + \phi$ where G^{∞} is the Green's function associated with the infinite domain (which is assumed to be known) and ϕ is the general solution of the corresponding homogeneous equation. Then determine the arbitrary coefficients in ϕ by requiring that $G^{\infty} + \phi$ satisfies the given boundary conditions on S. It is then clear that G satisfies both the differential equation and the boundary conditions.

3) Divide the domain Ω in two subdomains by a surface S' passing through the source point $\underline{r}'$. Then G in the interior of each subdomain satisfies an homogeneous equation. Find in each subdomain the general solution of the homogeneous equation subject to the given boundary conditions on S. Next match the two solutions on S' in a way obtained by integrating the differential equation for G around $\underline{r}'$. An elementary example of this technique was used in Sect.1.2.3. The interested reader may find a brief presentation of these techniques in the book by MATHEWS and WALKER [1.1]. A more comprehensive and rigorous presentation is given in the book by BYRON and FULLER [1.2] or by MORSE and FESHBACH [1.3]. Several books on electromagnetism such as those by SMYTHE [1.4] or JACKSON [1.5] contain several interesting examples of Green's functions.

Finally, it should be mentioned that for more complicated operators L (such as those describing the quantum-mechanical motion of a particle in an external field) the determination of G is a very complicated problem. More often than not one has to employ approximate techniques such as perturbation expansions. We will return to this very interesting subject in Chap.4.

1.3 Summary

Definition

The Green's function, corresponding to the linear, Hermitian, time-independent differential operator $L(\underline{r})$ and the complex variable $z = \lambda + is$, is defined as the solution of the equation[4]

$$[z - L(\underline{r})]G(\underline{r},\underline{r}';z) = \delta(\underline{r}' - \underline{r}) \qquad (1.1)$$

[4] The numbering of the equations appearing in each summary is that of the main text of the corresponding chapter.

subject to certain homogeneous boundary conditions on the surface S of the domain Ω of $\underline{r}$ and $\underline{r}'$. Equation (1.1) can be considered as the $\underline{r}$-representation of the operator equation

$$(z - L)G = 1 \quad . \tag{1.1'}$$

Basic Properties

1) If $\{|\phi_n\rangle\}$ is the complete orthonormal set of eigenfunctions (subject to the same boundary conditions on the surface S) of L and $\{\lambda_n\}$, the set of eigenvalues, one can write

$$G = (z - L)^{-1} \tag{1.10}$$

$$= \sum_n \frac{|\phi_n\rangle\langle\phi_n|}{z - \lambda_n} \quad . \tag{1.11}$$

2) From (1.11) one can see that $G(z)$ is uniquely defined if and only if $z \neq \{\lambda_n\}$. If z coincides with any of the discrete eigenvalues of L, G does not exist, since, as it can be seen from (1.11), $G(z)$ has simple (first-order) poles at the positions of the discrete eigenvalues. If z belongs to the continuum spectrum of L, then G usually exists, but it is not uniquely defined because one can add to any particular G the general solution of the homogeneous equation corresponding to (1.1). The continuous spectrum of L appears as a singular line (a branch cut and/or a natural boundary) of $G(z)$ as shown in Fig.1.1. Because L is Hermitian, all its eigenvalues are real;

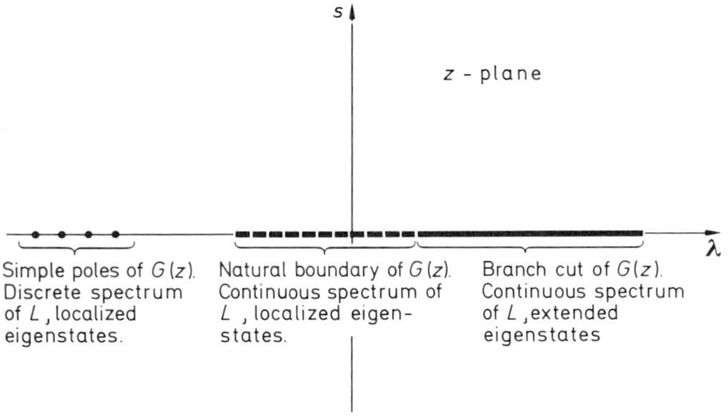

Fig.1.1. Analytic behavior of $G(z) \equiv (z - L)^{-1}$. The singular points or lines are on the real z-axis (when L is Hermitian) and provide information about the eigenvalues and eigenfunctions of L

hence, the singularities of $G(z)$ are on the real z-axis. For the branch cuts of $G(z)$ (which correspond to the continuous spectrum associated with extended eigenstates) we define the side limits

$$G^{\pm}(\lambda) = \lim_{s \to 0^+} G(\lambda \pm is) \quad . \tag{1.15'}$$

Methods of Calculation

$G(z)$ is calculated either by solving the defining equation (1.1) or by using (1.11).

Use

Once $G(z)$ is known one can:

1) Obtain information about the homogeneous equation corresponding to (1.1), i.e., about the eigenvalues and eigenfunctions of L. Thus the position of the poles of $G(z)$ give the discrete eigenvalues of L, and the residues at these poles provide information about the corresponding eigenfunctions. The branch cuts (or the natural boundaries, if any) give the location of the continuous spectrum and the discontinuity across the branch cut gives the density of states $N(\lambda)$

$$N(\lambda) = \mp \frac{1}{\pi} \operatorname{Im}\{\operatorname{Tr} G^{\pm}(\lambda)\} \quad . \tag{1.28}$$

2) Solve the inhomogeneous equation

$$[z - L(\underline{r})]u(\underline{r}) = f(\underline{r}) \quad , \tag{1.32}$$

where the unknown function $u(\underline{r})$ satisfies on S the same boundary conditions as $G(\underline{r},\underline{r}';z)$, and $f(\underline{r})$ is given. We have

$$u(\underline{r}) = \int G(\underline{r},\underline{r}';z)f(\underline{r}')d\underline{r}' \quad ; \quad z \neq \{\lambda_n\} \; , \tag{1.33a}$$

$$u(\underline{r}) = \int G^{\pm}(\underline{r},\underline{r}';\lambda)f(\underline{r}')d\underline{r}' + \phi(\underline{r}) \quad ; \quad z = \lambda \tag{1.33b}$$

where λ belongs to the branch cut of $G(z)$ and $\phi(\underline{r})$ is the general solution of the corresponding homogeneous equation for the given value of λ. For z coinciding with any of the discrete eigenvalues of L, let us say λ_n, there is no solution of (1.32) unless $f(\underline{r})$ is orthogonal to all eigenfunctions associated with λ_n.

3) Use $G_0(z)$ and L_1 to obtain information about the eigenfunctions and eigenvalues of $L = L_0 + L_1$; $G_0(z) \equiv (z - L_0)^{-1}$. Discussion of this third use of the Green's function formalism will be given in Chap.4.

Some very common applications of the formalism outlined above were discussed in Sect.1.2.

2. Time-Dependent Green's Functions

The Green's functions corresponding to linear partial differential equations of first and second order in time are defined; their main properties and uses are presented.

2.1 First-Order Case

The Green's function $g(\underline{r},\underline{r}',t - t')$ associated with the first-order (in time) partial differential equations

$$\left[\frac{i}{c}\frac{\partial}{\partial t} - L(\underline{r})\right]\phi(\underline{r},t) = 0 \tag{2.1}$$

$$\left[\frac{i}{c}\frac{\partial}{\partial t} - L(\underline{r})\right]\psi(\underline{r},t) = f(\underline{r},t) \tag{2.2}$$

is defined as the solution of

$$\left[\frac{i}{c}\frac{\partial}{\partial t} - L(\underline{r})\right] g(\underline{r},\underline{r}',t,t') = \delta(\underline{r} - \underline{r}')\delta(t - t') \tag{2.3}$$

subject to the same boundary conditions on the bounding surface S as $\phi(\underline{r},t)$ and $\psi(\underline{r},t)$. For the time being we will assume that c is a positive constant. The operator $L(\underline{r})$ is as in Chap.1. Expressing $g(\underline{r},\underline{r}',\tau)$, where $\tau = t - t'$, in terms of its Fourier transform

$$g(\tau) = \int_{-\infty}^{\infty} \frac{d\omega'}{2\pi}\, e^{-i\omega'\tau} g(\omega') \quad , \tag{2.4}$$

and substituting in (2.3) we obtain

$$(\frac{\omega}{c} - L)g(\omega) = \delta(\underline{r} - \underline{r}') \quad . \tag{2.5}$$

The $\underline{r}$, $\underline{r}'$ dependence of $g(\tau)$ and $g(\omega)$ is not displayed explicitly. Comparing with (1.1) we see that

$$g(\omega) = G(\frac{\omega}{c}) \quad , \tag{2.6}$$

where $G(z)$ is the Green's function associated with L and examined in detail in Chap.1.

Using the results of Chap.1, one concludes that $g(\omega)$ is an analytic function of the complex variable ω with singularities (poles and/or branch cuts) on the real ω-axis. Because of this property (2.4) is not well defined as it stands. One has to use a limiting procedure to define $g(\tau)$

$$g^C(\tau) = \lim_{C \to C_0} \int_C \frac{d\omega}{2\pi} \, G(\frac{\omega}{c}) \, e^{-i\omega\tau} \quad . \tag{2.7}$$

One can obtain infinitely many Green's functions depending on the way the path C approaches the real ω-axis C_0. However, there are only two choices of physical interest shown in Fig.2.1; they give

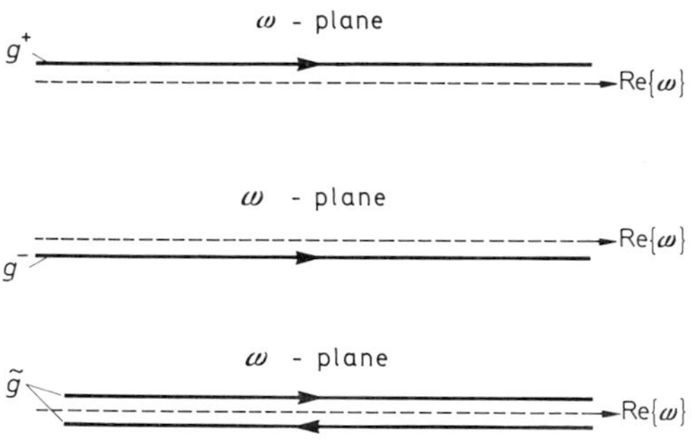

Fig.2.1. Integration paths in the ω-plane for obtaining $g^+(\tau)$, $g^-(\tau)$ and $\tilde{g}(\tau) = g^+(\tau) - g^-(\tau)$, where $g^\pm(\tau)$ satisfy a first-order (in time) differential equation. The singularities of the integrand lie on the real ω-axis

$$g^\pm(\tau) = \int_{-\infty}^{\infty} \frac{d\omega'}{2\pi} \, G^\pm(\omega'/c) \, e^{-i\omega'\tau} \quad . \tag{2.8}$$

It is useful to consider quantities denoted by the symbol $\tilde{g}^{CC'}$ and defined as the difference of two Green's function $g^C, g^{C'}$. Obviously the quantities $\tilde{g}^{CC'}$ can be expressed as integrals of $g(\omega)\exp(-i\omega\tau)/2\pi$ over closed contours enclosing part(s) of (or the whole) real ω-axis. Because each $\tilde{g}^{CC'}$ is the difference of two Green's functions, it satisfies the homogeneous equation (2.1) and not (2.3). Thus, strictly speaking, the various $\tilde{g}^{CC'}$'s are not Green's functions.

In the present case, because we introduce only two Green's functions of interest, there is only one $\tilde{g}^{CC'}$ of interest defined by

$$\tilde{g}(\tau) = g^{+}(\tau) - g^{-}(\tau) \quad . \tag{2.9}$$

By inspection of Fig.2.1 and taking into account (2.9) we see that $\tilde{g}(\tau)$ is given by integrating along the contour shown in Fig.2.1.

For $\tau>0(<0)$ the paths for $g^{\pm}$ can be closed by an infinite semicircle in the lower (upper)ω-half plane. Taking into account that all singularities of $g(\omega)$ or $G(\omega/c)$ are on the real axis we have from Fig.2.1 that

$$g^{\pm}(\tau) = \pm\,\theta(\pm\tau)\tilde{g}(\tau) \tag{2.10}$$

where $\theta(\tau)$ is defined as

$$\begin{aligned}\theta(\tau) &= 1 \quad ; \quad \tau>0\\ &= 0 \quad ; \quad \tau<0 \quad .\end{aligned} \tag{2.11}$$

Taking into account (1.16) and (2.8) we obtain

$$g^{-}(\underline{r},\underline{r}',\tau) = [g^{+}(\underline{r}',\underline{r},-\tau)]^{*} \quad . \tag{2.12}$$

From Fig.2.1, we can see that the quantity $\tilde{g}(\tau)$ can be written as

$$\begin{aligned}\tilde{g}(\tau) &= \int_{-\infty}^{\infty} \frac{d\omega'}{2\pi}\, e^{-i\omega'\tau}\tilde{G}\left(\frac{\omega'}{c}\right)\\ &= -2\pi i \int_{-\infty}^{\infty} \frac{d\omega'}{2\pi}\, e^{-i\omega'\tau} \sum_{n} \delta\left(\frac{\omega'}{c} - \lambda_n\right)\phi_n(\underline{r})\phi_n^{*}(\underline{r}')\\ &= -ic \sum_{n} e^{-ic\lambda_n\tau}\, \phi_n(\underline{r})\phi_n^{*}(\underline{r}') \quad .\end{aligned} \tag{2.13}$$

We have used (1.21,22) to express the difference $G^{+} - G^{-} \equiv \tilde{G}$ in terms of λ_n, $\phi_n(\underline{r})$.

The corresponding operator $\tilde{\mathcal{g}}(\tau)$ is then

$$\mathcal{g}(\tau) = - ic \sum_n e^{-ic\lambda_n\tau} |\phi_n><\phi_n| = - ice^{-ic\mathcal{L}\tau} \quad ; \tag{2.14}$$

the operator

$$\mathcal{U}(t - t') \equiv \exp[-ic\mathcal{L}(t - t')] \tag{2.15}$$

is a time-evolution operator or a propagator because

$$|\phi(t)> = \mathcal{U}(t - t')|\phi(t')> \quad , \tag{2.16}$$

where $|\phi(t)>$ satisfies (2.1). Thus the operator $\mathcal{U}(t - t')$ propagates $|\phi>$ from time t' to t. We see that

$$\mathcal{U}(t - t') = \frac{i}{c}\,\tilde{\mathcal{g}}(t - t') \quad ; \tag{2.17}$$

in the $\underline{r}$-representation (2.17) becomes when t = t'

$$\tilde{g}(\underline{r},\underline{r}',0) = - ic\delta(\underline{r} - \underline{r}') \quad . \tag{2.18}$$

Note that

$$\mathcal{U}(t_1 - t_2) = \mathcal{U}(t_1 - t_3)\mathcal{U}(t_3 - t_2) \quad ; \tag{2.19}$$

$\mathcal{g}(t_1 - t_2)$ obeys a similar relation. Combining (2.16,17) and transforming to the $\underline{r}$-representation we obtain

$$\phi(\underline{r},t) = \frac{i}{c} \int \tilde{g}(\underline{r},\underline{r}', t-t')\phi(\underline{r}',t')d\underline{r}' \tag{2.20}$$

which, with the help of (2.10), can be expressed in terms of $g^{\pm}(\tau)$.

The solution of the inhomogeneous equation (2.2), $\psi(\underline{r},t)$, can be expressed in terms of the general solution of the homogeneous equation (2.1), $\phi(\underline{r},t)$, and the Green's function $g^{+}(t - t')$ as follows

$$\psi(\underline{r},t) = \phi(\underline{r},t) + \int d\underline{r}'\, dt'\, g^{+}(\underline{r},\underline{r}',t - t')f(\underline{r}',t') \quad . \tag{2.21}$$

To prove this, substitute (2.21) in (2.2) and take into account (2.1) and (2.3). Note that if we had used $g^{-}(\tau)$ instead of $g^{+}(\tau)$, the resulting $\psi(\underline{r},t)$ would again have satisfied (2.2). We have excluded the solution corresponding to $g^{-}(\tau)$ on the basis of the physical argument that the response of a system at a time t depends only on what the source was in the past (t'<t). Using (2.10) we can rewrite (2.21) as

$$\psi(\underline{r},t) = \phi(\underline{r},t) + \int d\underline{r}' \int_{-\infty}^{t} dt'\ \tilde{g}(\underline{r},\underline{r}',t - t')f(\underline{r}',t') \quad . \tag{2.22}$$

2.2 Examples

Here we calculate the various Green's functions for the case $L = -\nabla^2$. It is enough to calculate $\tilde{g}(\tau)$; $g^{\pm}(\tau)$ can be obtained from (2.10). For the present case the most convenient way of calculating $\tilde{g}(\tau)$ is (2.13); $\phi_n(\underline{r}) = \exp(i\underline{k}\underline{r})/\sqrt{\Omega}$, and $\lambda_n = k^2$. We have

$$\tilde{g}(\underline{r},\underline{r}',\tau) = -\ ic \sum_{\underline{k}} \frac{e^{i\underline{k}(\underline{r} - \underline{r}')}}{\Omega} e^{-ick^2\tau}$$

$$= -\ ic \int \frac{d^dk}{(2\pi)^d}\ e^{i\underline{k}\underline{\rho}-ick^2\tau} \quad , \tag{2.23}$$

where $\rho \equiv \underline{r} - \underline{r}'$ and d is the dimensionality. Taking into account that $\underline{k}\underline{\rho} = \sum_{i=1}^{d} k_i\rho_i$ and $k^2 = \sum_{i=1}^{d} k_i^2$, where k_i,ρ_i are the cartesian coordinates of $\underline{k}$ and $\underline{\rho}$, respectively, we can rewrite (2.23) as

$$\tilde{g}(r,r',\tau) = -ic \prod_{i=1}^{d} \int_{-\infty}^{\infty} \frac{dk_i}{2\pi}\ e^{ik_i\rho_i - ick_i^2\tau}$$

$$= -\ ic \prod_{i=1}^{d} \int_{-\infty}^{\infty} \frac{dk}{2\pi}\ e^{i\rho_i^2/4c\tau}\ e^{-ic\tau k^2}$$

$$= -\ ic \prod_{i=1}^{d}\ e^{i\rho_i^2/4c\tau} \frac{1}{2\pi} \sqrt{\frac{\pi}{ic\tau}} \tag{2.24}$$

$$= -\ ic \left(\frac{1}{4\pi ic\tau}\right)^{d/2} \exp(i\rho^2/4c\tau) \quad ,$$

where $\sqrt{\pi/ic\tau}$ is the square root with a positive real part. Equation (2.24), together with (2.20), allow us to study how a free wave-mechanical packet evolves in time. Examples of particular interest are the evolution of the minimum uncertainty wave packet (see, e.g., [2.1]) and the spread of an initially completely localized quantum-mechanical particle.

Note that if we choose $c = -i\kappa$, where κ is a positive constant, and $L = -\nabla^2$, we obtain the diffusion equation $-\partial\phi/\kappa\partial t + \nabla^2\phi = 0$. For such an equation we must consider the evolution towards the future and as a result we should keep $g^+(\tau)$ only. Instead of the Fourier transform it is more convenient to consider the Laplace transform. The final result for $g^+(\tau)$ is obtained by substituting $c = -i\kappa$ in (2.24). We have

$$g^+(\underline{r},\underline{r}',t - t') = - \kappa\theta(t - t')\left(\frac{1}{4\pi\kappa(t-t')}\right)^{d/2} \exp[-(\underline{r} - \underline{r}')^2/4\kappa(t - t')] \quad . \tag{2.25}$$

Note that $g^+ \to -\kappa\delta(\underline{r} - \underline{r}')$ as $t - t' \to 0^+$. As t increases, g^+ describes the diffusion of this local initial pulse. Note further that the average diffusion range increases as the square root of $t - t'$. For further details see [Ref.1.3,I, p.862] or [Ref.1.2, p.447].

2.3 Second-Order Case

The Green's functions associated with the second-order (in time) differential equations

$$\left[-\frac{1}{c^2}\frac{\partial^2}{\partial t^2} - L(\underline{r})\right]\phi(\underline{r},t) = 0 \tag{2.26}$$

$$\left[-\frac{1}{c^2}\frac{\partial^2}{\partial t^2} - L(\underline{r})\right]\psi(\underline{r},t) = f(\underline{r},t) \tag{2.27}$$

are defined as the solution of

$$\left[-\frac{1}{c^2}\frac{\partial^2}{\partial t^2} - L(\underline{r})\right]g(\underline{r},\underline{r}',t - t') = \delta(\underline{r} - \underline{r}')\delta(t - t') \tag{2.28}$$

subject to the same boundary conditions on S as $\phi(\underline{r})$ and $\psi(\underline{r})$; c is a positive constant. Expressing the general solution $g(\tau)$ of (2.28) as

$$g(\tau) = \int_{-\infty}^{\infty} \frac{d\omega'}{2\pi} e^{-i\omega'\tau} g(\omega') \tag{2.29}$$

and substituting in (2.28) we obtain for $g(\omega)$

$$g(\omega) = G(\omega^2/c^2) \quad . \tag{2.30}$$

Since G(z) is analytic on the complex plane except on the real z-axis, it follows that $g(\omega)$ is analytic on the complex ω-plane except on the real or imaginary ω-axes. The singularities of $g(\omega)$ on the real (imaginary) ω-axis come from the singularities of G(z) on the positive (negative) real semi-axis. Here we will restrict ourselves on physical grounds to the case where there are no singularities off the real ω-axis (i.e., all eigenfrequencies ω_n are real) which means that the singularities of G(z) are located on the positive real z-semiaxis.

Since G(z) has singularities for z real and positive, $g(\omega)$ may not be well defined when ω is real. We again need to employ a limiting procedure:

$$g^C(\tau) = \lim_{C \to C_0} \int_C \frac{d\omega}{2\pi} g(\omega) e^{-i\omega\tau} = \lim_{C \to C_0} \int_C \frac{d\omega}{2\pi} G\left(\frac{\omega^2}{c^2}\right) e^{-i\omega\tau} \quad . \tag{2.31}$$

Of all the infinite choices for C, only three are of physical interest. These three choices are shown in Fig.2.2. Note that the fourth case shown in Fig.2.2 can be expressed in terms of the other three because

$$g^R + g^A = g + g^- \tag{2.32}$$

as can be seen from Fig.2.2. Usually we take g, g^R, and g^A as the three basic Green's functions; $g(\tau)$ is called the causal Green's function or simply the Green's function in many body and field theory; as we will see in Part III it is widely used. $g^R(\tau)$ and $g^A(\tau)$ are the retarded and advanced Green's function respectively; they are used in solutions of inhomogeneous equations; $g^R(\tau) = 0$ for $\tau<0$ and $g^A(\tau) = 0$ for $\tau>0$. The Fourier transforms of $g(\tau)$, $g^R(\tau)$, $g^A(\tau)$, $g^-(\tau)$ are given by

$$g(\omega') = \lim_{s \to 0^+} G\left(\frac{\omega'^2}{c^2} + is\right) = G^+\left(\frac{\omega'^2}{c^2}\right) = \lim_{s \to 0^+} g(\omega' + is\bar{\varepsilon}(\omega')) \tag{2.33}$$

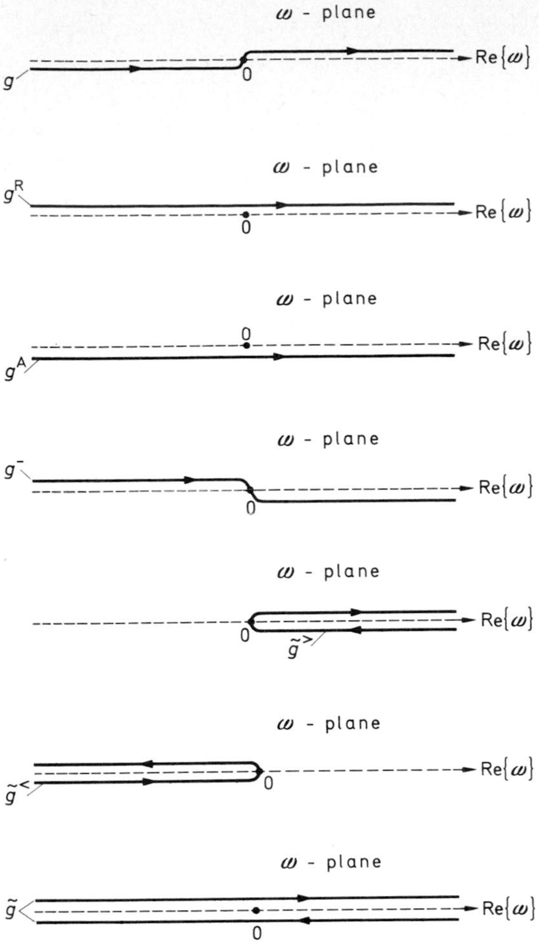

Fig.2.2. Integration paths in the ω-plane for obtaining various Green's functions satisfying a second-order (in time) differential equation. The singularities of the integrand are located on the real ω-axis. The quantities $\tilde{g}^{\gtrless}$, $\tilde{g}$ are differences of pairs of Green's function (see text)

$$g^R(\omega') = \lim_{s\to 0^+} G\left(\frac{\omega'^2}{c^2} + is\bar{\varepsilon}(\omega')\right) = \lim_{s\to 0^+} g(\omega' + is) \tag{2.34}$$

$$g^A(\omega') = \lim_{s\to 0^+} G\left(\frac{\omega'^2}{c^2} - is\bar{\varepsilon}(\omega')\right) = \lim_{s\to 0^+} g(\omega' - is) \tag{2.35}$$

$$g^-(\omega') = \lim_{s\to 0^+} G\left(\frac{\omega'^2}{c^2} - is\right) = G^-\left(\frac{\omega'^2}{c^2}\right) = \lim_{s\to 0^+} g(\omega' - is\bar{\varepsilon}(\omega')) \tag{2.36}$$

where

$$\bar{\varepsilon}(x) \equiv \theta(x) - \theta(-x) = 1 \ ; \quad x>0$$

$$= -1 \ ; \quad x<0 \tag{2.37}$$

and ω' is real.

Since there are three independent g's of interest we can define three $\tilde{g}$'s, each as a difference of a pair of g's. More explicitly we have

$$\tilde{g}^{>} = g - g^{A} \quad , \tag{2.38}$$

$$\tilde{g}^{<} = g - g^{R} \quad , \tag{2.39}$$

$$\tilde{g} = g^{R} - g^{A} = \tilde{g}^{>} - \tilde{g}^{<} \quad . \tag{2.40}$$

Obviously $\tilde{g}^{\gtrless}(\tau)$ and $\tilde{g}(\tau)$ can be obtained be integrating $g(\omega)\exp(-i\omega\tau)/2\pi$ along the contours shown in Fig.2.2. We remind the reader that the $\tilde{g}$'s satisfy the homogeneous equation (2.26) and not (2.28).

Taking into account that the path for the various g's can be closed in the lower (upper) ω-plane when τ is larger (smaller) than zero, we obtain by inspection of Fig.2.2

$$g(\tau) = \theta(\tau)\tilde{g}^{>}(\tau) + \theta(-\tau)\tilde{g}^{<}(\tau) \tag{2.41}$$

$$g^{R}(\tau) = \theta(\tau)\,\tilde{g}(\tau) \tag{2.42}$$

$$g^{A}(\tau) = -\theta(-\tau)\tilde{g}(\tau) \tag{2.43}$$

$$g^{-}(\tau) = -\theta(\tau)\tilde{g}^{<}(\tau) - \theta(-\tau)\tilde{g}^{>}(\tau) \quad . \tag{2.44}$$

As we can see from (2.40-44), knowledge of $\tilde{g}^{>}$, $\tilde{g}^{<}$ allows us to determine all g's and $\tilde{g}$.

From Fig.2.2 we can see that

$$\begin{aligned}\tilde{g}^{>}(\underline{r},\underline{r}',\tau) &= \int_0^\infty \frac{d\omega}{2\pi}\, e^{-i\omega\tau}\, \tilde{G}\left(\frac{\omega^2}{c^2}\right) \\ &= -\int_0^\infty \frac{d\omega}{2\pi}\, e^{-i\omega\tau}\, 2\pi i \sum_n \phi_n(\underline{r})\phi_n^*(\underline{r}')\,\delta\left(\frac{\omega^2}{c^2} - \lambda_n\right) \\ &= \frac{-ic}{2}\sum_n \frac{\phi_n(\underline{r})\phi_n^*(\underline{r}')}{\sqrt{\lambda_n}}\, e^{-ic\sqrt{\lambda_n}\tau}\end{aligned} \tag{2.45}$$

where $\sqrt{\lambda_n} \geq 0$. To obtain (2.45) we have used (1.21,22) and (2.30). Similarly, we have

$$\tilde{g}^{<}(\underline{r},\underline{r}',\tau) = \frac{-ic}{2} \sum_n \frac{\phi_n(\underline{r})\phi_n^*(\underline{r}')}{\sqrt{\lambda_n}} e^{ic\sqrt{\lambda_n}\tau} \quad . \tag{2.46}$$

From (2.45) and (2.46) it follows that

$$\tilde{g}^{<}(\underline{r},\underline{r}',\tau) = -[\tilde{g}^{>}(\underline{r}',\underline{r},\tau)]^* \tag{2.47}$$

$$\tilde{g}(\underline{r},\underline{r}',\tau) = -c\sum_n \frac{\phi_n(\underline{r})\phi_n^*(\underline{r}')}{\sqrt{\lambda_n}} \sin(c\sqrt{\lambda_n}\tau) \quad . \tag{2.48}$$

Equation (2.48) in operator form is

$$\tilde{\mathcal{g}}(\tau) = - c \, \frac{\sin(c\sqrt{L}\tau)}{\sqrt{L}} \; ; \; \tau = t - t' \quad . \tag{2.49}$$

Consider now the expression

$$|\phi(t)\rangle = - \frac{1}{c^2} [\tilde{\mathcal{g}}(t - t')|\dot{\phi}(t')\rangle + \dot{\tilde{\mathcal{g}}}(t - t')|\phi(t')\rangle] \tag{2.50}$$

where the dot denotes differentiation with respect to t and $\dot{\phi}(t')$ is $d\phi/dt$ for $t = t'$. Since $\tilde{g}(t - t')$ satisfies the homogeneous equation (2.26) so does the function $|\phi(t)\rangle$ given by (2.50). Furthermore, $\phi(t) \to \phi(t')$ and $\dot{\phi}(t) \to \dot{\phi}(t')$ as $t \to t'$. Thus (2.50) determines the solution of (2.26) for an arbitrary time t in terms of $\phi(t')$ and $\dot{\phi}(t')$ at a particular time t'. Rewriting (2.50) in the $\underline{r}$-representation we have

$$\phi(\underline{r},t) = - \frac{1}{c^2} \int d\underline{r}'\tilde{g}(\underline{r},\underline{r}',t - t')\dot{\phi}(\underline{r}',t') - \frac{1}{c^2} \int d\underline{r}'\dot{\tilde{g}}(\underline{r},\underline{r}',t-t')\phi(\underline{r}',t'). \tag{2.51}$$

It is easy to verify that

$$\begin{aligned} \psi(\underline{r},t) &= \phi(\underline{r},t) + \int d\underline{r}'dt'g^R(\underline{r},\underline{r}',t - t')f(\underline{r}',t') \\ &= \phi(\underline{r},t) + \int d\underline{r}' \int_{-\infty}^{t} dt'\tilde{g}(\underline{r},\underline{r}',t - t')f(\underline{r}',t') \end{aligned} \tag{2.52}$$

satisfies the inhomogeneous equation (2.27) where $\phi(\underline{r},t)$ is the general solution of (2.26). We have used g^R in (2.52) because of the physical argument that the response $\psi(\underline{r},t)$ at time t depends on the values of the source at times t' prior to t.

2.4 Examples

Consider first the case $L = -\nabla^2$ in 3-D space for which $G^{\pm}(\lambda)$ are given by (1.40). Substituting in (2.45) we obtain

$$\tilde{g}^{>}(\rho,\tau) = -\frac{1}{4\pi\rho}\int_0^{\infty}\frac{d\omega}{2\pi}\, e^{-i\omega\tau}(e^{i\omega\rho/c} - e^{-i\omega\rho/c}) \tag{2.53}$$

where $\underline{\rho} = \underline{r} - \underline{r}'$. Using (2.47) we have

$$\tilde{g}^{<}(\rho,\tau) = -\frac{1}{4\pi\rho}\int_0^{\infty}\frac{d\omega}{2\pi}\, e^{i\omega\tau}(e^{i\omega\rho/c} - e^{-i\omega\rho/c}) \quad . \tag{2.54}$$

Subtracting (2.54) from (2.53) we get

$$\begin{aligned}
\tilde{g}(\rho,\tau) &= \frac{1}{4\pi\rho}\int_0^{\infty}\frac{d\omega}{2\pi}\,(e^{i\omega\rho/c} - e^{-i\omega\rho/c})(e^{+i\omega\tau} - e^{-i\omega\tau}) \\
&= \frac{1}{4\pi\rho}\int_{-\infty}^{\infty}\frac{d\omega}{2\pi}\left(e^{i\omega(\frac{\rho}{c} + \tau)} - e^{i\omega(\frac{\rho}{c} - \tau)}\right) \\
&= \frac{1}{4\pi\rho}\left[\delta(\tfrac{\rho}{c} + \tau) - \delta(\tfrac{\rho}{c} - \tau)\right] \\
&= \frac{c}{4\pi\rho}\left[\delta(\rho + c\tau) - \delta(\rho - c\tau)\right]
\end{aligned} \tag{2.55}$$

where $\rho = |\underline{r} - \underline{r}'|$ and $\tau = t - t'$.

Using (2.55) and (2.42,43) we obtain

$$g^R(\underline{r},\underline{r}',t - t') = -\frac{c}{4\pi|\underline{r} - \underline{r}'|}\,\delta\Big(|\underline{r} - \underline{r}'| - c(t - t')\Big) \tag{2.56}$$

$$g^A(\underline{r},\underline{r}',t - t') = -\frac{c}{4\pi|\underline{r} - \underline{r}'|}\,\delta\Big(|\underline{r} - \underline{r}'| + c(t - t')\Big) \quad . \tag{2.57}$$

The solution of the inhomogeneous wave equation $(\nabla^2 - \partial^2/c^2\partial t^2)\psi(\underline{r},t) = f(\underline{r},t)$ is then according to (2.52)

$$\psi(\underline{r},t) = \phi(\underline{r},t) - \frac{1}{4\pi}\int d\underline{r}'dt'\delta\left(\frac{\rho}{c} - (t - t')\right)f(\underline{r}',t')/\rho$$

$$= \phi(\underline{r},t) - \frac{1}{4\pi}\int d\underline{r}' \frac{f(\underline{r}',t - |\underline{r} - \underline{r}'|/c)}{|\underline{r} - \underline{r}'|} \quad . \tag{2.58}$$

Equation (2.58) is the basic result in electromagnetic theory.

In order to obtain the Green's functions associated with the 2-D wave equation we will use the following argument [1.3]. A point source at the point (x_1', x_2') of a two-dimensional space is equivalent to a uniformly distributed source along a line parallel to the x_3-axis and passing through the point $(x_1', x_2', 0)$. Hence, the 2-D g can be obtained by integrating the 3-D g over the third component of $\underline{r}'$, x_3'. Writing $\underline{r} = \underline{R} + x_3\underline{i}_3$, $\underline{r}' = \underline{R}'+x_3'\underline{i}_3$ and $\rho^2 = (\underline{R} - \underline{R}')^2 + (x_3 - x_3')^2$, we obtain for the 2-D $\tilde{g}$

$$\tilde{g}(\underline{R},\underline{R}',\tau) = \int_{-\infty}^{\infty} dx_3' \, \tilde{g}(\rho,\tau) = \int_{-\infty}^{\infty} dy \, \tilde{g}(\sqrt{P^2 + y^2},\tau) \quad , \tag{2.59}$$

where $P = |\underline{R} - \underline{R}'|$ and $y = x_3 - x_3'$. Substituting (2.55) into (2.59) and performing the integration, we get the final result for the 2-D $\tilde{g}$

$$\tilde{g}(\underline{R},\underline{R}',\tau) = - \bar{\varepsilon}(\tau)\theta(c|\tau| - P)c/2\pi\sqrt{c^2\tau^2 - P^2} \quad . \tag{2.60}$$

The 2-D g^R is then

$$g^R(\underline{R},\underline{R}',\tau) = - \theta(c\tau - P)c/2\pi\sqrt{c^2\tau^2 - P^2} \quad . \tag{2.61}$$

The 1-D g^R can be obtained by integrating the 2-D g^R once more over x_2'. We find

$$g^R(x,x',\tau) = - \frac{c}{2}\,\theta(c\tau - |x - x'|) \quad . \tag{2.62}$$

A very interesting discussion of the physical significance of the results for g^R associated with the wave equation in 3-D, 2-D and 1-D is given in [Ref. 1.3,I, pp.842-848].

We will conclude this chapter by obtaining the various Green's functions associated with the 3-D Klein-Gordon equation, which corresponds to $L(\underline{r})$ being equal to $-\nabla^2 + \mu^2$ where μ is a positive constant. The eigenvalues are $\lambda_n = k^2 + \mu^2$ and the eigenfunctions $\langle \underline{r}|\phi_n\rangle = \exp(i\underline{k}\underline{r})/\sqrt{\Omega}$. Substituting in (2.46) we obtain

$$\tilde{g}^<(\underline{r},\underline{r}',\tau) = \frac{-ic}{2} \int \frac{d^3k}{(2\pi)^3} \frac{e^{i(\underline{k}\underline{\rho} + ck_0\tau)}}{k_0} \quad , \tag{2.63}$$

where $k_0 = \sqrt{\lambda_n} = \sqrt{k^2 + \mu^2}$ and $\underline{\rho} = \underline{r} - \underline{r}'$. After performing the integration over the angle variables, we obtain

$$\tilde{g}^<(\underline{r},\underline{r}',\tau) = - \frac{c}{4\pi\rho} \int_{-\infty}^{\infty} \frac{dk}{2\pi} \frac{k}{k_0} \; e^{i(k\rho + ck_0\tau)} = \frac{c}{4\pi\rho} \frac{\partial f}{\partial\rho} \quad , \tag{2.64}$$

where

$$f = i \int_{-\infty}^{\infty} \frac{dk}{2\pi} \frac{e^{i(k\rho + ck_0\tau)}}{k_0} \quad ; \tag{2.65}$$

changing the integration variable in (2.65) to ϕ where

$$k = \mu \sinh(\phi) \quad ; \quad k_0 = \mu \cosh(\phi)$$

we obtain

$$f = \frac{i}{2\pi} \int_{-\infty}^{\infty} d\phi \; e^{i\mu[\rho \sinh(\phi) + c\tau \cosh(\phi)]} \quad . \tag{2.66}$$

In evaluating the integral (2.66) one has to distinguish four cases according to the signs of $c^2\tau^2 - \rho^2$ and τ. For example, for $\tau>0$ and $c^2\tau^2 - \rho^2 > 0$ the quantity $\rho \sinh(\phi) + c\tau \cosh(\phi)$ can be written as $\sqrt{c^2\tau^2 - \rho^2} \cosh(\phi + \phi_0)$ where $\tanh(\phi_0) = \rho/c\tau$. Substituting in (2.66) we obtain

$$f = \frac{i}{2\pi} \int_{-\infty}^{\infty} d\phi \; \exp[i\mu\sqrt{c^2\tau^2 - \rho^2} \cosh(\phi + \phi_0)] = - \frac{1}{2} H_0^{(1)}(\mu\sqrt{c^2\tau^2 - \rho^2}) \quad . \tag{2.67}$$

Similar expressions are obtained for the other three cases. Performing the differentiation with respect to ρ(keeping in mind the discontinuities of f at $c^2\tau^2 = \rho^2$ which produce δ functions), we obtain finally for $\tilde{g}^<$ the following expression:

$$\tilde{g}^<(\underline{r},\underline{r}',t - t') = \bar{\varepsilon}(\tau)\delta(\nu)\,\frac{c}{4\pi} - \theta(\nu)\bar{\varepsilon}(\tau)\,\frac{\mu c}{8\pi\sqrt{\nu}}\,J_1(\mu\sqrt{\nu})$$

$$- \theta(\nu)\,\frac{i\mu c}{8\pi\sqrt{\nu}}\,Y_1(\mu\sqrt{\nu}) + \theta(-\nu)\,\frac{i\mu c}{4\pi^2\sqrt{-\nu}}\,K_1(\mu\sqrt{-\nu}) \qquad (2.68)$$

where

$$\nu = c^2\tau^2 - \rho^2 \quad ; \quad \tau = t - t' \quad ; \quad \underline{\rho} = \underline{r} - \underline{r}' \quad . \qquad (2.69)$$

From (2.47) we obtain for $\tilde{g}^>\,(\underline{r},\underline{r}',t - t')$

$$\tilde{g}^>(\underline{r},\underline{r}',t - t') = - \operatorname{Re}\{\tilde{g}^<(\underline{r},\underline{r}',t - t')\} + i\operatorname{Im}\{\tilde{g}^<(\underline{r},\underline{r}',t - t')\} \quad . \qquad (2.70)$$

The rest of the g's and $\tilde{g}$'s are then

$$\tilde{g}(\underline{r},\underline{r}',t - t') = - \bar{\varepsilon}(\tau)\delta(\nu)\,\frac{c}{2\pi} + \theta(\nu)\bar{\varepsilon}(\tau)\,\frac{\mu c}{4\pi\sqrt{\nu}}\,J_1(\mu\sqrt{\nu}) \qquad (2.71)$$

$$g(\underline{r},\underline{r}',t - t') = - \delta(\nu)\,\frac{c}{4\pi} + \theta(\nu)\,\frac{\mu c}{8\pi\sqrt{\nu}}\,J_1(\mu\sqrt{\nu}) + i\operatorname{Im}\{\tilde{g}^<(\underline{r},\underline{r}',t - t')\} \qquad (2.72)$$

$$g^R(\underline{r},\underline{r}',t - t') = - \theta(\tau)\,\frac{c}{2\pi}\left[\delta(\nu) - \frac{\mu}{2\sqrt{\nu}}\,\theta(\nu)\,J_1(\mu\sqrt{\nu})\right] \qquad (2.73)$$

$$g^A(\underline{r},\underline{r}',t - t') = - \theta(-\tau)\,\frac{c}{2\pi}\left[\delta(\nu) - \frac{\mu}{2\sqrt{\nu}}\,\theta(\nu)\,J_1(\mu\sqrt{\nu})\right] \quad . \qquad (2.74)$$

As $\mu \to 0^+$, the above expressions reduce to those obtained for the wave equation. To obtain this reduction, take into account that $\delta(\nu) = \delta(c^2\tau^2 - \rho^2) = [\delta(c\tau - \rho) + \delta(c\tau + \rho)]/2\rho$. More information about the g's and $\tilde{g}$'s associated with various relativistic equations can be found in [2.2]. Note the following correspondence of our notation with that of [2.2]: $g^R \to -D^{ret}$, $g^A \to -D^{adv}$, $g \to -D^c$, $\tilde{g} \to -D$, $\tilde{g}^> \to -D^-$, $\tilde{g}^< \to D^+$.

2.5 Summary

Definition

The Green's function g associated with a first-order (in time) partial differential equation of the form $i\partial\phi/c\partial t - L(\underline{r})\phi = 0$ is defined as solution of the equation

$$\left[\frac{i}{c}\frac{\partial}{\partial t} - L(\underline{r})\right] g(\underline{r},\underline{r}',t,t') = \delta(\underline{r} - \underline{r}')\delta(t - t') \tag{2.3}$$

subject to certain homogeneous boundary conditions on the surface S of the domain Ω of $\underline{r},\underline{r}'$. Here $L(\underline{r})$ is a linear, Hermitian, time-independent operator possessing a complete set of eigenfunctions as in Chap.1. The constant c is either real (in which case it will be taken as positive without loss of generality) or imaginary. The former case corresponds to a Schrödinger type equation while the latter to a diffusion type equation.

Basic Properties

The Green's function g is a function of the difference $\tau \equiv t - t'$. The Fourier transform of $g(\underline{r},\underline{r}',\tau)$ with respect to τ, $g(\underline{r},\underline{r}';\omega)$ is directly related with the time-independent Green's function examined in Chap.1. More explicitly $g(\underline{r},\underline{r}';\omega) = G(\underline{r},\underline{r}';\omega/c) \equiv \langle\underline{r}|(\omega/c - L)^{-1}|\underline{r}'\rangle$. However, for real ω, $G(\omega/c)$ may not be well defined; thus we must obtain $g(\tau)$ by a limiting procedure of the type

$$g^C(\tau) = \lim_{C\to C_0} \int_C \frac{d\omega}{2\pi} G\left(\frac{\omega}{c}\right) e^{-i\omega\tau} \tag{2.7}$$

as the integration path C approaches the real ω-axis C_0. One can obtain infinitely many $g(\tau)$'s depending on how C approaches the real axis. However, there are only two "natural" (i.e., corresponding to situations of physical interest) choices of C shown in Fig.2.1; the Green's functions corresponding to these two paths, $g^{\pm}(\tau)$, are then given by

$$g^{\pm}(\tau) = \int_{-\infty}^{\infty} \frac{d\omega'}{2\pi} G^{\pm}\left(\frac{\omega'}{c}\right) e^{-i\omega'\tau} \quad ; \tag{2.8}$$

For $\tau>0$ ($\tau<0$) we can close the integration paths with an infinite semicircle in the lower (upper) ω-half plane. Consequently, $g^+(\tau) = 0$ for $\tau<0$ and $g^-(\tau) = 0$ for $\tau>0$.

Some particular examples of $g(\tau)$ associated with $L = -\nabla^2$ were presented in Sect.2.2.

Definition

The Green's function $g(\tau)$ associated with a second-order (in time) differential equation is defined as the solution of

$$\left[-\frac{1}{c^2}\frac{\partial^2}{\partial t^2} - L(\underline{r})\right] g(\underline{r},\underline{r}',\tau) = \delta(\underline{r} - \underline{r}')\delta(\tau) \quad , \tag{2.28}$$

where $\tau \equiv t - t'$.

Basic Properties

The Fourier transform of $g(\tau)$, $g(\omega)$, is related to $G(z)$ defined in Chap.1 by

$$g(\omega) = G\left(\frac{\omega^2}{c^2}\right) \quad . \tag{2.30}$$

Because of the singularities of $G(z)$ on the real axis one needs to employ a limiting procedure:

$$g^C(\tau) = \lim_{C \to C_0} \int_C \frac{d\omega}{2\pi} G\left(\frac{\omega^2}{c^2}\right) e^{-i\omega\tau} \quad . \tag{2.31}$$

Again there are infinitely many $g^C(\tau)$ one can obtain depending on the way the path C approaches the real axis C_0. In the present case there are only three independent "natural" (i.e., of physical interest) choices shown in Fig.2.2 as g, g^R, g^A. In particular, $g(\tau)$ or its Fourier transform $g(\omega')$ is the so-called causal Green's function or simply Green's function in many body or field theory. It is given by the Fourier transform of (2.33)

$$g(\tau) = \int_{-\infty}^{\infty} \frac{d\omega'}{2\pi} G^+(\omega'^2/c^2) e^{-i\omega'\tau} \quad ;$$

$g^R(\tau)$ is the retarded Green's function which has the property $g^R(\tau) = 0$ for $\tau<0$; $g^A(\tau)$ is the advanced Green's function which satisfies the relation $g^A(\tau) = 0$ for $\tau>0$; $g^-(\tau) = g^R(\tau) + g^A(\tau) - g(\tau)$. All these Green's functions are interrelated.

Some particular examples associated with $L = -\nabla^2$ or $L = -\nabla^2 + \mu^2$ were presented in Sect.2.4.

Use

Having $g^{\pm}(\tau)$, which satisfies (2.3), we can:

1) Solve the homogeneous equation $i\partial\phi(\underline{r},t)/c\partial t = L(\underline{r})\phi(\underline{r},t)$ as

$$\phi(\underline{r},t) = \frac{i}{c} \int \tilde{g}(\underline{r},\underline{r}',t - t')\phi(\underline{r}'t')d\underline{r}' \tag{2.20}$$

where

$$\tilde{g}(\tau) = g^{+}(\tau) - g^{-}(\tau) \tag{2.9}$$

in terms of the initial value of $\phi, \phi(\underline{r}',t')$.

2) Solve the inhomogeneous equation $i\partial\psi(\underline{r},t)/c\partial t - L(\underline{r})\psi(\underline{r},t) = f(\underline{r},t)$. The function $\psi(\underline{r},t)$ is given by

$$\psi(\underline{r},t) = \phi(\underline{r},t) + \int d\underline{r}'\, dt'\, g^{+}(\underline{r},\underline{r}',t - t')f(\underline{r}',t') \tag{2.21}$$

where $\phi(\underline{r},t)$ is the solution of the homogeneous equation.

3) Use $g_0(\tau)$ and $L_1(t)$ to obtain information about the solution of $i\partial\psi/c\partial t - L_0\psi - L_1\psi = 0$ where g_0 corresponds to L_0. This aspect will be discussed in Chap.4.

4) Relate $g^{\pm}(\tau)$ with the commutators or anticommutators of field operators in quantum field theory. These relations will be given in Part III.

Similarly, the various Green's functions satisfying (2.28) can also be used for solving the corresponding homogeneous and inhomogeneous equations as well as for carrying out a perturbative approach; they are also related to the commutators and anticommutators of field operators.

Part II

Green's Functions in One-Body Quantum Problems

3. Physical Significance of G. Application to the Free Particle Case

The general theory developed in Chap.1 can be applied directly to the time-independent one-particle Schrödinger equation by making the substitutions $L(\underline{r}) \to H(\underline{r})$, $\lambda \to E$, where $H(\underline{r})$ is the Hamiltonian. The formalism presented in Chap.2 is applicable to the time-dependent one-particle Schrödinger equation.

3.1 General Relations

The one-particle time-independent Schrödinger equation has the form

$$[E - H(\underline{r})]\psi(\underline{r}) = 0 \quad , \tag{3.1}$$

and the corresponding Green's function satisfies the equation

$$[E - H(\underline{r})]G(\underline{r},\underline{r}';E) = \delta(\underline{r} - \underline{r}') \quad . \tag{3.2}$$

Here $H(\underline{r})$ is the Hamiltonian operator in the $\underline{r}$-representation and $G(\underline{r},\underline{r}';E)$ as a function of $\underline{r}$ or $\underline{r}'$ satisfies the same boundary conditions as the wave function $\psi(\underline{r})$. It is clear that the general formalism developed in Chap.1 is directly applicable to the present case with the substitutions

$$L(\underline{r}) \longrightarrow H(\underline{r}) \tag{3.3a}$$

$$\lambda \longrightarrow E \tag{3.3b}$$

$$\lambda + is = z \longrightarrow z = E + is \tag{3.3c}$$

$$\lambda_n \longrightarrow E_n \tag{3.3d}$$

$$\phi_n(\underline{r}) \longrightarrow \phi_n(\underline{r}) \quad . \tag{3.3e}$$

Thus, the basic relation expressing G in terms of the eigenvalues E_n and the complete set of orthonormal eigenfunctions ϕ_n of $\mathcal{H}$ is

$$G(\underline{r},\underline{r}';z) = \sum_n \frac{\phi_n(\underline{r})\phi_n^*(\underline{r}')}{z - E_n} \tag{3.4}$$

or, in the bra and ket notation,

$$\mathcal{G}(z) = \sum_n \frac{|\phi_n\rangle\langle\phi_n|}{z - E_n} = \frac{1}{z - \mathcal{H}} . \tag{3.5}$$

The singularities of G(z) versus z are on the real z-axis. They can be used as follows:

1) The position of the poles of G(z) coincide with the discrete eigenenergies corresponding to $\mathcal{H}$, and vice versa.

2) The residue at each pole E_n of $G(\underline{r},\underline{r}';z)$ equals $\sum_i \phi_i(\underline{r})\phi_i^*(\underline{r}')$ where the summation runs over the f_n degenerate eigenstates corresponding to the discrete eigenenergy E_n.

3) The degeneracy f_n can be found by integrating the residue (Res) of the diagonal matrix element $G(\underline{r},\underline{r};E_n)$ over $\underline{r}$, i.e.,

$$\begin{aligned} f_n &= \int d^3r \mathrm{Res}\{G(\underline{r},\underline{r};E_n)\} \\ &= \mathrm{Tr}\{\mathrm{Res}\{\mathcal{G}(E_n)\}\} \quad . \end{aligned} \tag{3.6}$$

For a nondegenerate eigenstate $f_n = 1$, and consequently

$$\phi_n(\underline{r})\phi_n^*(\underline{r}') = \mathrm{Res}\{G(\underline{r},\underline{r}';E_n)\} \quad . \tag{3.7}$$

From (3.7) we see that

$$|\phi_n(\underline{r})| = [\mathrm{Res}\{G(\underline{r},\underline{r};E_n)\}]^{\frac{1}{2}} \tag{3.8}$$

$$p_n(\underline{r}) = -i\ell n \left\{ \frac{\mathrm{Res}\{G(\underline{r},0;E_n)\}}{[\mathrm{Res}\{G(\underline{r},\underline{r};E_n)\} \cdot \mathrm{Res}\{G(0,0;E_n)\}]^{\frac{1}{2}}} \right\} \tag{3.9}$$

where $p_n(\underline{r})$ is the phase of $\phi_n(\underline{r})$ (assuming that the phase of $\phi_n(\underline{r})$ for $\underline{r} = 0$ is zero).

4) The branch cuts of $G(z)$ along the real z-axis coincide with the continuous spectrum of $\mathcal{H}$ and vice versa. (We assume that the continuous spectrum of $\mathcal{H}$ consists of extended or propagating eigenstates.)

5) The density of states per unit volume $\rho(\underline{r},E)$ is given by

$$\rho(\underline{r};E) = \mp \frac{1}{\pi} \operatorname{Im}\{G^{\pm}(\underline{r},\underline{r};E)\} \quad . \tag{3.10}$$

6) The density of states $N(E)$ is given by integrating $\rho(\underline{r};E)$ over $\underline{r}$, i.e.,

$$N(E) = \int d\underline{r}\rho(\underline{r};E)$$

$$= \mp \frac{1}{\pi} \operatorname{Tr}\{\operatorname{Im}\{G^{\pm}(E)\}\} \quad . \tag{3.11}$$

7) Using the results of Chap.2 we can write the solution of the time-dependent Schrödinger equation

$$\left(i\hbar \frac{\partial}{\partial t} - \mathcal{H}\right)|\psi(t)\rangle = 0^{1} \tag{3.12}$$

as follows

$$|\psi(t)\rangle = \mathcal{U}(t - t_0)|\psi(t_0)\rangle \tag{3.13}$$

where the time evolution operator

$$\mathcal{U}(t - t_0) = \exp[- i(t - t_0)\mathcal{H}/\hbar] \tag{3.14}$$

can be expressed simply in terms of the Green's function

$$\mathcal{U}(t - t_0) = i\hbar\tilde{g}(t - t_0) = i\hbar \int_{-\infty}^{\infty} \frac{d\omega}{2\pi} e^{-i\omega(t-t_0)} \tilde{G}(\hbar\omega) \quad . \tag{3.15}$$

Equation (3.13) can be written in the $\underline{r}$-representation as

$$\psi(\underline{r},t) = i\hbar \int \tilde{g}(\underline{r},\underline{r}',t - t_0)\, \psi(\underline{r}',t_0)d\underline{r}' \quad . \tag{3.16}$$

[1] $\hbar \equiv h/2\pi$, where h is Planck's constant.

3.2 The Free-Particle ($H_0 = p^2/2m$) Case

We denote by H_0 the free-particle Hamiltonian

$$H_0 = p^2/2m = -\frac{\hbar^2}{2m}\nabla^2 \tag{3.17}$$

and by $G_0(z)$ the corresponding Green's function

$$\left(z + \frac{\hbar^2}{2m}\nabla_r^2\right) G_0(\underline{r},\underline{r}';z) = \delta(\underline{r}-\underline{r}') \quad . \tag{3.18}$$

Equation (3.18) can be written as

$$\left(\frac{2mz}{\hbar^2} + \nabla_r^2\right)\left[\frac{\hbar^2}{2m} G_0(\underline{r},\underline{r}';z)\right] = \delta(\underline{r}-\underline{r}') \quad . \tag{3.19}$$

Comparing (3.19) with (1.36) we see that

$$G_0(\underline{r},\underline{r}';z) = \frac{2m}{\hbar^2} G\left(\underline{r},\underline{r}';\frac{2mz}{\hbar^2}\right) \quad , \tag{3.20}$$

where $G(\underline{r},\underline{r}';z)$ is the Green's function corresponding to the operator $L = -\nabla_r^2$; this Green's function has been calculated in Sect.1.2. Thus we have for G_0

3-D Case

$$G_0(\underline{r},\underline{r}';E) = -\frac{m}{2\pi\hbar^2}\,\frac{e^{-k_0|\underline{r}-\underline{r}'|}}{|\underline{r}-\underline{r}'|} \quad ; \qquad E \le 0 \tag{3.21}$$

$$G_0^{\pm}(\underline{r},\underline{r}';E) = -\frac{m}{2\pi\hbar^2}\,\frac{e^{\pm ik_0|\underline{r}-\underline{r}'|}}{|\underline{r}-\underline{r}'|} \quad ; \qquad E \ge 0 \tag{3.22}$$

where

$$k_0 = \sqrt{\frac{2m|E|}{\hbar^2}} \ge 0 \quad . \tag{3.23}$$

Here $G_0(z)$ has a branch cut along the positive E-axis; this branch cut corresponds to the continuous spectrum of H_0. The discontinuity of G across the branch cut, $\tilde{G}$, where

$$\tilde{G}(\underline{r},\underline{r}';E) = G^+(\underline{r},\underline{r}';E) - G^-(\underline{r},\underline{r}';E) \quad , \tag{3.24}$$

is given in the present case by

$$\tilde{G}_0(\underline{r},\underline{r}';E) = -2\pi i \frac{m}{2\pi^2\hbar^2} \frac{\sin(k_0|\underline{r}-\underline{r}'|)}{|\underline{r}-\underline{r}'|} \theta(E) \quad . \tag{3.25}$$

In particular the DOS per unit volume is

$$\rho_0(\underline{r};E) = \tilde{G}_0(\underline{r},\underline{r};E)/(-2\pi i) = \frac{mk_0}{2\pi^2\hbar^2} \theta(E) = \theta(E) \frac{m^{3/2}}{\sqrt{2}\pi^2\hbar^3} \sqrt{E} \quad . \tag{3.26}$$

Note that $\rho_0(\underline{r};E)$ does not depend on $\underline{r}$. The reason is the translational invariance of the Hamiltonian H_0.

2-D Case

$$G_0(\underline{r},\underline{r}';E) = -\frac{m}{\pi\hbar^2} K_0(k_0|\underline{r}-\underline{r}'|) \quad ; \quad E<0 \tag{3.27}$$

$$G_0^{\pm}(\underline{r},\underline{r}';E) = -\frac{im}{2\hbar^2} H_0^{(1)}(\pm k_0|\underline{r}-\underline{r}'|) \; ; \quad E>0 \tag{3.28}$$

where k_0 is given by (3.23), K_0 is the zeroth-order modified Bessel function and $H_0^{(1)}$ is the Hankel function of first kind of zero order. The quantity $\tilde{G}_0(\underline{r},\underline{r}';E)$ is then

$$\tilde{G}_0(\underline{r},\underline{r}';E) = -2\pi i\theta(E) \frac{m}{2\pi\hbar^2} J_0(k_0|\underline{r}-\underline{r}'|) \quad , \tag{3.29}$$

where $J_0(x)$ is the Bessel function of first kind of zero order, and k_0 is given by (3.23). To obtain (3.29) from (3.28) we have used the relations [3.1]

$$H_0^{(1)}(-x) = -H_0^{(2)}(x)$$

and

$$H_0^{(1)} = J_0 + iY_0 \quad ; \quad H_0^{(2)} = J_0 - iY_0 \quad ,$$

where Y_o is the Bessel function of second kind of zero order [3.1]. The DOS per unit area is

$$\rho(\underline{r};E) = \tilde{G}_o(\underline{r},\underline{r};E)/(-2\pi i) = \theta(E)\,\frac{m}{2\pi\hbar^2} \quad . \tag{3.30}$$

1-D Case

$$G_o(x,x';E) = -\frac{m}{\hbar^2 k_o}\, e^{-k_o|x-x'|} \quad ; \quad E<0 \tag{3.31}$$

$$G_o^{\pm}(x,x';E) = \mp\frac{im}{\hbar^2 k_o}\, e^{\pm ik_o|x-x'|} \quad ; \quad E>0 \tag{3.32}$$

where k_o is given by (3.23). The quantity $\tilde{G}_o$ is then

$$\tilde{G}_o(x,x';E) = -\,2\pi i\theta(E)\,\frac{m}{\pi\hbar^2 k_o}\,\cos(k_o|x-x'|) \quad . \tag{3.33}$$

The DOS per unit length is

$$\rho_o(x;E) = \tilde{G}_o(x,x;E)/(-2\pi i) = \theta(E)\,\frac{m}{\pi\hbar^2 k_o} = \theta(E)\,\frac{m^{1/2}}{\sqrt{2}\pi\hbar}\,\frac{1}{\sqrt{E}} \quad . \tag{3.34}$$

It must be stressed that the DOS is a quantity of great importance for most branches of physics. The reason is that most quantities of physical interest depend on the DOS. For example, the thermodynamic properties of a system of noninteracting particles can be expressed in terms of the DOS of each particle; transition probabilities, scattering amplitudes, etc., depend on the DOS of final and initial states.

In Fig.3.1 we plot the DOS versus E for a free particle whose motion in 3-D, 2-D and 1-D space is governed by the Schrödinger equation. In all cases the spectrum has a lower bound (at E = 0) below which the DOS is zero. Energies at which the continuous spectrum terminates are called band edges in solid-states physics. We will adopt this name here.

As we shall see later, the behavior of $\rho(E)$, $\tilde{G}(E)$ and $G(z)$ for E or z near a band edge E_B is of great physical interest. It should be noted that the analytic structures of $\rho_o(E)$ and $\tilde{G}_o(E)$ near the band edge $E_B = 0$ are identical [compare (3.25,29,33) with (3.26,30,34) respectively]. Furthermore, the analytic behavior of $\tilde{G}(E)$ or $\rho(E)$ around an arbitrary energy E_o determines the analytical properties of $G(z)$ for z around E_o. (See Appendix 1

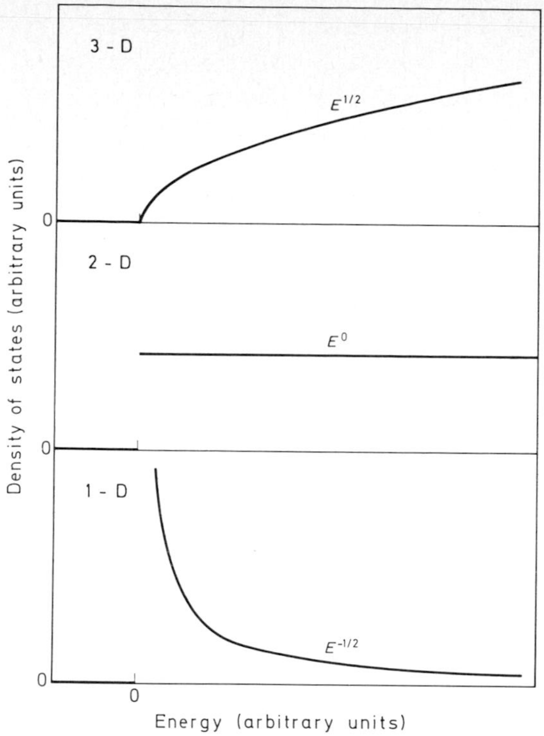

Fig.3.1. Density of states $N(E)$ ($N(E)dE$ gives the number of eigenstates in the energy interval $[E, E+dE]$) vs. E for a free particle obeying Schrödinger equation (energy momentum relation $E = \hbar^2\vec{k}^2/2m$) in a D-dimensional (D = 1,2,3) space

and [3.2].) Thus, if $\tilde{G}(E)$ or $\rho(E)$ is continuous around E_0, $G^{\pm}(E)$ are continuous functions of E for E in the vicinity of E_0; if $\tilde{G}(E)$ [or $\rho(E)$] has a discontinuity at E_0, $G^{\pm}(E)$ exhibit a logarithmic singularity; finally if $\tilde{G}(E)$ [or $\rho(E)$] blows up at E_0 as $(E-E_0)^{-\gamma}$ where $0<\gamma<1$, $G^{\pm}(E)$ also approach infinity as $(E-E_0)^{-\gamma}$.

The present results (see, e.g., Fig.3.1) show that the behavior of $\rho_0(E)$ [and hence $\tilde{G}_0(E)$ and $G_0^{\pm}(E)$] near the band edge $E_B = 0$ depends critically on the dimensionality of the system.

In the 3-D case the DOS and $\tilde{G}(E)$ are continuous functions of E approaching zero like $\sqrt{E - E_B}$ in the limit $E \to E_B^+$, [see (3.25,26)]. As a result of the general theorem stated in Appendix A, $G^{+}(E)$ are continuous functions of E as E crosses the band edge [compare (3.21) with (3.22)].

In the 2-D case the DOS and $\tilde{G}(E)$ are discontinuous at $E = 0$, see (3.29, 30). As a result the Green's function $G(z)$ develops a logarithmic singularity as $z \to 0$ [see (3.27,28) and take into account that $K_0(x) \to -\ln(x)$ and $H_0^{(1)}(\pm x) \to (2i/\pi)\ln(x)$ as $x \to 0^+$].

Finally, in the 1-D case the DOS and $\tilde{G}(E)$ approach infinity as $1/\sqrt{E}$ in the limit $E \to 0^+$. As a result of the general theorems stated in Appendix A, $G(z)$ blows up like $1/\sqrt{z}$ as $z \to 0$, see (3.31,32).

The above statements connecting the dimensionality with the behavior of $G(z)$, $\tilde{G}(E)$ and the DOS near a band edge were based upon the results for the free-particle case. We shall see in Chap.5 that the same connection exists for almost all cases where the Hamiltonian H is periodic.

3.3 The Free-Particle Klein-Gordon Case

In this section we calculate the density of states (i.e., the number of eigenstates in the energy interval [E, E + dE] divided by dE) for a free particle of rest mass m obeying the time-independent Klein-Gordon equation which has the form

$$\left(\frac{E^2}{\hbar^2 c^2} - \mu^2 + \nabla_r^2\right) \psi(\underline{r}) = 0 \tag{3.35}$$

with $\mu = cm/\hbar$. The formalism developed in Chap.1 is directly applicable to the present case if one makes the substitutions

$$L(\underline{r}) \rightarrow -\nabla_r^2 \tag{3.36}$$

$$\lambda \rightarrow \frac{E^2}{\hbar^2 c^2} - \mu^2 \quad . \tag{3.37}$$

The DOS $N(E)$ and $\rho(E)$ with respect to the energy variable E can be obtained from the DOS $N_\lambda(\lambda)$ and $\rho_\lambda(\lambda)$ with respect to the variable λ, by a simple change of variables according to (3.37), i.e.,

$$N(E) = N_\lambda(\lambda(E)) \frac{d\lambda}{dE} = N_\lambda(E^2/\hbar^2 c^2 - \mu^2) \frac{2E}{\hbar^2 c^2} \tag{3.38}$$

$$\rho(E) = \rho_\lambda(\lambda) \frac{d\lambda}{dE} = \rho_\lambda(E^2/\hbar^2 c^2 - \mu^2) \frac{2E}{\hbar^2 c^2} \quad , \tag{3.39}$$

where $N_\lambda(\lambda)$, $\rho_\lambda(\lambda)$ can be obtained immediately by use of (1.27,28,40,49,56). After some simple algebra we obtain for the DOS of a free particle of rest mass m obeying the Klein-Gordon equation

$$\rho(E) = \theta(E - mc^2) \frac{E\sqrt{E^2 - m^2 c^4}}{2\pi^2 \hbar^3 c^3} \quad ; \quad \text{3-D} \quad , \tag{3.40}$$

$$\rho(E) = \theta(E - mc^2) \frac{E}{2\pi\hbar^2 c^2} \quad ; \quad 2\text{-D} \quad , \text{ and} \tag{3.41}$$

$$\rho(E) = \theta(E - mc^2) \frac{E}{\pi\hbar c\sqrt{E^2 - m^2c^4}} \quad ; \quad 1\text{-D} \quad . \tag{3.42}$$

In (3.40-42) we have kept only the positive energy solutions. If we want to keep the negative energy solutions as well, we must replace E by $|E|$ in the above equations.

Note first that (3.40-42) reduce to (3.26,30,34), respectively, by introducing $E' = E - mc^2$ and assuming that $E' \ll mc^2$. Secondly, when $m = 0$, we obtain the case of a free particle obeying the wave equation. Examples of such particles are those resulting by quantizing classical wave equations such as the electromagnetic equations (the corresponding particles are of course the photons and c is the velocity of light) or the equation describing the propagation of sound waves in a continuous medium (the corresponding particles are called phonons and c is the velocity of sound).

Thus, for particles whose energy E is related with their momentum $\hbar|\underline{k}|$ by the relation

$$E = \hbar c|\underline{k}| \quad , \tag{3.43}$$

the DOS is obtained from (3.40-42) by putting $m = 0$. We have explicitly

$$\rho(E) = \theta(E) \frac{E^2}{2\pi^2\hbar^3c^3} \quad ; \quad 3\text{-D} \quad , \tag{3.44}$$

$$\rho(E) = \theta(E) \frac{E}{2\pi\hbar^2c^2} \quad ; \quad 2\text{-D} \quad , \tag{3.45}$$

$$\rho(E) = \frac{\theta(E)}{\pi\hbar c} \quad ; \quad 1\text{-D} \quad . \tag{3.46}$$

It should be noted that the DOS given in (3.40-42) and (3.44-46) are not related with the corresponding Green's functions in the way described in Sect.3.2. The reason is that the relation between λ and E is not the same as in the Schrödinger case but is given by (3.37).

In Fig.3.2 we plot the DOS versus E for the case where the energy versus $\underline{k}$ is given by (3.43).

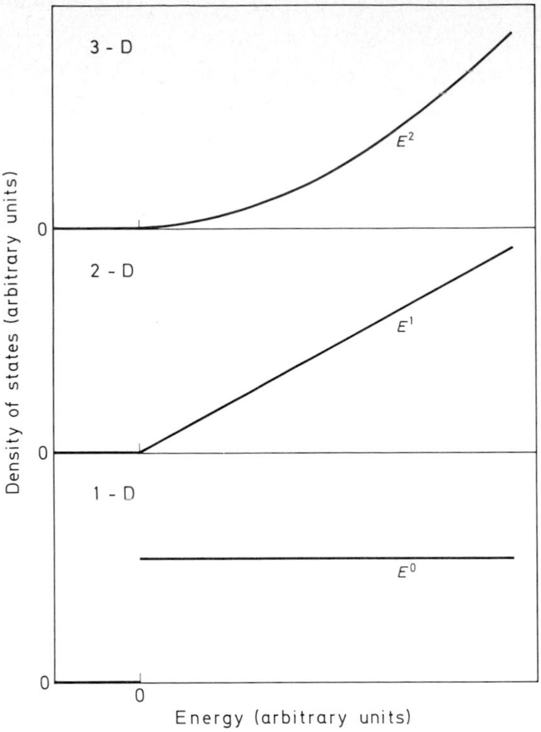

Fig.3.2. Density of states N(E) (N(E)dE gives the number of eigenstates in the energy interval [E, E+dE]) vs. E for a free particle obeying the wave equation (energy momentum relation $E = \hbar c|\vec{k}|$) in a D-dimensional (D = 1,2,3) space

3.4 Summary

Knowledge of $G(z) \equiv (z - H)^{-1}$ permits us to obtain the discrete eigenenergies, the corresponding eigenfunctions, and the density of states in the continuous parts of the spectrum of H. Knowledge of $\tilde{g}(\tau)$ allows us to calculate the time development of the wave function.

For the simple case where $H = H_0 \equiv p^2/2m$, we obtain the density of states per unit volume (area, or length) as follows:

$$\rho_0(E) = \theta(E) \frac{m^{3/2}}{\sqrt{2}\pi^2\hbar^3} \sqrt{E} \quad ; \quad \text{3-D case} \tag{3.26}$$

$$= \theta(E) \frac{m}{2\pi\hbar^2} \quad ; \quad \text{2-D case} \tag{3.30}$$

$$= \theta(E) \frac{m^{1/2}}{\sqrt{2}\pi\hbar} \frac{1}{\sqrt{E}} \quad ; \quad \text{1-D case} \quad . \tag{3.34}$$

The behavior of $\rho_0(E)$ near the bound of the spectrum (E = 0) depends strongly on the dimensionality. The behavior of $\rho(E)$ around an energy E_0 determines the analytical structure of $G^{\pm}(E)$ around E_0. Thus, continuity of $\rho(E)$ (as in the 3-D case) implies continuity of $G^{\pm}(E)$; discontinuity of $\rho(E)$ (as in the 2-D case) implies a logarithmic singularity in $G^{\pm}(E)$; and divergence of $\rho(E)$ (as in the 1-D case) implies divergence of $G^{\pm}(E)$. The quantity $\rho_0(E)$ is plotted in Fig.3.1.

We calculate also the density of states for a free particle of mass m obeying the Klein-Gordon equation. In the particular case where m = 0 we have the wave equation which implies an energy momentum relation of the form $E = \hbar c|\underline{k}|$ and for which the density of states per unit volume (area or length) is

$$\rho(E) = \theta(E)\,\frac{E^2}{2\pi^2\hbar^3c^3}\ ; \quad \text{3-D case} \tag{3.44}$$

$$= \theta(E)\,\frac{E}{2\pi\hbar^2c^2}\ ; \quad \text{2-D case} \tag{3.45}$$

$$= \theta(E)/\pi\hbar c\ ; \quad \text{1-D case}\ ; \tag{3.46}$$

$\rho(E)$ is plotted in Fig.3.2.

4. Green's Functions and Perturbation Theory

The problem of finding the eigenvalues and eigenfunctions of a Hamiltonian $H = H_0 + H_1$ can be solved in three steps: 1) calculate the Green's function $G_0(z)$ corresponding to H_0; 2) express $G(z)$ as a perturbation series in terms of $G_0(z)$ and H_1, where $G(z)$ is the Green's function associated with H; and 3) extract from $G(z)$ information about the eigenvalues and eigenfuctions of H.

4.1 Formalism

4.1.1 Time-Independent Case

In this chapter we consider the very important and common case where the one-particle Hamiltonian H can be separated into an unperturbed part H_0, and a perturbation H_1

$$H = H_0 + H_1 \quad . \tag{4.1}$$

It is implicitly assumed that H_0 is such that its eigenvalues and eigenfunctions can be easily obtained. The question is to determine the eigenvalues and eigenfunctions of H. Very often this goal is achieved by taking the following indirect path:

1) Determine first the Green's function G_0 associated with the unperturbed part H_0;

2) Express the Green's function G associated with the total Hamiltonian H in terms of G_0 and H_1;

3) Obtain information about the eigenvalues and eigenfunctions of H from G.

Step 3) above has been examined in detail in Chap.1 and Sect.3.1. The implementation of step 1) depends on H_0. For the very common and important case where $H_0 = p^2/2m$, G_0 has been obtained in Sect.3.2. In the next chapter a whole class of H_0's will be introduced, and the corresponding G_0's will be calculated. In the present section we examine in some detail step 2), i.e., how G can be expressed in terms of G_0 and H_1.

The Green's functions $G_0(z)$ and $G(z)$ corresponding to H_0 and H, respectively, are

$$G_0(z) = (z - H_0)^{-1} \quad \text{and} \tag{4.2}$$

$$G(z) = (z - H)^{-1} \quad . \tag{4.3}$$

Using (4.1,2) we can rewrite (4.3) as follows

$$\begin{aligned} G(z) &= (z - H_0 - H_1)^{-1} = \{(z - H_0)[1 - (z - H_0)^{-1}H_1]\}^{-1} \\ &= [1 - (z - H_0)^{-1}H_1]^{-1}(z - H_0)^{-1} \\ &= [1 - G_0(z)H_1]^{-1}G_0(z) \quad . \end{aligned} \tag{4.4}$$

Expanding the operator $(1 - G_0H_1)^{-1}$ in power series we obtain

$$G = G_0 + G_0H_1G_0 + G_0H_1G_0H_1G_0 + \dots \quad . \tag{4.5}$$

Equation (4.5) can be written in a compact form

$$\begin{aligned} G &= G_0 + G_0H_1(G_0 + G_0H_1G_0 + \dots) \\ &= G_0 + G_0H_1G \end{aligned} \tag{4.6}$$

or

$$\begin{aligned} G &= G_0 + (G_0 + G_0H_1G_0 + \dots)H_1G_0 \\ &= G_0 + GH_1G_0 \quad . \end{aligned} \tag{4.7}$$

In the $\underline{r}$-representation (4.6) becomes

$$\mathrm{G}(\underline{r},\underline{r}';z) = \mathrm{G}_0(\underline{r},\underline{r}';z) + \int d\underline{r}_1 d\underline{r}_2 \mathrm{G}_0(\underline{r},\underline{r}_1;z)\mathrm{H}_1(\underline{r}_1,\underline{r}_2)\mathrm{G}(\underline{r}_2,\underline{r}';z) \quad . \tag{4.6'}$$

Usually $H_1(\underline{r}_1,\underline{r}_2)$ has the form $\delta(\underline{r}_1 - \underline{r}_2)V(\underline{r}_1)$; then (4.6') becomes

$$G(\underline{r},\underline{r}';z) = G_0(\underline{r},\underline{r}';z) + \int d\underline{r}_1 G_0(\underline{r},\underline{r}_1;z)V(\underline{r}_1)G(\underline{r}_1,\underline{r}';z) \quad , \tag{4.8}$$

i.e., $G(\underline{r},\underline{r}';z)$ satisfies a linear inhomogeneous integral equation with a kernel $G_0(\underline{r},\underline{r}_1;z)V(\underline{r}_1)$. Equation (4.7) can be also written in a similar form. If we use the $\underline{k}$-representation, we can rewrite (4.6) as follows

$$G(\underline{k},\underline{k}';z) = G_0(\underline{k},\underline{k}';z) + \sum_{\underline{k}_1\underline{k}_2} G_0(\underline{k},\underline{k}_1;z)H_1(\underline{k}_1,\underline{k}_2)G(\underline{k}_2,\underline{k}';z) \quad . \tag{4.9}$$

Taking into account that $<\underline{r}|\underline{k}> = \exp(i\underline{k}\underline{r})/\sqrt{\Omega}$ and that

$$\sum_{\underline{k}} = \Omega \int \frac{d\underline{k}}{(2\pi)^d} \quad , \tag{4.10}$$

where d is the dimensionality, one can easily show that (4.9) is the Fourier transform of (4.6'), as it should be.

At this point we introduce the so-called t-matrix, which is of central importance in scattering theory and is directly related with G and G_0. The t-matrix $T(z)$, corresponding to the *unperturbed Hamiltonian* H_0, the *perturbation part* H_1 and the *parameter* z, is defined by the operator equation

$$T(z) \equiv H_1 G(z)(z - H_0) \quad . \tag{4.11}$$

The above definition of $T(z)$ is valid for $z \neq \{E_n\}$ where $\{E_n\}$ are the eigenvalues of H. If z = E where E belongs to the continuous spectrum of H, we define

$$T^{\pm}(E) \equiv H_1 G^{\pm}(E)(E - H_0) \quad . \tag{4.12}$$

Finally, if z coincides with one of the discrete eigenvalues of H, let us say E_n, $T(E_n)$ is not defined because $G(z)$ [and hence $T(z)$] has a simple pole at E_n. This statement is correct except in the pathological case where the eigenvalue E_n and the corresponding eigenfunction $|\psi_n>$ of H satisfies the relation $H_0|\psi_n> = E_n|\psi_n>$. In this case, the pole of $G(z)$ at E_n is cancelled by the zero of $z - H_0$ at E_n, and $T(z)$ is analytic around E_n.

The analytic structure of $T(z)$ is quite similar to $G(z)$: $T(z)$ is analytic in the complex z-plane; it has singularities on the real z-axis. The positions of the poles of $T(z)$ on the real axis give the discrete eigenvalues of H and vice versa. The continuous spectrum of H produces a branch cut in $T(z)$. Note that the analytic continuation of $T(z)$ across the branch cut does not coincide with $T(z)$, and it may develop singularities for complex values of z.

Using (4.5 and 11), we obtain the following expansion for $T(z)$

$$T(z) = H_1 + H_1G_0(z)H_1 + H_1G_0(z)H_1G_0(z)H_1 + \ldots \quad . \tag{4.13}$$

The summation in (4.13) can be performed to give

$$T(z) = H_1 + H_1(G_0 + G_0H_1G_0 + \ldots)H_1 = H_1 + H_1GH_1 \quad , \tag{4.14}$$

$$= H_1 + H_1G_0(H_1 + H_1G_0H_1 + \ldots) = H_1 + H_1G_0T \quad , \tag{4.15}$$

$$= H_1 + (H_1 + H_1G_oH_1 + \ldots)G_oH_1 = H_1 + TG_0H_1 \quad . \tag{4.16}$$

With the help of T the basic equation (4.5) can be rewritten as

$$G(z) = G_0(z) + G_0(z)T(z)G_0(z) \quad , \tag{4.17}$$

which means that knowledge of T allows immediate determination of G.

Equations (4.15,16) in the $\underline{r}$- or $\underline{k}$-representation will become linear inhomogeneous integral equations for the unknown quantity $T(\underline{r},\underline{r}';z)$ or $T(\underline{k},\underline{k}';z)$; e.g., (4.15) in the $\underline{k}$-representation is

$$T(\underline{k},\underline{k}';z) = H_1(\underline{k},\underline{k}') + \sum_{\underline{k}_1\underline{k}_2} H_1(\underline{k},\underline{k}_1)G_0(\underline{k}_1,\underline{k}_2;z)T(\underline{k}_2,\underline{k}';z) \quad , \tag{4.18}$$

where

$$H_1(\underline{k},\underline{k}') \equiv \langle\underline{k}|H_1|\underline{k}'\rangle = \frac{1}{\Omega}\int d\underline{r}\, d\underline{r}'\, e^{-i\underline{k}\underline{r}+i\underline{k}'\underline{r}'} H_1(\underline{r},\underline{r}') \tag{4.19}$$

$$G_0(\underline{k}_1,\underline{k}_2;z) \equiv \langle\underline{k}_1|G_0(z)|\underline{k}_2\rangle = \frac{1}{\Omega}\int d\underline{r}_1 d\underline{r}_2\, \exp(-i\underline{k}_1\underline{r}_1 + i\underline{k}_2\underline{r}_2)G_0(\underline{r}_1,\underline{r}_2;z) \tag{4.20}$$

$$T(\underline{k},\underline{k}';z) \equiv \langle\underline{k}|T(z)|\underline{k}'\rangle = \frac{1}{\Omega}\int d\underline{r}\, d\underline{r}'\, e^{-i\underline{k}\underline{r}+i\underline{k}'\underline{r}'}\, T(\underline{r},\underline{r}';z) \quad . \tag{4.21}$$

When $H_1(\underline{r},\underline{r}') = \delta(\underline{r} - \underline{r}')V(\underline{r})$, $H_1(\underline{k},\underline{k}')$ reduces to

$$H_1(\underline{k},\underline{k}') = V(\underline{k} - \underline{k}')/\Omega \quad , \tag{4.22}$$

where $V(\underline{q})$ is the Fourier transform of $V(\underline{r})$, i.e.,

$$V(\underline{q}) = \int d\underline{r}\, V(\underline{r})\, e^{-i\underline{q}\underline{r}} \quad . \tag{4.23}$$

In the usual case, where $G_0(\underline{r}_1,\underline{r}_2)$ is a function of the difference $\underline{r}_1 - \underline{r}_2$ only, we have from (4.20) that

$$G_0(\underline{k}_1\underline{k}_2;z) = \delta_{\underline{k}_1\underline{k}_2}\, G_0(\underline{k}_1;z) \tag{4.24}$$

where $G_0(\underline{k};z)$ is the Fourier transform of $G(\underline{r}_1,\underline{r}_2;z)$ with respect to the variable $\underline{\rho} = \underline{r}_1 - \underline{r}_2$. Under these conditions (4.18) can be written as

$$T'(\underline{k},\underline{k}';z) = V(\underline{k} - \underline{k}') + \int \frac{d\underline{k}_1}{(2\pi)^d}\, V(\underline{k} - \underline{k}_1)G_0(\underline{k}_1;z)T'(\underline{k}_1,\underline{k}';z) \tag{4.25}$$

where $T'(\underline{k},\underline{k}';z) = \Omega T(\underline{k},\underline{k}';z)$.

As was mentioned before, knowledge of $G(z)$ [or equivalently $T(z)$] allows us to determine the *discrete* eigenvalues and the corresponding eigenfunctions of H; it permits us also to obtain the DOS of the continuous part of the spectrum of H. Now we would like to examine how the eigenstates associated with the continuous spectrum of H can be obtained. The time-independent Schrödinger equation $(E - H)|\psi\rangle = 0$ can be written as

$$(E - H_0)|\psi\rangle = H_1|\psi\rangle \quad ; \tag{4.26}$$

E belongs to the continuous spectrum of H. Equation (4.26) can be considered as an inhomogeneous equation whose general solution is [according to 1.33b)]

$$|\psi^{\pm}\rangle = |\phi\rangle + G_0^{\pm}(E)H_1|\psi^{\pm}\rangle \quad , \tag{4.27}$$

where $|\phi\rangle$ is the general solution of $(E - H_0)|\phi\rangle = 0$; the superscripts ± have been introduced in order to distinguish the solutions associated with G_0^+ from those associated with G_0^-. Equation (4.27) is an integral equation for the unknown functions $|\psi^{\pm}\rangle$; in the $\underline{r}$-representation we can rewrite (4.27) as

$$\psi^{\pm}(\underline{r}) = \phi(\underline{r}) + \int d\underline{r}_1 d\underline{r}_2 G_0^{\pm}(\underline{r},\underline{r}_1;E)H_1(\underline{r}_1,\underline{r}_2)\psi^{\pm}(\underline{r}_2) \quad , \tag{4.27'}$$

or, in the usual case where $H_1(\underline{r}_1,\underline{r}_2) = \delta(\underline{r}_1 - \underline{r}_2)V(\underline{r}_1)$,

$$\psi^{\pm}(\underline{r}) = \phi(\underline{r}) + \int d\underline{r}_1 G_0^{\pm}(\underline{r},\underline{r}_1;E)V(\underline{r}_1)\psi^{\pm}(\underline{r}_1) \quad . \tag{4.28}$$

Equation (4.27') or (4.28) is the Lippman-Schwinger equation. If E does not belong to the spectrum of H_0, (4.28) becomes homogeneous, because $\phi(\underline{r}) = 0$

$$\psi(\underline{r}) = \int d\underline{r}_1 G_0(\underline{r},\underline{r}_1;E)V(\underline{r}_1)\psi(\underline{r}_1) \quad . \tag{4.29}$$

Usually, H_0 and H_1 are chosen in such a way that the *continuous* spectra of H_0 and H coincide. Then (4.27,28) are appropriate for finding the eigenfunctions of H associated with the continuous spectrum (of either H_0 or H) while (4.29) gives the discrete eigenfunctions and eigenvalues of H. It should be recalled that once $G(z)$ [or $T(z)$] is known, one does not need to solve (4.29) to obtain the discrete spectrum, since the latter can be determined directly from G or T.

By iterating (4.27) we obtain

$$|\psi^{\pm}\rangle = |\phi\rangle + G_0^{\pm}H_1|\phi\rangle + G_0^{\pm}H_1G_0^{\pm}H_1|\phi\rangle + \dots \quad . \tag{4.30}$$

Using (4.13), we can express (4.30) in terms of $T^{\pm}$,

$$|\psi^{\pm}\rangle = |\phi\rangle + G_0^{\pm}T^{\pm}|\phi\rangle \quad . \tag{4.31}$$

By multiplying (4.5) from the left or from the right by H_1 and using (4.13), we have

$$H_1G = TG_0 \qquad \text{or} \qquad GH_1 = G_0T \quad . \tag{4.32}$$

Substituting (4.32) into (4.31) we have

$$|\psi^{\pm}\rangle = |\phi\rangle + G^{\pm}H_1|\phi\rangle \quad . \tag{4.33}$$

Equations (4.31,33) are important because they express the eigenfunctions $|\psi^{\pm}\rangle$ in terms of $T^{\pm}$ or $G^{\pm}$ in a closed form. Comparing (4.31) with (4.27) we obtain

$$T^{\pm}|\phi\rangle = H_1|\psi^{\pm}\rangle \quad . \tag{4.34}$$

4.1.2 Time-Dependent Case

Similar expressions can be obtained for the solution of the time-dependent Schrödinger equation, which is written as

$$\left(i\hbar\frac{\partial}{\partial t} - H_0\right)|\psi(t)\rangle = H_1(t)|\psi(t)\rangle \quad , \tag{4.35}$$

where H_1 can be time dependent. From the general relation (2.21) we have

$$|\psi^{\pm}(t)\rangle = |\phi(t)\rangle + \int_{-\infty}^{\infty} dt'\, g_0^{\pm}(t - t')H_1(t')|\psi^{\pm}(t')\rangle \quad . \tag{4.36}$$

On physical grounds one keeps the $|\psi^{+}(t)\rangle$ solution because it is causal, i.e., the effects of H_1 on the eigenfunction appear after H_1 has been applied. Equation (4.36) can be iterated to give

$$|\psi^{+}(t)\rangle = |\phi(t)\rangle + \int dt_1 g_0^{+}(t - t_1)H_1(t_1)|\phi(t_1)\rangle + \int dt_1 dt_2 g_0^{+}(t - t_1)$$

$$H_1(t_1)g_0^{+}(t_1 - t_2)\, H_1(t_2)|\phi(t_2)\rangle + \dots \, . \tag{4.37}$$

Let us assume that $H_1(t) = 0$ for $t \leq t_0$ and that $|\phi(t_0)\rangle$ is an eigenfunction of H_0, let us say, $|\phi_n\rangle$. From (2.16,17) we have that

$$|\phi(t)\rangle = e^{-iE_n(t-t_0)/\hbar}|\phi_n\rangle = e^{-iH_0(t-t_0)/\hbar}|\phi_n\rangle = i\hbar\tilde{g}_0(t - t_0)|\phi_n\rangle . \tag{4.38}$$

Under these initial conditions (4.37) can be written as

$$|\psi^{+}(t)\rangle = A(t,t_0)|\phi_n\rangle \tag{4.39}$$

where

$$A(t,t_0) = i\hbar\tilde{g}(t - t_0) + i\hbar\int_{t_0}^{t} dt_1\tilde{g}_0(t - t_1)H_1(t_1)\tilde{g}_0(t_1 - t_0)$$

$$+ i\hbar\int_{t_0}^{t} dt_1 dt_2\tilde{g}_0(t - t_1)H_1(t_1)g_0^{+}(t_1 - t_2)H_1(t_2)\tilde{g}_0(t_2 - t_0) + \dots \, . \tag{4.40}$$

In obtaining (4.39,40) we have taken into account (4.38) and the relation $g^+(\tau) = \theta(\tau)\tilde{g}(\tau)$.

The probability amplitude for a transition from the state $|\phi_n\rangle$ to the state $|\phi_m\rangle$ as a result of $H_1(t)$ acting during the time interval $[t_0,t]$ can be calculated from (4.38-40) as

$$\langle\phi_m|A(t,t_0)|\phi_n\rangle = e^{(-iE_mt+iE_nt_0)/\hbar}[\langle\phi_m|\phi_n\rangle$$

$$+ \frac{-i}{\hbar}\int_{t_0}^{t} dt_1\langle\phi_m|\, e^{iH_0t_1/\hbar}H_1(t_1)\, e^{-iH_0t_1/\hbar}|\phi_n\rangle$$

$$+ \frac{-i}{\hbar}\int_{t_0}^{t} dt_1dt_2\langle\phi_m|\, e^{iH_0t_1/\hbar}\, H_1(t_1)g_0^+(t_1 - t_2)H_1(t_2)$$

$$e^{-iH_0t_2/\hbar}|\phi_n\rangle + \ldots] \quad . \tag{4.41}$$

In order to get rid of the unimportant phase factor in (4.41) we define an operator $S(t,t_0)$ as follows:

$$S(t,t_0) \equiv e^{iH_0t/\hbar}\, A(t,t_0)\, e^{-iH_0t_0/\hbar} \quad . \tag{4.42}$$

The matrix element $\langle\phi_m|S(t,t_0)|\phi_n\rangle$ is the quantity in parentheses in (4.41) which can be rewritten as

$$\langle\phi_m|S(t,t_0)|\phi_n\rangle = \delta_{nm} + \frac{-i}{\hbar}\int_{t_0}^{t} dt_1\, e^{i\omega_{mn}t_1}\langle\phi_m|H_1(t_1)|\phi_n\rangle$$

$$+ \frac{-i}{\hbar}\int_{t_0}^{t} dt_1dt_2 \int\frac{d\omega}{2\pi}\, e^{it_1(\omega_m-\omega)}\, e^{it_2(\omega-\omega_n)}$$

$$\langle\phi_m|H_1(t_1)G_0^+(\hbar\omega)H_1(t_2)|\phi_n\rangle + \ldots \tag{4.43}$$

where $\omega_n = E_n/\hbar$, $\omega_m = E_m/\hbar$ and $\omega_{mn} = \omega_m - \omega_n$.

Equation (4.43) implies that the probability amplitude (apart from unimportant phase factors) for a transition between two *different* states $|\phi_n\rangle$ and $|\phi_m\rangle$ as a result of $H_1(t)$ acting during the infinite period from $-\infty$ to $+\infty$ is

$$<\phi_m|S|\phi_n> = \frac{-i}{\hbar}\int_{-\infty}^{\infty} dt_1\ e^{i\omega_{mn}t_1} <\phi_m|H_1(t_1)|\phi_n> + \ldots \quad , \tag{4.44}$$

where

$$S \equiv \lim_{\substack{t \to +\infty \\ t_0 \to -\infty}} S(t,t_0) \tag{4.45}$$

is the so-called S-matrix.

Equation (4.44) is the basic result in time-dependent perturbation theory.

For the particular case where H_1 is time independent we obtain for $<\phi_m|S|\phi_n>$

$$\begin{aligned}<\phi_m|S|\phi_n> &= \delta_{nm} + <\phi_m|H_1|\phi_n> \frac{-i}{\hbar}\int_{-\infty}^{\infty} dt_1\ e^{i\omega_{mn}t_1} \\ &\quad + \frac{-i}{\hbar}\int \frac{d\omega}{2\pi} <\phi_m|H_1 G_0^+(\hbar\omega)H_1|\phi_n> \int_{-\infty}^{\infty} dt_1 \int_{-\infty}^{\infty} dt_2\ e^{it_1(\omega_m-\omega)} \\ &\quad e^{it_2(\omega-\omega_n)} + \ldots \\ &= \delta_{nm} - 2\pi i\ \delta(E_n - E_m)[<\phi_m|H_1|\phi_n> + <\phi_m|H_1 G_0^+(E_n)H_1|\phi_n> + \ldots] \\ &= \delta_{nm} - 2\pi i\ \delta(E_n - E_m)<\phi_m|T^+(E_n)|\phi_n> \quad ;\end{aligned} \tag{4.46}$$

to arrive at the last result we employ the relation

$$\int_{-\infty}^{\infty} dt\ e^{iEt/\hbar} = 2\pi\hbar\ \delta(E) \quad . \tag{4.47}$$

The transition probability for the case of time-independent H_1 is

$$\begin{aligned}|<\phi_m|S|\phi_n>|^2 &= \frac{1}{\hbar^2}|<\phi_m|H_1|\phi_n>|^2 \int dt_1 dt_2\ e^{i\omega_{mn}t_1 - i\omega_{mn}t_2} + \\ &= \frac{1}{\hbar^2}|<\phi_m|H_1|\phi_n>|^2 \int_{\infty}^{\infty} dt \int_{-\infty}^{\infty} dt'\ e^{i\omega_{mn}t'} + \ldots \\ &= \frac{2\pi}{\hbar}|<\phi_m|H_1|\phi_n>|^2\ \delta(E_n - E_m)\int_{-\infty}^{\infty} dt + \ldots \quad .\end{aligned}$$

Thus, the transition probability per unit time, W_{mn}, is

$$\begin{aligned}W_{mn} &= \frac{2\pi}{\hbar}|<\phi_m|H_1|\phi_n>|^2\ \delta(E_n - E_m) + \ldots \\ &= \frac{2\pi}{\hbar}|<\phi_m|T^+(E_n)|\phi_n>|^2\ \delta(E_n - E_m) \quad .\end{aligned} \tag{4.48}$$

The last relation follows by observing that the inclusion of higher-order terms in the probability amplitude is achieved by replacing H_1 by the t-matrix $T^+(E_n)$ [see (4.46)]. Equation (4.48) is the well-known Fermi's "golden rule".

Since $|\phi_n\rangle$ is assumed normalized, we have that $|\psi^+(t)\rangle$ is normalized, i.e.,

$$1 = \langle\psi^+(t)|\psi^+(t)\rangle = \langle\phi_n|A^+(t,t_o)A(t,t_o)|\phi_n\rangle \quad ,$$

which implies that $A(t,t_o)$ is unitary. Using the definition (4.42) it is easy to show then that $S(t,t_o)$ is unitary, too. Thus for the S-matrix we have

$$S^+S = SS^+ = 1 \quad . \tag{4.49}$$

Equation (4.49) can be expressed in terms of the t-matrix using (4.46):

$$\langle\phi_n|T^+(E_n)|\phi_\ell\rangle - \langle\phi_n|T^-(E_n)|\phi_\ell\rangle = -2\pi i \sum_m \langle\phi_n|T^-(E_n)|\phi_m\rangle\langle\phi_m|T^+(E_n)|\phi_\ell\rangle \\ \delta(E_m - E_n) \quad . \tag{4.50}$$

Equation (4.50) can also be derived from $T = H_1 + H_1GH_1$ which implies that $T^+ - T^- = H_1(G^+ - G^-)H_1 = -2\pi i H_1\delta(E - H)H_1$. The rest of the proof is left to the reader. As we shall see in the next section, (4.50) is equivalent to the so-called optical theorem in scattering theory; in other words, the optical theorem stems from the unitarity of S.

Before we conclude the present discussion we will recast the expression for the S-matrix in a form which is convenient for future manipulations:

$$S = 1 + \frac{-i}{\hbar}\int_{-\infty}^{\infty} dt_1 H_1^I(t_1) + \left(\frac{-i}{\hbar}\right)^2 \int_{-\infty}^{\infty} dt_1 H_1^I(t_1) \int_{-\infty}^{t_1} dt_2 H_1^I(t_2) + \dots \tag{4.51}$$

where

$$H_1^I(t) \equiv e^{iH_0t/\hbar}\, H_1(t)\, e^{-iH_0t/\hbar} \quad . \tag{4.52}$$

To obtain (4.51) we have used the relation

$$g_0^+(t_1 - t_2) = \theta(t_1 - t_2)\tilde{g}_0(t_1 - t_2) = \theta(t_1 - t_2) \frac{-i}{\hbar} e^{-iH_0(t_1-t_2)/\hbar} \quad .(4.53)$$

The restrictions in the intermediate integrations in (4.51) can be relaxed if at the same time we divide the nth term ($n = 1,2,3, \ldots$) by $n!$ [4.1]. However, in writing products of H_1^I's, one must make certain that the original ordering of the operators is preserved, i.e., the product is ordered in such a way that earlier times appear to the right. To make this chronological ordering explicit we define the time-ordered product of operators H_1^I as

$$T[H_1^I(t_i) \ldots H_1^I(t_j) \ldots] = H_1^I(t_1)H_1^I(t_2) \ldots H_1^I(t_n) \quad , \tag{4.54}$$

where $t_1, t_2, \ldots, t_n$ satisfy the relations $t_1 > t_2 > \ldots > t_n$ and $t_i \ldots t_j \ldots$ is any permutation of $t_1, t_2, \ldots, t_n$. With this definition (4.51) becomes

$$S = \sum_{n=0}^{\infty} \left(\frac{-i}{\hbar}\right)^n \frac{1}{n!} \int dt_1 dt_2 \ldots dt_n \, T[H_1^I(t_1) \ldots H_1^I(t_n)]$$

$$= T \exp\left[-\frac{i}{\hbar} \int dt \, H_1^I(t)\right] \quad . \tag{4.55}$$

The last expression in (4.55) is simply a compact way of writing the sum.

4.2 Application: Scattering Theory (E>0)

In this and the next sections we take $H_0 = p^2/2m = -\hbar^2\nabla^2/2m$; the perturbation $H_1(\underline{r},\underline{r}')$ is of the form $\delta(\underline{r} - \underline{r}')V(\underline{r})$ where $V(\underline{r})$ is of finite extent [i.e., $V(\underline{r})$ decays fast enough as $r \to \infty$]. Here H_0 has a continuous spectrum which extends from zero to $+\infty$. The continuous spectrum of H coincides with that of H_0; H, however, may develop discrete levels of negative eigenenergies if $V(\underline{r})$ is negative in some region(s). In the present section we will examine questions related to the continuous spectrum (E>0) while in the next section we examine the question of a discrete level for very shallow $V(\underline{r})$.

The problem of physical interest for E>0 is scattering: An incident particle of energy $E = \hbar^2k^2/2m$, described by the unperturbed wave function $\exp(i\underline{k}\underline{r})/\sqrt{\Omega}$, comes under the influence of the perturbation $V(\underline{r})$ and as a result its wave function is modified. The question is to find the asymptotic behavior (as $r \to \infty$) of this modification.

The solution of the scattering problem can be obtained immediately from (4.31), which in the $\underline{r}$-representation becomes

$$\psi^{\pm}(\underline{r}) = \frac{1}{\sqrt{\Omega}} e^{i\underline{k}\underline{r}} + \frac{1}{\sqrt{\Omega}} \int d\underline{r}_1 d\underline{r}_2 \, G_0^{\pm}(\underline{r},\underline{r}_1) T^{\pm}(\underline{r}_1,\underline{r}_2) \, e^{i\underline{k}\underline{r}_2} \quad . \tag{4.56}$$

Using the expression for the 3-D $G_0^{\pm}(\underline{r},\underline{r}')$, which we obtained in Sect.3.2, we can rewrite (4.56) as

$$\sqrt{\Omega}\, \psi^{\pm}(\underline{r}) = e^{i\underline{k}\underline{r}} - \frac{2m}{4\pi\hbar^2} \int d^3r_1 d^3r_2 \, \frac{e^{\pm ik|\underline{r}-\underline{r}_1|}}{|\underline{r}-\underline{r}_1|} \, T^{\pm}(\underline{r}_1,\underline{r}_2) \, e^{i\underline{k}\underline{r}_2} \quad , \tag{4.57}$$

where $k = \sqrt{2mE/\hbar^2}$. As was mentioned before we are interested in the asymptotic behavior of $\psi^{\pm}(\underline{r})$ as $r \to \infty$. In this limit we can omit $\underline{r}_1$ in the denominator of the right-hand side of (4.57); in the exponent we can write $|\underline{r} - \underline{r}_1| \approx r - r_1 \cos\theta + O(1/r)$. Thus $k|\underline{r} - \underline{r}_1| \approx kr - kr_1 \cos\theta = kr - \underline{k}_f\underline{r}_1$, where $\underline{k}_f$ is a vector of magnitude k in the direction of $\underline{r}$. For large r we can then write (4.57) as

$$\sqrt{\Omega}\, \psi^{\pm}(\underline{r}) \xrightarrow[r \to \infty]{} e^{i\underline{k}\underline{r}} - \frac{m}{2\pi\hbar^2} \frac{e^{\pm ikr}}{r} \int d^3r_1 d^3r_2 \, e^{\mp i\underline{k}_f\underline{r}_1} \langle \underline{r}_1|\mathcal{T}^{\pm}(E)|\underline{r}_2\rangle \cdot e^{i\underline{k}\underline{r}_2}$$

$$= e^{i\underline{k}\underline{r}} - \frac{m}{2\pi\hbar^2} \frac{e^{\pm ikr}}{r} \langle \pm\underline{k}_f|\mathcal{T}'^{\pm}(E)|\underline{k}\rangle \quad . \tag{4.58}$$

To obtain (4.58) we have used the relations: $\langle \underline{r}|\underline{k}\rangle = \exp(i\underline{k}\underline{r})/\sqrt{\Omega}$, $T' = \Omega T$ and $\int d^3r |r\rangle\langle \underline{r}| = 1$.

Equation (4.58) shows that the solution ψ^- must be excluded because it produces a physically unacceptable ingoing spherical wave asymptotically. The quantity of physical importance is the scattering amplitude $f(\underline{k}_f,\underline{k})$ which is defined by the relation

$$\psi(\underline{r}) \xrightarrow[r \to \infty]{} \text{const.} \left[e^{i\underline{k}\underline{r}} + f(\underline{k}_f\underline{k}) \frac{e^{i\underline{k}\underline{r}}}{r} \right] \quad . \tag{4.59}$$

Comparing (4.58) with (4.59) we obtain for the scattering amplitude

$$f(\underline{k}_f,\underline{k}) = - \frac{m}{2\pi\hbar^2} \langle \underline{k}_f|\mathcal{T}'^{+}(E)|\underline{k}\rangle \tag{4.60}$$

where $E = \hbar^2 k_f^2/2m = \hbar^2 k^2/2m$. Thus the t-matrix in the $\underline{k}$-representation is essentially the scattering amplitude, which is directly related with the differential cross section, $d\sigma/dO$

$$d\sigma/dO = |f|^2 = \frac{m^2}{4\pi^2\hbar^4} |\langle \underline{k}_f | T'^{+}(E) | \underline{k} \rangle|^2 \quad . \tag{4.61}$$

Substituting (4.60) in (4.25) we obtain the following integral equation for the scattering amplitude f

$$f(\underline{k}_f,\underline{k}) = -\frac{m}{2\pi\hbar^2} V(\underline{k}_f - \underline{k}) + \int \frac{d^3k_1}{(2\pi)^3} \frac{V(\underline{k}_f - \underline{k}_1)}{E - \hbar^2 k_1^2/2m + is} f(\underline{k}_1,\underline{k}) \quad ; \tag{4.62}$$

we have taken into account that $G_0^+(k_1;E) = \lim\{E - \hbar^2 k_1^2/2m + is\}^{-1}$ as $s \to 0^+$. Thus, to the first order in the scattering potential

$$f(\underline{k}_f\underline{k}) \approx -\frac{m}{2\pi\hbar^2} V(\underline{k}_f - \underline{k}) \quad , \tag{4.63}$$

where $V(\underline{q})$ is the Fourier transform of $V(\underline{r})$ [see (4.23)]. Equation (4.63) is the Born approximation for the scattering amplitude.

Equation (4.61) can be derived in an alternative way. The differential cross section is defined as the probability per unit time for the transition $\underline{k} \to \underline{k}_f$, $W_{\underline{k}_f\underline{k}}$, times the number of final states divided by the solid angle 4π and by the flux $j = v/\Omega$ of the incoming particle.

$$d\sigma/dO = \frac{\Omega}{4\pi v} \int dE_f N(E_f) W_{\underline{k}_f\underline{k}} \quad . \tag{4.64}$$

Substituting $W_{\underline{k}_f\underline{k}}$ from (4.48) and $N(E_f)$ from (3.26) we obtain (4.61).

The total cross section

$$\sigma = \int dO d\sigma/dO \tag{4.65}$$

can be written in view of (4.64,48,50) as

$$\sigma = \frac{\Omega}{v} \sum_{\underline{k}_f} W_{\underline{k}_f\underline{k}} = \frac{\Omega}{v} \frac{2\pi}{\hbar} \sum_{\underline{k}_f} |\langle \underline{k}_f | T^{+}(E) | \underline{k} \rangle|^2 \, \delta(E_f - E)$$

$$= \frac{2\pi\Omega}{\hbar v} \sum_{\underline{k}_f} \langle\underline{k}|\mathcal{T}^-(E)|\underline{k}_f\rangle\langle\underline{k}_f|\mathcal{T}^+(E)|\underline{k}\rangle\, \delta(E_f - E)$$

$$= \frac{2\pi}{\hbar v}\frac{\Omega i}{2\pi} [\langle\underline{k}|\mathcal{T}^+(E)|\underline{k}\rangle - \langle k|\mathcal{T}^-(E)|\underline{k}\rangle] = -\frac{2\Omega}{\hbar v} \mathrm{Im}\{\langle\underline{k}|\mathcal{T}^+(E)|k\rangle\} \quad . \qquad (4.66)$$

Taking (4.60) into account we can rewrite (4.66) as

$$\sigma = \frac{4\pi}{k} \mathrm{Im}\{f(\underline{k},\underline{k})\} \quad . \qquad (4.67)$$

Equation (4.67) is the so-called optical theorem which connects the total cross section with the forward scattering amplitude.

As was already mentioned, the scattering amplitude f for positive energies is directly related with an observable quantity of great physical importance, namely the differential cross section. The behavior of f for negative energies is also of physical significance because the poles of f(E) [which coincide with the poles of $\mathcal{T}(E)$ as can be seen from (4.60)] give the discrete eigenenergies of the system. In other words, if the scattering problem has been solved and f versus E has been obtained, one only needs to find the position of the poles of f in order to find the discrete levels of the system. Of course, these poles are on the negative E-semiaxis. We should mention also that many times f versus E (or T^+ versus E) exhibits sharp peaks at certain positive energies. The states associated with such peaks in f are called resonances; their physical significance will be discussed in Chap.6.

An elementary example of the above comments is provided by the case where $V(\underline{r})$ is an attractive Coulomb potential $V(\underline{r}) = -e^2/r$. Then the scattering amplitude is [4.2], [4.3]

$$f = \frac{-t\Gamma(1 - it)}{\Gamma(1 + it)} \, \frac{\exp\{2it\ln[\sin(\theta/2)]\}}{2k\,\sin^2(\theta/2)} \quad , \qquad (4.68)$$

where

$$k = \sqrt{2mE/\hbar^2} \quad ; \quad \mathrm{Im}\{k\} \geq 0 \qquad (4.69)$$

$$t = me^2/\hbar^2 k \qquad (4.70)$$

and θ is the angle between $\underline{k}$ and $\underline{k}_f$. The poles of f occur when the argument of $\Gamma(1 - it)$ is a nonpositive integer, i.e., when $1 - it = -p$, where $p = 0,1,2, \ldots$, i.e., when

$$it = 1 + p = n \quad ; \quad n = 1,2,3, \ldots \quad . \tag{4.71}$$

Substituting (4.69,70) in (4.71) we obtain for the discrete eigenenergies of the Coulomb potential

$$E_n = -\frac{e^4 m}{2\hbar^2}\frac{1}{n^2} \quad ; \quad n = 1,2, \ldots \tag{4.72}$$

which is the standard result. If the potential was repulsive the scattering amplitude would be given by (4.68) with i replaced by - i. In this case the argument in the gamma function in the numerator in (4.68) cannot become a nonpositive integer; thus, f has no poles, which means that there are no discrete levels. This result was expected since the scattering potential is repulsive.

4.3 Application: Bound State in Shallow Potential Wells (E<0)

Here we assume that

$$\begin{aligned} V(\underline{r}) &= -V_0 \quad \text{for } \underline{r} \text{ inside } \Omega_0 \\ &= 0 \quad \text{for } \underline{r} \text{ outside } \Omega_0 \quad , \end{aligned} \tag{4.73}$$

where Ω_0 is a finite region in real space and V_0 is positive and very small: $V_0 \to 0^+$. We are interested in finding whether or not a discrete level E_0 appears and how it varies with V_0. To answer this question, it is enough to find the position of the pole of G(E), if any, for E in the range $[-V_0, 0]$. The basic equation (4.5) has the following form in the present case

$$\begin{aligned} G(\underline{r},\underline{r}';z) &= G_0(\underline{r},\underline{r}';z) \\ &- V_0 \int_{\Omega_0} d\underline{r}_1 G_0(\underline{r},\underline{r}_1;z)G_0(\underline{r}_1,\underline{r}';z) + V_0^2 \int_{\Omega_0} d\underline{r}_1 \int_{\Omega_0} d\underline{r}_2 G_0(\underline{r},\underline{r}_1;z) \\ &\quad G_0(\underline{r}_1,\underline{r}_2;z)G_0(\underline{r}_2,\underline{r}';z) + \ldots \quad . \end{aligned} \tag{4.74}$$

3-D Case

$$G_0(\underline{r},\underline{r}_1;E) = -\frac{m}{2\pi\hbar^2}\frac{e^{-k_0|\underline{r}-\underline{r}_1|}}{|\underline{r}-\underline{r}_1|} \quad ; \quad k_0 = \sqrt{2m|E|/\hbar^2} \quad ,$$

[see (3.21)]. In this case $G_0(E)$ remains finite for E around 0. As a result the various integrals in (4.74) are finite. Thus for sufficiently small V_0 the power series expansion in (4.74) converges, which means that the difference $G - G_0$ approaches zero as $V_0 \to 0^+$. This in turn means that G remains finite for small V_0 and E in the range $[-V_0, 0]$ and consequently has no poles in this range. The conclusion is that in 3-D sufficiently shallow potential wells do not produce discrete levels; the product $\Omega_0 V_0$ has to exceed a critical value before the first discrete level is formed.

2-D Case

$$G_0(\underline{r};\underline{r}_1;E) = -\frac{m}{\pi\hbar^2} K_0(k_0|\underline{r}-\underline{r}_1|) \quad ; \quad k_0 = \sqrt{2m|E|/\hbar^2} \quad ,$$

[see (3.27)]. Since $V_0 \to 0^+$, it follows that $|E| \to 0^+$ and $k_0 \to 0^+$. Using the small argument expansion for K_0 we obtain

$$G_0(\underline{r},\underline{r}_1;E) = \frac{m}{\pi\hbar^2}\ln(k_0|\underline{r}-\underline{r}_1|) + c_1 + O(k_0|\underline{r}-\underline{r}_1|) \tag{4.75}$$

where c_1 is a known constant. Substituting (4.75) in (4.74) and keeping the leading terms only, we obtain

$$G(\underline{r},\underline{r}';E) \approx G_0(\underline{r},\underline{r}';E)\sum_{n=0}^{\infty}\left[-\frac{V_0\Omega_0 m}{\pi\hbar^2}\ln(k_0 S^{\frac{1}{2}})\right]^n$$

$$= \frac{G_0(\underline{r},\underline{r}';E)}{1+\frac{V_0\Omega_0 m}{\pi\hbar^2}\ln(k_0 S^{\frac{1}{2}})} \quad , \tag{4.76}$$

where S is a constant of the order of the extension of the potential well Ω_0; its exact value requires explicit evaluation of integrals of the type $\int_{\Omega_0} d^2r_1 \ldots \int_{\Omega_0} d^2r_n \ln|\underline{r}-\underline{r}_1|\cdot\ln|\underline{r}_1-\underline{r}_2|\ldots\ln|\underline{r}_n-\underline{r}'|$. From (4.76) one finds for the pole of $G(\underline{r},\underline{r}';E)$, E_0, the following result

$$E_o \xrightarrow[V_o \to 0^+]{} -\frac{\hbar^2}{2mS} \exp\left(-\frac{2\pi\hbar^2}{mV_0\Omega_0}\right) = -\frac{\hbar^2}{2mS} \exp\left(-\frac{1}{\rho_0 V_0 \Omega_0}\right) \quad , \tag{4.77}$$

where $\rho_0 = m/2\pi\hbar^2$ is the unperturbed DOS per unit area, see (3.30). The conclusion is that in the 2-D case a discrete level is always formed no matter how shallow the potential well is. This property stems from the fact that $G_0(E)$ blows up as E approaches the band edges E = 0. As was discussed in Chap.3, the logarithmic divergence of $G_0(E)$ is linked to the discontinuity of the unperturbed DOS at the band edge. The logarithmic singularity of $G_0(E)$ at E = 0 produces a discrete level which depends on V_0 as $\exp(-1/\rho_0 V_0 \Omega_0)$ where ρ_0 is the discontinuity in the unperturbed DOS per unit area, V_0 is the depth of the potential well and Ω_0 its 2-D extent. As we will see in Chap.6 this result is valid in all cases where the unperturbed DOS exhibits a discontinuity at the band edge. The connection of this result with the theory of superconductivity will be discussed in Chap.6.

For the particular case of a circular potential well, the problem can be solved directly from Schrödinger's equation, and the result for the discrete level is

$$E_0 \xrightarrow[V_0 \to 0^+]{} -\frac{2\pi}{e^{2\gamma}} \frac{\hbar^2}{m\Omega_0} \exp\left(-\frac{2\pi\hbar^2}{mV_0\Omega_0}\right) \tag{4.78}$$

where Ω_0 is the area of the circular potential well and $\gamma = 0.577...$ is Euler's constant. For other shapes of the potential well (even simple ones like the square potential well) it becomes extremely complicated to solve the problem directly from Schrödinger's equation. Thus one can appreciate the power of the Green's function approach, which allowed us to obtain (4.77) for every shape of the 2-D potential well. Note that the shape of the potential well influences the preexponential factor S in (4.77) but not the dominant exponential factor.

1-D Case

$$G_0(x,x';E) = -\frac{m}{\hbar^2 k_0} e^{-k_0|x - x'|} ; \qquad k_0 = \sqrt{-2mE/\hbar^2} \quad ,$$

[see (3.31)]. In the limit $V_0 \to 0^+$, $E_0 \to 0^-$ and $k_0 \to 0^+$. Thus G_0 can be approximated by

$$G_o(x,x';E) \xrightarrow[E \to 0^-]{} -\sqrt{\frac{m}{-2\hbar^2 E}} \quad . \qquad (4.79)$$

Substituting in (4.74) we obtain

$$G \approx G_o \sum_{n=0}^{\infty} (-G_o V_o \Omega_o)^n = \frac{G_o}{1 + G_o V_o \Omega_o} = \frac{G_o}{1 - \Omega_o V_o \sqrt{-m/2\hbar^2 E}} \quad . \qquad (4.80)$$

Thus the discrete level, E_o, as given by the pole of G(E), is

$$E_o = -\frac{m\Omega_o^2 V_o^2}{2\hbar^2} \quad ; \qquad (4.81)$$

Ω_o is the linear extent of the 1-D potential well. In the 1-D case, as in the 2-D case, a potential well no matter how shallow always creates a discrete level. In contrast to the 2-D case the level E_o is an analytic function of $V_o\Omega_o$ [$E_o \sim -(V_o\Omega_o)^2$] as $V_o\Omega_o \to 0^+$. This behavior is a consequence of the square root singularity of $G_o(E)$ or $\rho_o(E)$ at the band edge.

The material of this section will be discussed again in detail in Chap.6.

4.4 Summary

In this chapter we were interested in finding the eigenvalues and eigenfunctions of a Hamiltonian H which can be decomposed as

$$H = H_o + H_1 \quad ; \qquad (4.1)$$

H_o is such that its eigenvalues and eigenfunctions can be easily determined. This problem was solved by: 1) calculating $G_o(z)$ corresponding to H_o; 2) expressing $G(z)$ in terms of $G_o(z)$ and H_1 where $G(z)$ is the Green's function associated with H; and 3) extracting from $G(z)$ information about the eigenvalues and eigenfunctions of H.

Here $G(z)$ is related to $G_o(z)$ and H_1 as follows

$$G(z) = G_o(z) + G_o(z)H_1G_o(z) + G_o(z)H_1G_o(z)H_1G_o(z) + \ldots \qquad (4.5)$$

$$= G_o(z) + G_o(z)H_1G(z) \qquad (4.6)$$

$$= G_0(z) + G(z)H_1G_0(z) \quad . \tag{4.7}$$

Equations (4.6,7) are inhomogeneous integral equations for G(z). It is very helpful to introduce an auxiliary quantity, which is called t-matrix and is given by

$$T(z) = H_1 + H_1G_0(z)H_1 + H_1G_0(z)H_1G_0(z)H_1 + \dots \tag{4.13}$$

$$= H_1 + H_1G(z)H_1 \tag{4.14}$$

$$= H_1 + H_1G_0(z)T(z) \tag{4.15}$$

$$= H_1 + T(z)G_0(z)H_1 \quad . \tag{4.16}$$

$G(z)$ can be easily expressed in terms of $T(z)$

$$G(z) = G_0(z) + G_0(z)T(z)G_0(z) \quad . \tag{4.17}$$

Here $T(z)$ has the same analytic structures as $G(z)$, i.e., it may have poles and/or branch cuts on the real z-axis. For E belonging to a branch cut we define the side limits $T^{\pm}(E) = \lim T(E \pm is)$ as $s \to 0^+$.

Information about the eigenvalues and eigenfunctions of H are extracted as follows:

1) The poles of $T(z)$ or $G(z)$ give the discrete eigenenergies of H.

2) The residue of $G(z)$ [or $T(z)$] at each pole determines the corresponding eigenfunction.

3) The eigenfunction(s) of H associated with E belonging to the continuous spectrum of H is given by

$$|\psi^{\pm}(E)\rangle = |\phi(E)\rangle + G_0^{\pm}(E)T^{\pm}(E)|\phi(E)\rangle \tag{4.31}$$

$$= |\phi(E)\rangle + G^{\pm}(E)H_1|\phi(E)\rangle \tag{4.33}$$

where $\phi(E)$ is any eigenfunction of H_0 corresponding to the same eigenvalue E. In most physical application the solution $|\psi^-(E)\rangle$ is excluded on physical grounds.

4) The discontinuity of $G(z)$ across the branch cut gives the density of states as was discussed before.

In the time-dependent case the time evolution of the system can be described in terms of an infinite series involving $g_0(\tau)$ and H_1. The main quantity of physical interest in this case is the probability amplitude for a transition from the eigenstate (of H_0) $|\phi_n\rangle$ to the eigenstate $|\phi_m\rangle$ as a result of H_1 acting during the infinite time interval $[-\infty, +\infty]$. This probability amplitude is expressed as the $\langle\phi_m|, |\phi_n\rangle$ matrix element of the so-called S-matrix which can be expressed as an infinite series involving $g_0(\tau)$ and H_1. This series for the S-matrix can be written as

$$S = 1 + \frac{-i}{\hbar}\int_{\infty}^{\infty} dt_1 H_1^I(t_1) + \left(\frac{-i}{\hbar}\right)^2 \int_{-\infty}^{\infty} dt_1 H_1^I(t_1) \int_{-\infty}^{t_1} dt_2 H_1^I(t_2) + \ldots \tag{4.51}$$

where $H_1^I(t) \equiv \exp(iH_0 t/\hbar) H_1(t) \exp(-iH_0 t/\hbar)$. The S-matrix is unitary;

$$SS^+ = S^+S = 1 \quad . \tag{4.49}$$

If H_1 is time independent, one finds that S is related to T^+ as follows

$$\langle\phi_m|S|\phi_n\rangle = \delta_{nm} - 2\pi i\delta(E_n - E_m)\langle\phi_m|T^+(E_n)|\phi_n\rangle \quad , \tag{4.46}$$

where $H_0|\phi_n\rangle = E_n|\phi_n\rangle$ and $H_0|\phi_m\rangle = E_m|\phi_m\rangle$. The probability per unit time for the transition $|\phi_n\rangle \rightarrow |\phi_m\rangle$ is

$$W_{nm} = \frac{2\pi}{\hbar} |\langle\phi_m|T^+(E_n)|\phi_n\rangle|^2 \delta(E_n - E_m) \quad . \tag{4.48}$$

Knowledge of $T(z)$ permits us to obtain immediately the scattering amplitude $f(\underline{k}_f, \underline{k})$ because the latter is directly related to T by

$$f(\underline{k}_f, \underline{k}) = -\frac{m\Omega}{2\pi\hbar^2} \langle\underline{k}_f|T^+(E)|\underline{k}\rangle \tag{4.60}$$

where $E = \hbar^2k^2/2m = \hbar^2k_f^2/2m$. From the unitarity of S and (4.46,60) the optical theorem follows,

$$\sigma = \frac{4\pi}{k} \operatorname{Im}\{f(\underline{k}, \underline{k})\} \quad , \tag{4.67}$$

where σ is the total cross section. Since f is proportional to T, it follows that the poles of scattering amplitude considered as a function of energy give the discrete levels of the system.

The formalism just outlined is applied to the question of the existence of a discrete level in very shallow potential wells. It is found that:

1) In the 3-D case the product of the extent times the depth of the well must exceed a critical value for a discrete level to appear.

2) For the 2-D case a discrete level at E_0 always exists no matter how shallow the potential well is; E_0 is given by

$$E_0 \sim - \exp\left(- \frac{1}{\rho_0 V_0 \Omega_0}\right) \quad \text{as } V_0\Omega_0 \to 0^+ \ , \tag{4.77}$$

where ρ_0 is the unperturbed DOS per unit area, V_0 is the depth and Ω_0 is the 2-D extent of the potential well.

3) For the 1-D case a discrete level is always present; it is given by

$$E_0 \xrightarrow[V_0\Omega_0 \to 0^+]{} - \frac{m\Omega_0^2 V_0^2}{2\hbar^2} \tag{4.81}$$

where Ω_0 is the linear extent of the well. The characteristic behavior above can be directly related with the analytic structure of the unperturbed DOS at the band edge.

5. Green's Functions for Tight-Binding Hamiltonians

We introduce the so-called tight-binding Hamiltonians, which have the form

$$H = \sum_{\underline{\ell}} |\underline{\ell}\rangle \varepsilon_{\ell} \langle \underline{\ell}| + \sum_{\underline{\ell}\underline{m}} |\underline{\ell}\rangle V_{\ell m} \langle \underline{m}| \quad ,$$

where each state $|\underline{\ell}\rangle$ is an atomic-like orbital centered at the site $\underline{\ell}$; the sites $\{\underline{\ell}\}$ form a lattice. Such Hamiltonians are very important in solid-state physics. Here we calculate the Green's functions associated with the TBH for various simple lattices. We also review briefly some applications in solid-state physics.

5.1 Introductory Remarks

In this chapter we examine the Green's functions associated with a class of *periodic* Hamiltonians, i.e., Hamiltonians remaining invariant under a translation by any vector $\underline{\ell}$, where $\{\underline{\ell}\}$ form a regular lattice

$$\underline{\ell} = \sum_{\alpha=1}^{d} \ell_{\alpha} \underline{r}_{\alpha} \quad ; \qquad \ell_{\alpha} = 0, \ \pm 1, \ \pm 2, \ \dots \ . \tag{5.1}$$

The reasons for considering here periodic Hamiltonians are the following:

1) They produce continuous spectra which possess not only a lower bound, as in the free particle case, but *upper* bound(s) as well. Thus the physics is not only richer but more symmetric and in some sense more satisfying.

2) They are of central importance for understanding the behavior of perfectly crystalline solids.

3) They provide the basis for understanding the properties of real imperfect crystalline solids since the imperfections can be treated as a perturbation H_1 using techniques developed in Chap.4.

4) The class of simplified periodic Hamiltonians we consider here allow us to obtain simple closed-form results for certain perturbation problems. Thus the physics presented in Sects.4.2 and 3 can be better appreciated without the burden of a complicated algebra. This point will be examined in detail in the next chapter.

Before we introduce the class of Hamiltonians under consideration, we remind the reader some of the basic properties of the eigenfunctions and eigenvalues of a periodic Hamiltonian.

The eigenfunctions, which are called Bloch functions, are plane waves modulated by a periodic factor $u_{n\underline{k}}(\underline{r})$, i.e.,

$$\psi_{n\underline{k}}(\underline{r}) = \frac{1}{\sqrt{\Omega}} e^{i\underline{k}\underline{r}} u_{n\underline{k}}(\underline{r}) \quad ; \tag{5.2}$$

the quantum number $\underline{k}$, determines how much the phase changes if we propagate by a lattice vector $\underline{\ell}$.

$$\psi_{n\underline{k}}(\underline{r} + \underline{\ell}) = e^{i\underline{k}\underline{\ell}} \psi_{n\underline{k}}(\underline{r}) \quad . \tag{5.3}$$

Note that the quantity $\underline{k}$ is restricted to a finite region in $\underline{k}$ - space, called the first Brillouin zone [5.1]. The other quantum number n, which is called the band index, takes integer values; the presence of n compensates somehow the restriction of $\underline{k}$ in the first Brillouin zone. It is worthwhile to stress that eigenstates of the type (5.2) imply propagation without any resistance as in the free-particle case.

Periodicity has more profound and characterisitc effects on the energy spectrum, which consists of continua (called bands) separated by forbidden energy regions called gaps. The points in energy where the bands terminate are called band edges. The eigenenergies $E_n(\underline{k})$ are continuous functions of $\underline{k}$ within each band n. The band edges correspond to absolute minima or maxima of E versus k for a given n.

For many purposes it is useful to introduce the Wannier functions. The Wannier function associated with the band index n and the lattice site $\underline{\ell}$ is defined as [5.2,3]

$$w_n(\underline{r} - \underline{\ell}) = \frac{1}{\sqrt{N}} \sum_{\underline{k}} e^{-i\underline{k}\underline{\ell}} \psi_{n\underline{k}}(\underline{r}) \tag{5.4}$$

where $\psi_{n\underline{k}}(\underline{r})$ are the eigenfunctions given by (5.2). The Wannier functions $w_n(\underline{r} - \underline{\ell})$ are localized around the lattice site $\underline{\ell}$.

The functions $w_n(\underline{r} - \underline{\ell})$ for all n and $\underline{\ell}$ form a complete set; thus any operator, e.g., the Hamiltonian, can be expressed in the Wannier representation. It may happen that the eigenenergies associated with a particular band index n_0 are well separated from all the other eigenenergies. In this case the matrix elements of the Hamiltonian H between $w_{n_0}(\underline{r} - \underline{\ell})$ and $w_n(\underline{r} - \underline{m})$, where $n \neq n_0$, may be smaller than $|\varepsilon_n - \varepsilon_{n_0}|$. Then these small matrix elements to a first approximation can be omitted, and the subspace spanned by the states $w_{n_0}(\underline{r} - \underline{\ell})$, where $\underline{\ell}$ runs over all lattice vectors, becomes decoupled from the rest. In what follows we work within this subspace; for simplicity we omit the subscript n_0 and we write for the basic vectors of the subspace

$$w_{n_0}(\underline{r} - \underline{\ell}) = \langle\underline{r}|\underline{\ell}\rangle \quad . \tag{5.5}$$

The matrix elements of the Hamiltonian within this subspace are

$$\langle\underline{\ell}|H|\underline{m}\rangle = \varepsilon_{\underline{\ell}}\delta_{\underline{\ell}\underline{m}} + V_{\underline{\ell}\underline{m}} \quad . \tag{5.6}$$

Following the usual notation we have denoted the diagonal matrix elements by $\varepsilon_{\underline{\ell}}$ and the off-diagonal matrix elements by $V_{\underline{\ell}\underline{m}}$ ($V_{\underline{\ell}\underline{\ell}} \equiv 0$).

The periodicity of the Hamiltonian, i.e., its invariance under translations by a lattice vector $\underline{\ell}$, implies that

$$\varepsilon_{\underline{\ell}} = \varepsilon_0 \quad \text{for all } \underline{\ell} \tag{5.7a}$$

$$V_{\underline{\ell}\underline{m}} = V_{\underline{\ell}-\underline{m}} \quad . \tag{5.7b}$$

It should be stressed that the Hamiltonian, which describes a periodic solid, has matrix elements outside the subspace spanned by the vectors $|\underline{\ell}\rangle$ and that this subspace is coupled with the rest of the Hilbert space. Nevertheless we restrict ourselves to this subspace for the sake of simplicity. The price for this approximation can be considered reasonable, since many important *qualitative* features are retained in spite of this drastic simplification. Furthermore, bands arising from atomic orbitals weakly overlapping with their neighbors (i.e., tightly bound to their atoms) can be described rather accurately by working within the above defined subspace or its straightforward generalization [5.1,4]. For this reason, the Hamiltonian (5.6), which is confined within the subspace spanned by $\{|\underline{\ell}\rangle\}$, where $\underline{\ell}$ runs over all lattice sites, is called the tight-binding Hamiltonian (TBH) or the tight-binding model (TBM).

5.2 The Tight-Binding Hamiltonian (TBH)

The TBH is defined as

$$H = \sum_{\underline{i}} |\underline{i}\rangle \varepsilon_{\underline{i}} \langle \underline{i}| + \sum_{\underline{i}\underline{j}} |\underline{i}\rangle V_{\underline{i}\underline{j}} \langle \underline{j}| \tag{5.8}$$

where each $\langle \underline{r}|\underline{i}\rangle$ is centered around the corresponding lattice site $\underline{i}$ and decays far away from it. If the Hamiltonian H possesses the periodicity of the lattice, the quantities $\{\varepsilon_{\underline{i}}\}$ and $\{V_{\underline{i}\underline{j}}\}$ satisfy (5.7). We shall consider also the case where the lattice can be divided into two interpenetrating sublattices such that each point of sublattice 1 is surrounded by points belonging to sublattice 2; the Hamiltonian remains invariant under translation by vectors of sublattice 1 or sublattice 2. In this case

$$\varepsilon_{\underline{i}} = \varepsilon_1 \text{ if } \underline{i} \text{ belongs to sublattice 1} \tag{5.9a}$$

$$= \varepsilon_2 \text{ if } \underline{i} \text{ belongs to sublattice 2} \quad . \tag{5.9b}$$

For the sake of simplicity we assume in our explicit results that

$$V_{\underline{i}\underline{j}} = V \quad , \quad \underline{i}, \underline{j} \text{ nearest neighbors} \tag{5.10}$$

$$= 0 \quad , \quad \text{otherwise} \quad .$$

This last assumption, although not neccesary, simplifies the calculational effort. The set $\{|\underline{i}\rangle\}$ is assumed orthonormal;

$$\langle \underline{i}|\underline{j}\rangle = \delta_{\underline{i}\underline{j}} \quad . \tag{5.11}$$

Thus a TBH is characterized by: 1) The lattice structure associated with the points $\{\underline{i}\}$. 2) The values of the diagonal matrix element $\{\varepsilon_{\underline{i}}\}$; in the simple periodic case, where (5.7a) is satisfied there is only one common value which can be taken as zero by a proper redefinition of the origin of energy; in the double spacing periodic case, where (5.9) is satisfied, the quantity of physical significance is the difference $\varepsilon_1 - \varepsilon_2 > 0$. 3) The off-diagonal matrix elements $V_{\underline{i}\underline{j}}$ which in the periodic case depend only on the difference $\underline{i} - \underline{j}$; if the simplifying assumption (5.10) is made, there is only one quantity V which, following the usual practice in the literature, can be taken as positive. However, it must be pointed out that V is in general negative; moreover, a negative V, in contrast to a positive V, preserves the well-known property that as the energy of real eigenfunctions increases so does the number of their sign alternation. In any case one can obtain the negative V Green's functions from those calculated here by employing the relation

$$G(\underline{\ell},\underline{m};\ E+is,\{\varepsilon_{\underline{i}}\},V) = -G(\underline{\ell},\underline{m};\ -E-is,\{-\varepsilon_{\underline{i}}\},-V) \quad .$$

The first term on the right-hand side of (5.8) describes a particle which can be trapped around any particular lattice site $\underline{i}$ with an eigen-energy $\varepsilon_{\underline{i}}$. The second term allows the particle to hop from the site $\underline{i}$ to the site $\underline{j}$ with a transfer matrix element $V_{\underline{i}\underline{j}}$. The quantum motion associated with the Hamiltonian (5.8) is equivalent to the wave motion of the coupled pendulums shown in Fig.5.1. This can be easily seen by writting the time-independent Schrödinger equation $H|\psi\rangle = E|\psi\rangle$ as

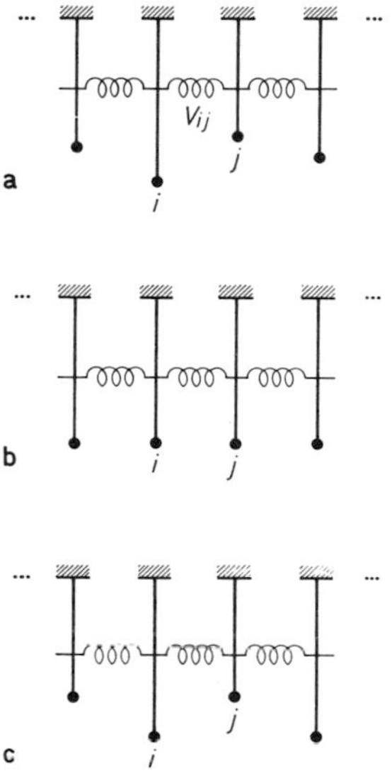

Fig.5.1. One-dimensional coupled pendulum analog of the tight-binding Hamiltonian. Only nearest neighbor couplings are shown (a). In the periodic case, examined in this chapter, all pendulums and all nearest neighbor couplings are identical (b). The double spacing periodic case is also shown (c)

$$(\varepsilon_{\underline{i}} - E)c_{\underline{i}} + \sum_{\underline{j}} V_{\underline{i}\underline{j}}c_{\underline{j}} = 0 \quad , \tag{5.12}$$

where $|\psi\rangle = \sum_{\underline{i}} c_{\underline{i}}|\underline{i}\rangle$ and (5.8,11) are used. The equations of motion of the coupled pendulums are

$$\left(m_{\underline{i}}\Omega_{\underline{i}}^2 + \sum_{\underline{j}} \kappa_{\underline{i}\underline{j}} - m_{\underline{i}}\Omega^2\right)u_{\underline{i}} - \sum_{\underline{j}} \kappa_{\underline{i}\underline{j}}u_{\underline{j}} = 0 \tag{5.13}$$

where u_i is the displacement of the pendulum located at the i site, Ω_i is its eigenfrequency in the absence of coupling, and $-\sum_j \kappa_{ij}(u_i - u_j)$ is the force exercised on the pendulum at the site i as a result of the couplings with all the other pendulums; m_i is the mass at i. The correspondence between the electronic and the pendulum case is illustrated in Table 5.1.

By direct substitution in the equation $H|\psi\rangle = E|\psi\rangle$, it is easy to show that the eigenfunctions and eigenvalues of H given by (5.8), where $\varepsilon_i = \varepsilon_0$ for all i, are

$$|\underline{k}) = \frac{1}{\sqrt{N}} \sum_{\underline{i}} e^{i\underline{k}\underline{i}} |\underline{i}\rangle \tag{5.14}$$

$$E(\underline{k}) = \varepsilon_0 + \sum_{\underline{\ell}} V_{0\underline{\ell}} \, e^{i\underline{k}\underline{\ell}} \quad . \tag{5.15}$$

Table 5.1. Analogy between the tight-binding Hamiltonian and a system of coupled pendulums

Electronic case	Pendulum case
c_i: component of the eigenfunction at the site i	u_i: displacement of the pendulum located at the site i
$-V_{ij}$: minus transfer matrix element	κ_{ij}: spring constant coupling sites i and j
E: eigenenergy	$m_i\Omega^2$: square of the eigenfrequency times mass at i
ε_i: site energy	$m_i\Omega_i^2 + \sum_j \kappa_{ij}$: square of the uncoupled eigenfrequency times mass at i plus sum of spring constants at i

Equation (5.14) means that the eigenmodes are propagating waves such that the amplitude at each site is the same and the phase changes in a regular way: $\phi_\ell = \underline{k}\underline{\ell}$. To obtain explicit results we employ the simplifying equation (5.10) so that (5.15) becomes

$$E(\underline{k}) = \varepsilon_0 + V {\sum_{\underline{\ell}}}' e^{i\underline{k}\underline{\ell}} \tag{5.16}$$

where the summation extends over the sites neighboring the origin. For the 1-D case we have

$$E(k) = \varepsilon_0 + 2V\cos(ka) \quad ; \quad \text{1-D} \quad , \tag{5.17}$$

where a is the lattice constant. For a 2-D square lattice

$$E(\underline{k}) = \varepsilon_0 + 2V[\cos(k_1a) + \cos(k_2a)] \quad ; \quad \text{2-D square} \quad . \tag{5.18}$$

For the 3-D simple cubic we have

$$E(\underline{k}) = \varepsilon_0 + 2V[\cos(k_1a) + \cos(k_2a) + \cos(k_3a)] \quad ; \quad \text{3-D simple cubic} \quad . \tag{5.19}$$

In Fig.5.2 we plot E versus k for the 1-D case; k is restricted within the first Brillouin zone, which for the 1-D case extends from $-\pi/a$ to π/a. The function E(k) has an absolute maximum (which corresponds to the upper band edge) for $k = 0$ with a value $E_{max} = \varepsilon_0 + 2V$; it has an absolute minimum (which corresponds to a lower band edge) for $k = \pm\pi/a$ with a value $E_{min} = \varepsilon_0 - 2V$. Thus the spectrum is a continuum (a band) extending from $\varepsilon_0 - 2V$ to $\varepsilon_0 + 2V$. The band width is 4V. One can easily show that cases (5.17-19) produce a single band extending from $\varepsilon_0 - ZV$ to $\varepsilon_0 + ZV$ where Z is the number of nearest neighbors. The quantity ZV, which is equal to half the bandwidth, is usually symbolized by B. For 2-D and 3-D cases the functions $E(\underline{k})$ are presented either by plotting E versus $\underline{k}$ as $\underline{k}$ varies along chosen directions or by plotting the lines (2-D case) or surfaces (3-D case) of constant energy. In all cases $\underline{k}$ is restricted within the first Brillouin zone.

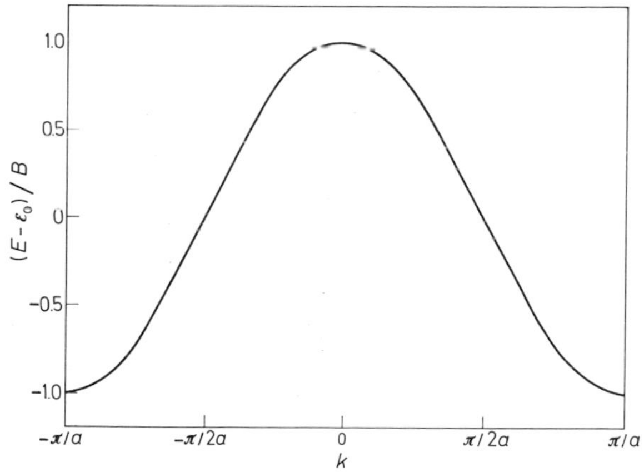

Fig.5.2. $(E-\varepsilon_0)/B$ vs. k for the 1-D tight-binding case with nearest-neighbor coupling $V(>0)$. $B = 2V$ is half the bandwidth

5.3 Green's Functions

The Green's function for the tight-binding Hamiltonian examined in the previous section is

$$G(z) = \sum_{\underline{k}} \frac{|\underline{k})(\underline{k}|}{z-E(\underline{k})} \tag{5.20}$$

where $|\underline{k})$ is given by (5.14) and $E(\underline{k})$ by (5.16). The matrix elements of $G(z)$ are

$$G(\underline{\ell},\underline{m};z) \equiv \langle\underline{\ell}|G(z)|\underline{m}\rangle = \sum_{\underline{k}} \frac{\langle\underline{\ell}|\underline{k})(\underline{k}|\underline{m}\rangle}{z-E(\underline{k})}$$

$$= \frac{\Omega}{N(2\pi)^d} \int_{1BZ} d\underline{k}\, \frac{e^{i\underline{k}(\underline{\ell}-\underline{m})}}{z-E(\underline{k})} \tag{5.21}$$

where the symbol "1BZ" denotes that the integration must be restricted within the first Brillouin zone. In particular, all diagonal matrix elements are equal to each other and are given by

$$G(\underline{\ell},\underline{\ell};z) = \frac{\Omega}{N(2\pi)^d} \int_{1BZ} \frac{d\underline{k}}{z-E(\underline{k})} \quad . \tag{5.22}$$

For large z one can omit $E(\underline{k})$ in the denominator of the integrand in (5.22) so that

$$G(\underline{\ell},\underline{\ell};z) \xrightarrow[z\to\infty]{} \frac{1}{z}\frac{\Omega}{N(2\pi)^d} \int_{1BZ} d\underline{k} \quad .$$

The volume of the first Brillouin zone equals $(2\pi)^d/\Omega_0$ where $\Omega_0 = \Omega/N$ is the volume of the unit cell of the lattice. Hence,

$$G(\underline{\ell},\underline{\ell};z) \xrightarrow[z\to\infty]{} \frac{1}{z} \quad . \tag{5.23}$$

This behavior can be understood if one expresses $G(\underline{\ell},\underline{\ell};z)$ in terms of the density of states per site $\rho(E)$

$$G(\underline{\ell},\underline{\ell};z) = \int \frac{dE\rho(E)}{z-E} \xrightarrow[z\to\infty]{} \frac{\int\rho(E)dE}{z} \quad ;$$

but $\int\rho(E)dE = 1$ since there is one state per site. Below we give some explicit results for the various matrix elements of G for several lattices.

5.3.1 One-Dimensional Lattice

Substituting (5.17) in (5.21) we obtain

$$G(\ell,m;z) = \frac{L}{N(2\pi)} \int_{-\pi/a}^{\pi/a} dk \frac{e^{ika(\ell-m)}}{z-\varepsilon_0-2V\cos(ka)}$$

$$= \frac{1}{2\pi} \int_{-\pi}^{\pi} d\phi \frac{e^{i\phi(\ell-m)}}{z-\varepsilon_0-2V\cos\phi} \quad . \tag{5.24}$$

To evaluate the integral, we observe first that it depends on the absolute value $|\ell - m|$. Next we transform it to an integral over the complex variable w along the unit circle. Thus we have

$$G(\ell,m;z) = \frac{-1}{2\pi i V} \oint dw \frac{w^{|\ell-m|}}{w^2-2xw+1} \tag{5.25}$$

where

$$x = \frac{z-\varepsilon_0}{B} \quad ; \quad B = 2V \tag{5.26}$$

the two roots of $w^2-2xw+1 = 0$ are given by

$$\rho_1 = x - \sqrt{x^2 - 1} \tag{5.27a}$$

$$\rho_2 = x + \sqrt{x^2 - 1} \tag{5.27b}$$

where by $\sqrt{x^2 - 1}$ we denote the square root whose imaginary part has the same sign as the Im{x}. (For real x one has to follow a limiting procedure.) It follows that $\rho_1\rho_2 = 1$. One can easily show that $|\rho_1|<1$ and $|\rho_2|>1$ unless x is real and satisfies the relation $-1\leq x\leq 1$. In the latter case both roots lie on the unit circle, and the integral (5.25) is not well defined. Hence, this condition gives the continuous spectrum of H which lies in the real E-axis between $\varepsilon_0 - 2V$ and $\varepsilon_0 + 2V$. For z not coinciding with this singular line, we obtain for $G(\ell,m;z)$ by the method of residues

$$G(\ell,m;z) = \frac{-1}{V} \frac{\rho_1^{|\ell-m|}}{\rho_1 - \rho_2}$$

$$= \frac{1}{\sqrt{(z-\varepsilon_0)^2-B^2}}\, \rho_1^{|\ell-m|} \qquad (5.28)$$

where ρ_1 is given by (5.27a). For z coinciding with the spectrum we have

$$G^{\pm}(\ell,m;E) = \frac{\mp i}{\sqrt{B^2-(E-\varepsilon_0)^2}} \left(x \mp i\sqrt{1-x^2}\right)^{|\ell-m|} \quad , \qquad (5.29)$$

where $\varepsilon_0 - B \leq E \leq \varepsilon_0 + B$, $x = (E-\varepsilon_0)/B$ and the symbol $\sqrt{1-x^2}$ denotes the positive square root.

The density of states per site is given by

$$\rho(E) = \mp \frac{1}{\pi} \operatorname{Im}\{G^{\pm}(\ell,\ell;E)\} = \frac{\theta(B-|E-\varepsilon_0|)}{\pi\sqrt{B^2-(E-\varepsilon_0)^2}} \quad . \qquad (5.30)$$

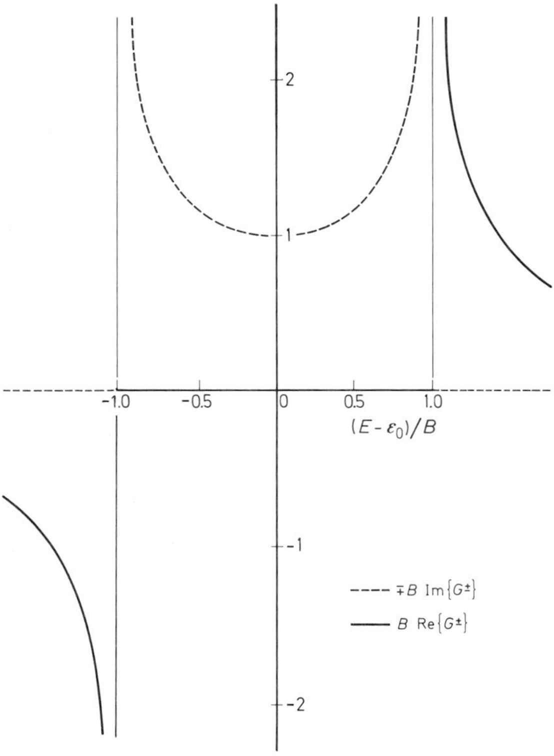

Fig.5.3. The diagonal matrix element $G^{\pm}(\ell,\ell;E)$ vs. E for the 1-D lattice. B = 2V is half the bandwidth

In Fig.5.3 we plot the real and the imaginary part of the diagonal matrix element $G(\ell,\ell;E)$ versus E. Note the square root singularities at both band edges. As was mentioned in Chap.3 this behavior is characteristic of the one-dimensionality. It is worthwhile to observe that the off-diagonal matrix elements $G(\ell,m;z)$ decay exponentially with the distance $|\ell-m|$ when z does not coincide with the spectrum. On the other hand, when z belongs to the spectrum, $|\rho_1| = 1$, and the matrix elements $G(\ell,m;z)$ do not decay with the distance $|\ell-m|$.

5.3.2 Square Lattice

For the square lattice we have, by substituting (5.18) in (5.21),

$$G(\underline{\ell},\underline{m};z) = \frac{a^2}{(2\pi)^2} \int_{1BZ} d^2k \frac{e^{i\underline{k}(\underline{\ell} - \underline{m})}}{z-\varepsilon_0-2V[\cos(k_1a) + \cos(k_2a)]} \tag{5.31}$$

where

$$\underline{k}(\underline{\ell} - \underline{m}) = a[k_1(\ell_1 - m_1) + k_2(\ell_2 - m_2)] \quad . \tag{5.32}$$

Here ℓ_1, ℓ_2, m_1 and m_2 are integers, a is the lattice constant; in the present case the first Brillouin zone is the square

$$-\pi/a \le k_1 < \pi/a$$

$$-\pi/a \le k_2 < \pi/a \quad .$$

Thus (5.31) can be rewritten as

$$G(\underline{\ell},\underline{m};z) = \frac{1}{(2\pi)^2} \int_{-\pi}^{\pi} d\phi_1 \int_{-\pi}^{\pi} d\phi_2 \frac{e^{i\phi_1(\ell_1-m_1)+i\phi_2(\ell_2-m_2)}}{z-\varepsilon_0-2V[\cos\phi_1 + \cos\phi_2]}$$

$$= \frac{1}{\pi^2} \int_0^{\pi} d\phi_1 \int_0^{\pi} d\phi_2 \frac{[\cos\phi_1(\ell_1-m_1)][\cos\phi_2(\ell_2-m_2)]}{z-\varepsilon_0-2V[\cos\phi_1 + \cos\phi_2]} \tag{5.33a}$$

$$= \frac{1}{\pi^2} \int_0^{\pi} d\phi_1 \int_0^{\pi} d\phi_2 \frac{[\cos(\ell_1-m_1+\ell_2-m_2)\phi_1][\cos(\ell_1-m_1-\ell_2+m_2)\phi_2]}{z-\varepsilon_0-4V \cos\phi_1 \cos\phi_2} \quad . \tag{5.33b}$$

For a derivation of the last expression see [5.5]. By taking the matrix elements of the equation $(z - \mathcal{H})G = 1$, one obtains recurrence relations which allow to express the arbitrary matrix element $G(\underline{\ell},\underline{m};z)$ in terms of the matrix elements $G(\underline{\ell},\underline{m};z)$ with $\ell_1 - m_1 = \ell_2 - m_2$ [5.5]. Furthermore, the matrix elements $G(\underline{\ell},\underline{m};z)$ with $\ell_1 - m_1 = \ell_2 - m_2$ can be expressed by recurrence relations in terms of the diagonal matrix element $G(\underline{\ell},\underline{\ell};z)$ and the matrix element $G(\underline{\ell},\underline{m};z)$ with $\ell_1 - m_1 = \ell_2 - m_2 = 1$ which will be denoted by $G(1;z)$. For the diagonal matrix element $G(\underline{\ell},\underline{\ell};z)$ we have

$$G(\underline{\ell},\underline{\ell};z) = \frac{1}{2\pi}\int_{-\pi}^{\pi} d\phi_1 \frac{1}{2\pi}\int_{-\pi}^{\pi} d\phi_2 \frac{1}{z-\varepsilon_0-B\cos\phi_1\cos\phi_2}$$

$$= \frac{1}{2\pi}\int_{-\pi}^{\pi} d\phi_1 \frac{1}{[(z-\varepsilon_0)^2-B^2\cos^2\phi_1]^{\frac{1}{2}}}$$

$$= \frac{1}{\pi(z-\varepsilon_0)}\int_{0}^{\pi} \frac{d\phi}{(1-\lambda^2\cos^2\phi)^{\frac{1}{2}}} \tag{5.34}$$

where

$$\lambda = B/(z - \varepsilon_0) \quad ; \quad B = 4V \quad , \tag{5.35}$$

hence,

$$G(\underline{\ell},\underline{\ell};z) = \frac{2}{\pi(z-\varepsilon_0)} \mathbb{K}(\lambda) \tag{5.36}$$

where $\mathbb{K}$ is the complete elliptic integral of the first kind. In a similar way, by performing the integration over ϕ_2, we obtain from (5.33b)

$$G(1;z) = \frac{1}{\pi}\int_{0}^{\pi} d\phi_1 \frac{\cos(2\phi_1)}{[(z-\varepsilon_0)^2-B^2\cos^2\phi_1]^{\frac{1}{2}}} \tag{5.37a}$$

$$= \frac{2}{\pi(z-\varepsilon_0)}\left[\left(\frac{2}{\lambda^2} - 1\right)\mathbb{K}(\lambda) - \frac{2}{\lambda^2}\cdot \mathbb{E}(\lambda)\right] \tag{5.37b}$$

where $\mathbb{E}(\lambda)$ is the complete elliptic integral of the second kind.

For $z = E+is$, $s \to 0^{\pm}$, and E within the band one may use the analytic continuation of $\mathbb{K}(\lambda)$ and $\mathbb{E}(\lambda)$ [3.1,5.5] to obtain explicit expressions for $G^{\pm}(\underline{\ell},\underline{\ell};E)$ and $G^{\pm}(1;E)$; e.g., for $G^{\pm}(\underline{\ell},\underline{\ell};E)$ we have

$$G(\underline{\ell},\underline{\ell};E) = \frac{2}{\pi(E-\varepsilon_0)}\,\mathbb{K}\left(\frac{B}{E-\varepsilon_0}\right) \;, \qquad |E-\varepsilon_0|>4V = B$$

$$\mathrm{Re}\{G^{\pm}(\underline{\ell},\underline{\ell};E)\} = -\frac{2}{\pi B}\,\mathbb{K}\left(\frac{E-\varepsilon_0}{B}\right) \;, \qquad -4V<E-\varepsilon_0<0$$

$$\mathrm{Re}\{G^{\pm}(\underline{\ell},\underline{\ell};E)\} = \frac{2}{\pi B}\,\mathbb{K}\left(\frac{E-\varepsilon_0}{B}\right) \;, \qquad 0<E-\varepsilon_0<4V$$

$$\mathrm{Im}\{G^{\pm}(\underline{\ell},\underline{\ell};E)\} = \mp\frac{2}{\pi B}\,\mathbb{K}\left(\sqrt{1-(E-\varepsilon_0)^2/B^2}\right) \;, \qquad |E-\varepsilon_0|<4V \;. \tag{5.38}$$

The density of states per site is given by

$$\rho(E) = \mp\frac{1}{\pi}\,\mathrm{Im}\{G^{\pm}(\underline{\ell},\underline{\ell};E)\} = \frac{2}{\pi^2 B}\,\theta(B-|E-\varepsilon_0|)\;\mathbb{K}\left(\sqrt{1-(E-\varepsilon_0)^2/B^2}\right) \;. \tag{5.39}$$

These functions are plotted in Fig.5.4. Note that the DOS exhibits at both band edges a discontinuity which produces the logarithmic singularities of the Re{G} at the band edges. As was discussed before, this behavior is characteristic of the 2-dimensionality of the system. Note also the singu-

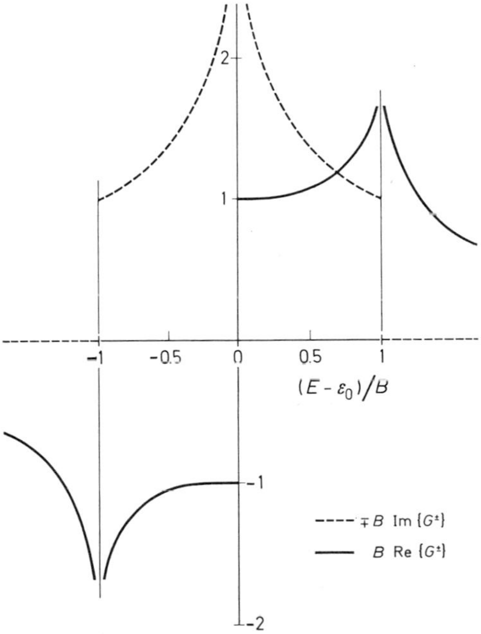

Fig.5.4. The diagonal matrix elements $G^{\pm}(\underline{\ell},\underline{\ell};E)$ vs. E for the 2-D square lattice. B = 4V is half the bandwidth

larity at the interior of the band (at $E = \varepsilon_0$); the $\mathrm{Re}\{G^\pm\}$ is discontinuous there and the $\mathrm{Im}\{G^\pm\}$ has a logarithmic singularity. The singularities of $G^\pm$ within the band are associated with saddle points in the function $E(\underline{k})$. A minimum number of such saddle points exists and depends on the number of independent variables $k_1 \ldots k_d$ [5.6].

As it was mentioned before, recurrence relations allow one to express all $G(\underline{\ell},\underline{m};z)$ in terms of $G(0;z) \equiv G(\underline{\ell},\underline{\ell};z)$, (5.36), and $G(1;z)$, (5.37). These recurrence relations develop numerical instabilities for $|E - \varepsilon_0| \gtrsim B$, especially along the direction of the x and y axes. The reason is that the recurrence relations are satisfied not only by the Green's functions, which decay to zero as $|\underline{\ell} - \underline{m}| \to \infty$, but by an independent set as well, which blows up as $|\underline{\ell} - \underline{m}| \to \infty$. Because of numerical errors, a very small component of this divergent set is present in $G(0;z)$ and $G(1;z)$ and is magnified as $|\underline{\ell} - \underline{m}|$ increases, until it eventually dominates the solution. To avoid this difficulty one has to use asymptotic expansions of $G(\underline{m},0;z)$, which are valid for $|\underline{m}| \gg R_0$, and which can be obtained by using the method of steepest descent in evaluating the integral for $G(\underline{m},0;z)$ [5.7]. The result for $|E| \leq B$ and for $|\underline{m}| \gg R_0$ is

$$2VG^+(\underline{m},0;E) = \frac{\pm 1 - i}{2\sqrt{\pi}} \, \frac{\exp[i(m_1\phi_1 + m_2\phi_2)]}{(|\underline{m}|/R_0)^{1/2}} \quad , \tag{5.40}$$

where the upper (lower) sign corresponds to $E > 0$ ($E < 0$); m_1, m_2 are the cartesian components of $\underline{m}$; ϕ_1, ϕ_2 are solutions of the following relations:

$$(E - \varepsilon_0)/2V = \cos\phi_1 + \cos\phi_2 \quad , \tag{5.41}$$

$$m_2 \sin\phi_1 = m_1 \sin\phi_2 \quad , \tag{5.42}$$

$$m_1 \sin\phi_1 \leq 0 \quad . \tag{5.43}$$

The characteristic length R_0 is (in units of the lattice constant a)

$$R_0 = \frac{(\sin^2\phi_1 + \sin^2\phi_2)^{1/2}}{|\sin^2\phi_1 \cos\phi_2 + \sin^2\phi_2 \cos\phi_1|} \quad . \tag{5.44}$$

It follows from (5.41-44) that the length R_0 depends on the direction m_1/m_2 as well as the energy E. In particular, as we approach the singular points $E = \varepsilon_0$ and $E = \varepsilon_0 + B$, the quantity $R_0 \to \infty$.

In the limit $E \to \varepsilon_0$, we obtain that $R_0 \to [4V(m_1^2 + m_2^2)/|E - \varepsilon_0|\,|m_1^2 - m_2^2|]^{1/2}$ when $m_2 \neq m_1$ and $R_0 \to \sqrt{2}\, 2V/|E - \varepsilon_0|$ when $m_2 = m_1$. In the same limit

$G^{\pm}(\underline{m},0;E) \to (-1)^{m_2} G^{\pm}(0;E)$ when $m_1 - m_2$ is a nonnegative even integer; and $G^{\pm}(\underline{m},0;E) \to (-i)^{m_1}/4V$ when $m_1 - m_2$ is a positive odd integer. All the other cases can be obtained by symmetry.

In the limit $E \to \varepsilon_0 + B$, the quantity $R_0 \to (V/|E - \varepsilon_0 - B|)^{1/2}$. In the same limit the differences $\delta G(\underline{m},0;E) \equiv G(\underline{m},0;E) - G(0;E)$ are finite numbers [5.8], which obey the same recurrence relations as the G's; the starting values are $\delta G(0;E) = 0$ and $\delta G(1;E) = -1/\pi V$. These recurrence relations eventually develop numerical instabilities and, in order to face this problem, one may use approximate asymptotic expansions which are of the form

$$\delta G(\underline{m},0;\ \varepsilon_0 + B) = -\frac{1}{\pi V}(\ln 2 + \frac{\gamma}{2}) - \frac{1}{2\pi V}\ln\left(\frac{m_1^2 + m_2^2}{2}\right)^{1/2} + O\left(\frac{1}{m_1^2 + m_2^2}\right) \quad . \tag{5.45}$$

For E outside the band the quantities ϕ_1, ϕ_2 in (5.40) become imaginary and consequently $G(\underline{m},0;E)$ decays exponentially with the distance $|\underline{m}|$. In particular for $E = \varepsilon_0 + B + \delta E$, $\delta E > 0$ and $\delta E/B << 1$, one obtains for large $|m|$

$$G(\underline{m},0;\ \varepsilon_0 + B + \delta E) = \frac{\exp(-|\underline{m}|\sqrt{\delta E/V})}{\sqrt{2\pi}(|\underline{m}|^2 \delta E/V)^{1/4}} \quad . \tag{5.46}$$

Several other 2-D lattices have been studied. Thus, HORIGUCHI [5.9] has expressed the Green's functions for the triangular and honeycomb lattices in terms of the complete elliptic integrals of first and second kind. HORIGUCHI and CHEN [5.10] obtained the Green's function for the diced lattice. For all these lattices the DOS exhibits the characteristic discontinuity at the band edges and the Re{G} has a logarithmic singularity. There are singular point(s) within the band where the Re{G} exhibits a discontinuity and the DOS has a logarithmic singularity.

If we are not interested in quantitative details we can approximate the Green's functions for 2-D lattices by a simple function which retains the correct analytic behavior near the band edges and which gives one state per site. This simple approximation is

$$G(\underline{\ell},\underline{\ell};z) = \frac{1}{2B}\ln\left(\frac{z-\varepsilon_0+B}{z-\varepsilon_0-B}\right) \tag{5.47}$$

which gives the following DOS

$$\rho(E) = \frac{1}{2B}\,\theta(B-|E-\varepsilon_0|) \quad . \tag{5.48}$$

We must stress that the simple expression (5.47) does not correspond to a real 2-D lattice and that it does not possess any Van Hove singularity in the interior of the band. In Fig.5.5 we plot $G(\underline{\ell},\underline{\ell};E)$ as given by (5.47) versus E.

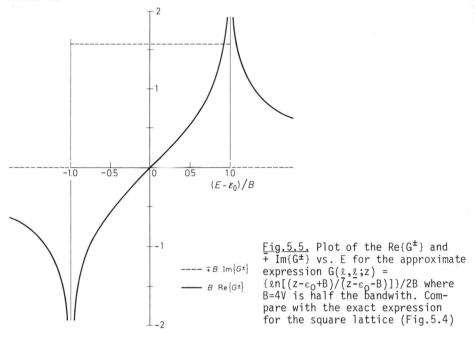

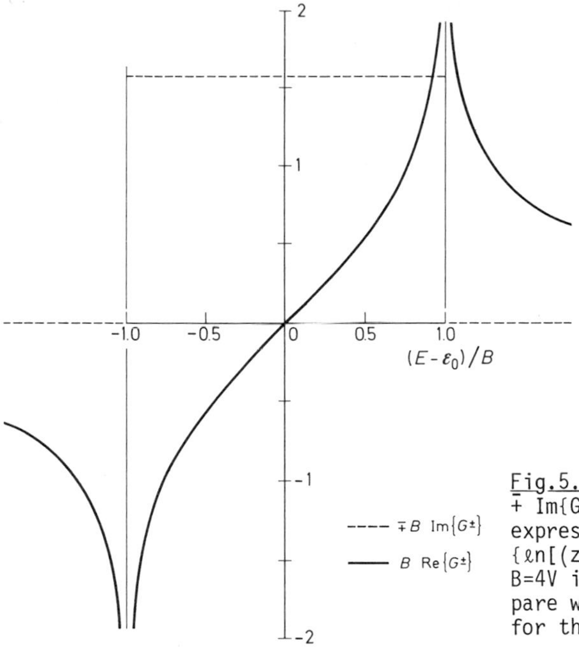

Fig.5.5. Plot of the $\mathrm{Re}\{G^\pm\}$ and $\mp\mathrm{Im}\{G^\pm\}$ vs. E for the approximate expression $G(\underline{\ell},\underline{\ell};z) = \{\ell n[(z-\varepsilon_0+B)/(z-\varepsilon_0-B)]\}/2B$ where B=4V is half the bandwith. Compare with the exact expression for the square lattice (Fig.5.4)

5.3.3 Simple Cubic Lattice

The first Brillouin zone for the simple cubic lattice is the cube

$$-\pi/a \leq k_1,\ k_2,\ k_3 < \pi/a$$

where a is the lattice constant. Substituting (5.19) in (5.21) and introducing the variables $\phi_i = k_i a$ (i = 1,2,3) we obtain

$$G(\underline{\ell},\underline{m};z) = \frac{1}{(2\pi)^3}\int_{-\pi}^{\pi} d\phi_1 \int_{-\pi}^{\pi} d\phi_2 \int_{-\pi}^{\pi} d\phi_3 \frac{\cos[(\ell_1-m_1)\phi_1+(\ell_2-m_2)\phi_2+(\ell_3-m_3)\phi_3]}{z-\varepsilon_0-2V(\cos\phi_1+\cos\phi_2+\cos\phi_3)} . \tag{5.49}$$

In particular the diagonal matrix element $G(\underline{\ell},\underline{\ell};z)$ is

$$G(\underline{\ell},\underline{\ell};z) = \frac{1}{(2\pi)^3}\int_{-\pi}^{\pi} d\phi_1 \int_{-\pi}^{\pi} d\phi_2 \int_{-\pi}^{\pi} d\phi_3 \frac{1}{z-\varepsilon_0-2V(\cos\phi_1+\cos\phi_2+\cos\phi_3)} ; \tag{5.50}$$

the integration over ϕ_2 and ϕ_3 in (5.50) can be done as in the square lattice case yielding $t\mathbb{K}(t)/2\pi V$ where

$$t = 4V/(z - \varepsilon_0 - 2V\cos\phi_1) \qquad (5.51)$$

thus,

$$G(\underline{\ell},\underline{\ell};z) = \frac{1}{2\pi^2 V}\int_0^\pi d\phi_1\; t\mathbb{K}(t) \quad . \qquad (5.52)$$

The last integral can be calculated numerically [5.4,11]. The $\mathrm{Re}\{G^\pm(\underline{\ell},\underline{\ell};E)\}$ and $\mathrm{Im}\{G^\pm(\underline{\ell},\underline{\ell};E)\}$ for real E are plotted in Fig.5.6. The behavior is typical for a 3-D system: near both band edges the DOS $-\mathrm{Im}\{G^+\}/\pi$ approaches zero continuously as $\sqrt{|\Delta E|}$, where $|\Delta E|$ is the difference of E and the corresponding band edge. The Re{G} remains finite around the band edges although its derivative with respect to E blows up as we approach the band edges from outside the band. Within the band there are two Van Hove singularities where both Im{G} and Re{G} are continuous while their derivatives are discontinuous and blow up.

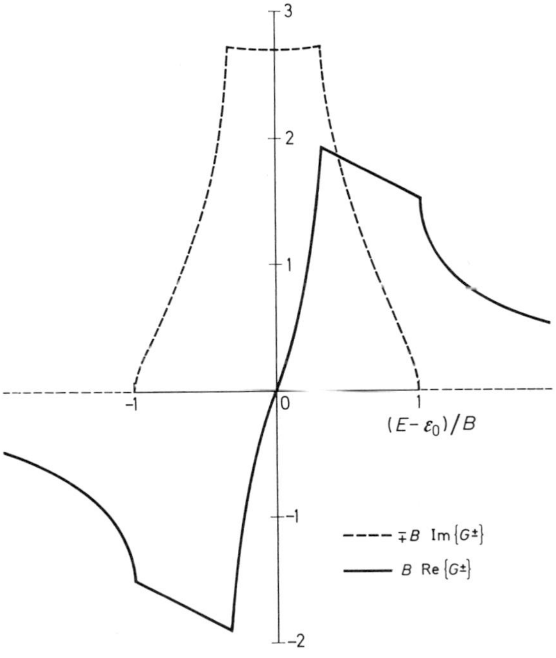

Fig.5.6. The diagonal matrix elements $G(\underline{\ell},\underline{\ell};E)$ vs. E for the simple cubic lattice. B = 6V is half the bandwidth

When $\underline{\ell}-\underline{m}$ is along an axis (e.g., the x-axis), one can perform the integration over ϕ_2 and ϕ_3 in (5.49) as in the square lattice yielding again $t\mathbb{K}(t)/2\pi V$; thus

$$G(\ell_1-m_1,0,0;z) = \frac{1}{2\pi^2 V}\int_0^\pi d\phi_1 \cos(\ell_1-m_1)\phi_1 \; t\mathbb{K}(t) \quad . \tag{5.53}$$

Details concerning the evaluation of the integral (5.53) are given in [5.12]; in the same reference some recurrence relations[1] are obtained which allow the computation of $G(\underline{\ell},\underline{m};z)$ for small $|\underline{\ell}-\underline{m}|$ in terms of $G(\ell_1-m_1,0,0;z)$. Using these techniques Table 5.2 at the end of this chapter was constructed. JOYCE [5.13] was able to prove that $G(\underline{\ell},\underline{\ell};z)$ can be expressed as a product of two complete elliptic integrals of the first kind. MORITA [5.14] obtained recurrence relations for the simple cubic lattice which allow the evaluation of all $G(\underline{\ell},\underline{m})$ in terms of $G(0,0,0) \equiv G(\underline{\ell},\underline{\ell})$, $G(2,0,0)$, and $G(3,0,0)$. The last two Green's functions were expressed by HORIGUCHI and MORITA [5.15] in closed forms in terms of complete elliptic integrals. The reader is also referred to the papers by AUSTEN and LOLY [5.16], by HIOE [5.17], by RASHID [5.18], by INAWASHIRO et al. [5.19], by KATSURA et al. [5.20], and by JOYCE [5.21]. Asymptotic expansions valid for large values of $|\underline{\ell}-\underline{m}|$ were obtained [5.7]; within the band $G(\underline{\ell},\underline{m})$ decays slowly as $|\underline{\ell}-\underline{m}|^{-1}$; outside the band it decays exponentially.

The Green's functions for other 3-D lattices, such as face- and body-centered cubic, have been calculated [5.11,14,16-18,22-26]. It is worthwhile to point out that the Green's functions for the fcc and bcc lattices blow up at the lower band edge and in the interior of the band, respectively [5.11]. This behavior is atypical for a 3-D system. A small perturbation such as the inclusion of second-nearest-neighbor transfer matrix elements will eliminate these pathological infinities.

In many cases, where quantitative details are not important, it is very useful to have a simple approximate expression (for G) which can be considered as typical for a 3-D lattice. Such an expression must exhibit the correct analytic behavior near the band edges and must give one state per site. Such a simple expression, widely used and known as the Hubbard Green's function, is the following

$$G(\underline{\ell},\underline{\ell};z) = \frac{2}{z-\varepsilon_0+\sqrt{(z-\varepsilon_0)^2-B^2}} \tag{5.54}$$

---

[1] There is a typographical error in Eqs.(3.8,9) of Ref.[5.12]: $G_{sc}(t;1,0,0)$ must be multiplied by t^3 instead of t^2.

where the sign of the $\mathrm{Im}\{\sqrt{(z-\varepsilon_0)^2 - B^2}\}$ is the same as the sign of the $\mathrm{Im}\{z\}$. The DOS corresponding to (5.54) is

$$\rho(E) = \frac{2\theta(B-|E-\varepsilon_0|)}{\pi B^2}\sqrt{B^2-(E-\varepsilon_0)^2} \quad . \tag{5.55}$$

The real and imaginary parts of G given by (5.54) are shown in Fig.5.7. Note that the simple approximation (5.54) does not reproduce any Van Hove singularity within the band.

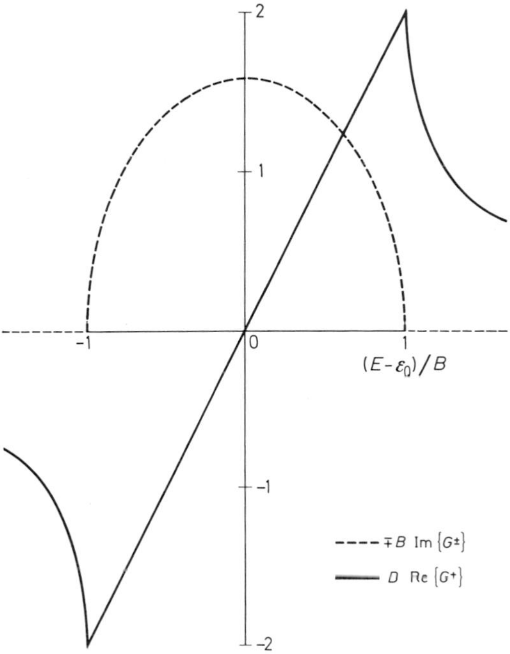

Fig.5.7. $\mathrm{Re}\{G^\pm\}$ and $\mp\,\mathrm{Im}\{G^\pm\}$ vs. E for the approximate 3-D $G(\ell,\ell;z) = 2/\{(z-\varepsilon_0)+[(z-\varepsilon_0)^2 - B^2]^{1/2}\}$ where B is half the bandwidth. Compare with the exact $G(\ell,\ell;z)$ for the simple cubic lattice (Fig.5.6)

5.3.4 Green's Functions for Bethe Lattices (Cayley Trees)

Bethe lattices or Cayley trees are lattices which have no closed loops and are completely characterized by the number of nearest neighbors Z or the connectivity K = Z - 1. In Fig.5.8 we show a portion of a Bethe lattice for K = 2 (Z = 3). The 1-D lattice is a Cayley tree with K = 1 (Z = 2).

Calculation of Green's functions in Bethe lattices can be done by using the renormalized perturbation expansion (RPE) as explained in Appendix B.

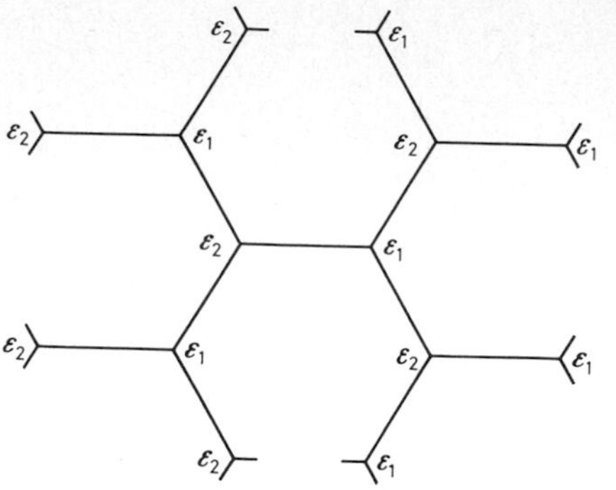

Fig.5.8. Part of a Bethe lattice (Cayley tree) with Z=3 nearest neighbors. The site energies $\{\varepsilon_\ell\}$ are arranged in an alternating periodic manner

For the double spacing periodic case shown in Fig.5.8 the result for $G(\ell,\ell;z)$ is (see Appendix B)

$$G(\ell,\ell;z) = G(1;z) \equiv \frac{2K(z-\varepsilon_2)}{D} \quad ; \quad \varepsilon_\ell = \varepsilon_1$$

$$G(2;z) \equiv \frac{2K(z-\varepsilon_1)}{D} \quad ; \quad \varepsilon_\ell = \varepsilon_2 \quad . \tag{5.56}$$

Here D is given by

$$D = (K-1)(z-\varepsilon_1)(z-\varepsilon_2)+(K+1)\{(z-\varepsilon_1)(z-\varepsilon_2)[(z-\varepsilon_1)(z-\varepsilon_2)-4KV^2]\}^{\frac{1}{2}} \tag{5.57}$$

where the sign of the imaginary part of the square root in (5.57) is the same as the sign of $\mathrm{Im}\{(z-\varepsilon_1)(z-\varepsilon_2)\}$. For the off-diagonal matrix element $G(\ell,m;z)$ we have (see Appendix B)

$$G(\ell,m) = G(\ell,\ell)V^{|m-\ell|}\left[G(\ell+1,\ell+1[\ell])\right]^{|m-\ell|/2}\left[G(\ell,\ell[\ell+1])\right]^{|m-\ell|/2}, \quad |m-\ell| \text{ even}$$

$$G(\ell,m) = G(\ell,\ell)V^{|m-\ell|}\left[G(\ell+1,\ell+1[\ell])\right]^{(|m-\ell|+1)/2}\left[G(\ell,\ell[\ell+1])\right]^{(|m-\ell|-1)/2}, \quad |m-\ell| \text{ odd} \quad . \tag{5.58}$$

where

$$G(\ell+1,\ell+1[\ell];z) = \frac{K+1}{z-\varepsilon_{\ell+1}+K[G(\ell+1,\ell+1;z)]^{-1}} \tag{5.59}$$

$$G(\ell,\ell[\ell+1];z) = \frac{K+1}{z-\varepsilon_\ell+K[G(\ell,\ell;z)]^{-1}} \quad . \tag{5.60}$$

As can be seen from (5.57) the spectrum consists of two subbands, the lower one extending from $(\varepsilon_1 + \varepsilon_2)/2 - \sqrt{(\varepsilon_1 - \varepsilon_2)^2/4 + 4KV^2}$ to ε_2 and the upper one from ε_1 to $(\varepsilon_1 + \varepsilon_2)/2 + \sqrt{(\varepsilon_1 - \varepsilon_2)^2/4 + 4KV^2}$. The DOS at sites with ε_1 (ε_2) is ρ_1 (ρ_2) where

$$\rho_i(E) = -\frac{1}{\pi} \mathrm{Im}\{G^+(i;E)\} \ ; \quad i = 1,2 \quad . \tag{5.61}$$

A plot of G(1;E) is shown in Fig.5.9. Note the analytic behavior at the band edges which is typically three dimensional except at one of the interior band edges where a square root divergence appears.

The simple periodic case is obtained by putting $\varepsilon_1 = \varepsilon_2 = \varepsilon_0$. We have then

$$G(\ell,\ell;z) = \frac{2K}{(K-1)(z-\varepsilon_0)+(K+1)\sqrt{(z-\varepsilon_0)^2-4KV^2}} \tag{5.62}$$

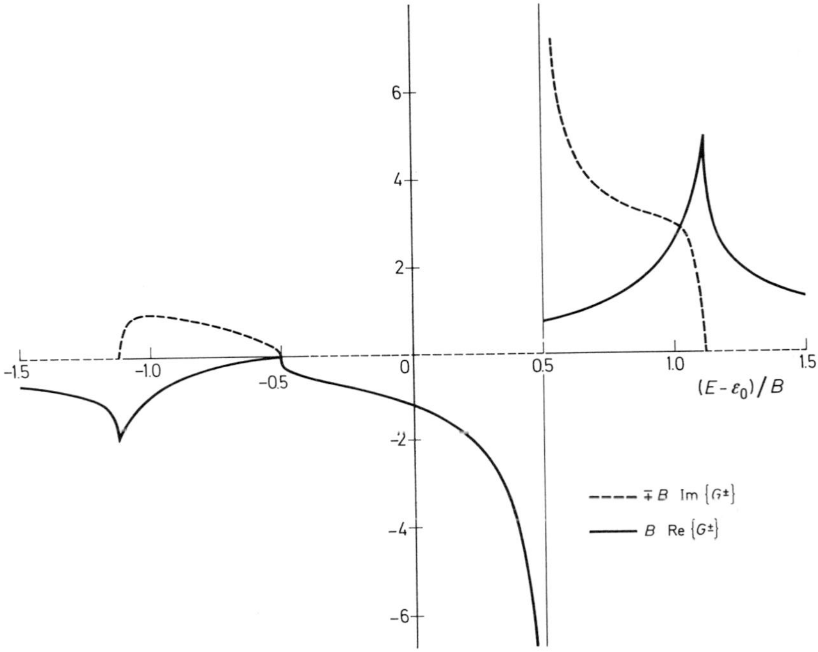

Fig.5.9. $\mathrm{Re}\{G^\pm(\ell,\ell)\}$ and $\mp\,\mathrm{Im}\{G^\pm(\ell,\ell)\}$ vs. E for a Bethe lattice with four nearest neighbors (K=3); $\varepsilon_{\ell+2n} = \varepsilon_1$, $\varepsilon_{\ell+2n+1} = \varepsilon_2$ where n integer and $\varepsilon_1 \geq \varepsilon_2$; $\varepsilon_0 = (\varepsilon_1+\varepsilon_2)/2$, $B = 2\sqrt{KV}$ and the difference $\varepsilon_1-\varepsilon_2$ has been taken equal to B. $\mathrm{Re}\{G^\pm(\ell+1,\ell+1;E)\} = -\mathrm{Re}\{G^\pm(\ell,\ell;-E+2\varepsilon_0)\}$; $\mathrm{Im}\{G^\pm(\ell+1,\ell+1;E)\} = \mathrm{Im}\{G^\pm(\ell,\ell;-E+2\varepsilon_0)\}$

where the sign of the imaginary part of the square root is the same as the sign of Im{z}. We obtain thus a single band centered around ε_0 with a half bandwidth $B = 2V\sqrt{K}$. Note that (5.62) reduces to the Hubbard Green's function in the limit $K \to \infty$ with $B = 2V\sqrt{K}$ remaining constant. For the quantity $G(\ell,\ell[\ell])$ we have when $\varepsilon_1 = \varepsilon_2 = \varepsilon_0$

$$G(\ell,\ell[\ell+1]) = \frac{2}{(z-\varepsilon_0)+\sqrt{(z-\varepsilon_0)^2-4KV^2}} \quad . \tag{5.63}$$

Hence the off-diagonal matrix elements are

$$G(\ell,m) = \frac{2K}{(K-1)(z-\varepsilon_0)+(K+1)\sqrt{(z-\varepsilon_0)^2-4KV^2}} \left[\frac{2V}{(z-\varepsilon_0)+\sqrt{(z-\varepsilon_0)^2-4KV^2}}\right]^{|\ell-m|} \quad . \tag{5.64}$$

For K = 1 the Bethe lattice is identical with the 1-D lattice, and consequently (5.64) reduces to (5.28) when K = 1. Similarly (5.56) and (5.58) give the double spacing periodic 1-D results when K = 1.

5.4 Brief Review of Applications in Solid-State Physics

There are two simple diametrically opposite approaches to the problem of electronic motion in solids. One, the nearly free electron (NFE) model, takes the point of view that in a solid the total *effective* potential felt by each electron is weak enough to be treated by perturbation methods. Developments based on this approach led to the pseudopotential method, which was proven to be very fruitful indeed, mainly for simple metals and semiconductors [5.27-34].

The other approach views the solids as made up of atoms brought together from infinite relative distance. It is then natural (following the usual practice for molecules) to try to express the unknown electronic wave function as a linear combination of atomic orbitals (LCAO). Thus one is led to a more complicated version of the simple tight-binding model examined here.

The complications are the following:

1) Usually one needs more than one orbital per site, e.g., tetrahedral solids (C, Si, Ge, etc.) require at least four orbitals per site (one s-like and three p-like), while transition metals require in addition five d-like orbitals.

2) One may have more than one atom per primitive crystal cell.

3) The matrix elements between orbitals at different sites may not decay fast enough so that more than nearest-neighbors matrix elements are needed.

4) The atomic-like orbitals at different sites may not be orthogonal to each other. Indeed, true atomic orbitals are not orthogonal.

Note that there is freedom in choosing the atomic-like orbitals. The Wannier functions, mentioned before, are one possible choice, which has the advantage of orthogonality and completeness and the disadvantage of weak decay (and hence many appreciable matrix elements). One may employ this freedom in choosing the basis in order to simplify the problem, i.e., to reduce the number of appreciable matrix elements and facilitate their computation.

The direct calculation of the matrix elements is in general a very difficult task. To obtain them indirectly one fits the LCAO results to either alternative calculations (see, e.g., [5.35]) or to experimental data (see, e.g., [5.36-38]). Recently, HARRISON [5.29,39] obtained very simple, general, and rather satisfactory expressions for the matrix elements by requiring that the NFE and the LCAO methods produce consistent results. SLATER and KOSTER [5.4] were the first to study in detail the tight-binding (or LCAO) method for various lattices. As is shown in HARRISON's book [5.29], the LCAO method is a very useful tool for analyzing many classes of materials such as covalent solids [5.35,40-45], ionic solids [5.46], simple metals [5.47], transition metals [5.48-50], transition metal compounds [5.51], and the A15 (such as Nb_3Sn) compounds [5.52,53].

It is worthwhile to mention that the Green's functions for Bethe lattices have several applications in solid-state physics, especially for analyzing amorphous solids [5.54,55] by the method of "cluster-Bethe-lattice". For a recent review the reader is referred to a paper by THORPE [5.56].

5.5 Summary

In this chapter we introduced the tight-binding Hamiltonian (TBH)

$$H = \sum_{\underline{\ell}} |\underline{\ell}\rangle \varepsilon_{\underline{\ell}} \langle \underline{\ell}| + \sum_{\underline{\ell}\underline{m}} |\underline{\ell}\rangle V_{\underline{\ell}\underline{m}} \langle \underline{m}| \tag{5.8}$$

where each state $|\underline{\ell}\rangle$ is an atomic like orbital centered at the site $\underline{\ell}$. The sites $\{\underline{\ell}\}$ form a regular lattice. The quantity $\varepsilon_{\underline{\ell}}$ is the energy of an electron located at the site $\underline{\ell}$ in the absence of $V_{\underline{\ell}\underline{m}}$. The quantity $V_{\underline{\ell}\underline{m}}$ is the amplitude for transfering an electron from the site $\underline{\ell}$ to the site $\underline{m}$. The electronic motion governed by the TBH (5.8) is mathematically equivalent to the motion of a coupled set of pendulums (see Table 5.1).

We examined the case of a periodic TBH. Periodicity implies that

$$\varepsilon_{\underline{\ell}} = \varepsilon_0 \quad \text{for all } \underline{\ell} \tag{5.7a}$$

$$V_{\underline{\ell}\underline{m}} = V_{\underline{\ell}-\underline{m}} \quad . \tag{5.7b}$$

The eigenfunctions and eigenvalues of the periodic TBH are of the following form

$$|\underline{k}) = \frac{1}{\sqrt{N}} \sum_{\underline{\ell}} e^{i\underline{k}\underline{\ell}} \, |\underline{\ell}> \tag{5.14}$$

$$E(\underline{k}) = \varepsilon_0 + \sum_{\underline{\ell}} V_{0\underline{\ell}} \, e^{i\underline{k}\underline{\ell}} \tag{5.15}$$

where $\underline{k}$ is restricted to a finite region called the first Brillouin zone. Using the general formula $G(z) = \sum_{\underline{k}} (|\underline{k})(\underline{k}|)/[z - E(\underline{k})]$ together with (5.14, 15), we can calculate $G(z)$. In the case where

$$\begin{aligned} V_{\underline{\ell}\underline{m}} &= V \quad ; \quad \underline{\ell},\underline{m} \text{ nearest neighbors} \\ &= 0 \quad ; \quad \text{otherwise} \end{aligned} \tag{5.10}$$

we obtain explicit expressions for G(z) for various lattices such as the 1-D, the 2-D square, the 3-D simple cubic and the Bethe lattice (see Figs. 5.3,4,6,9). The analytic behavior of these Green's functions near the band edges depends on the dimensionality as in the free particle case.

Finally, we discussed briefly the applications of the TBH [or the linear combination of atomic orbitals (LCAO) method] to the problem of the electronic structure of solids.

Table 5.2. Numerical values for the simple cubic Green's functions $G(\ell_1-m_1, \ell_2-m_2, \ell_3-m_3)$ [in units of $(2V)^{-1}$]

E/2V	Re{G(0,0,0)}	Im{G(0,0,0)}	Re{G(1,0,0)}	Im{G(1,0,0)}	Re{G(1,1,0)}	Im{G(1,1,0)}
0.0	0.0	0.896 441	-0.333 333	0.0	0.0	-0.185 788
0.2	0.074 176	0.896 927	-0.328 388	0.059 795	-0.036 758	-0.183 040
0.4	0.152 660	0.898 396	-0.312 979	0.119 786	-0.073 595	-0.174 773
0.5	0.195 323	0.899 508	-0.300 780	0.149 918	-0.092 168	-0.168 550
0.6	0.241 798	0.900 881	-0.284 974	0.180 176	-0.111 043	-0.160 914
0.8	0.356 091	0.904 444	-0.238 376	0.241 185	-0.151 676	-0.141 336
1.0	0.642 882	0.909 173	-0.119 039	0.303 058	-0.239 523	-0.115 850
1.2	0.623 924	0.617 641	-0.083 764	0.247 056	-0.210 658	0.000 621
1.4	0.606 577	0.506 449	-0.050 264	0.236 343	-0.180 223	0.049 291
1.5	0.598 434	0.463 545	-0.034 116	0.231 772	-0.164 446	0.067 430
1.6	0.590 611	0.425 657	-0.018 341	0.227 017	-0.148 310	0.082 687
1.8	0.575 841	0.360 233	0.012 171	0.216 140	-0.115 001	0.106 135
2.0	0.562 116	0.303 994	0.041 411	0.202 662	-0.080 371	0.121 426
2.2	0.549 312	0.253 400	0.069 496	0.185 826	-0.044 490	0.129 014
2.4	0.537 326	0.205 800	0.096 528	0.164 640	-0.007 418	0.128 408
2.5	0.531 612	0.182 285	0.109 677	0.151 904	0.011 546	0.124 557
2.6	0.526 070	0.158 374	0.122 594	0.137 257	0.030 785	0.117 828
2.8	0.515 470	0.105 986	0.147 772	0.098 920	0.070 069	0.092 137
3.0	0.505 462	0.0	0.172 129	0.0	0.110 383	0.0
3.2	0.400 104	0.0	0.093 444	0.0	0.042 757	0.0
3.4	0.357 493	0.0	0.071 825	0.0	0.028 498	0.0
3.5	0.341 048	0.0	0.064 556	0.0	0.024 187	0.0
3.6	0.326 621	0.0	0.058 612	0.0	0.020 855	0.0
3.8	0.302 143	0.0	0.049 381	0.0	0.016 041	0.0
4.5	0.242 855	0.0	0.030 949	0.0	0.007 868	0.0

E/2V	Re{G(1,1,1)}	Im{G(1,1,1)}	Re{G(2,0,0)}	Im{G(2,0,0)}	Re{G(2,1,0)}	Im{G(2,1,0)}
0.0	0.275 664	0.0	0.0	-0.153 291	0.057 669	0.0
0.2	0.265 774	-0.082 250	-0.058 501	-0.140 847	0.055 263	-0.014 153
0.4	0.234 932	-0.163 723	-0.108 662	-0.103 475	0.048 608	-0.025 972
0.5	0.210 461	-0.204 031	-0.127 430	-0.075 392	0.044 234	-0.030 162
0.6	0.178 622	-0.243 895	-0.139 596	-0.041 012	0.039 726	-0.032 830
0.8	0.083 045	-0.321 774	-0.130 786	0.046 798	0.033 989	-0.032 481
1.0	-0.201 680	-0.396 661	0.077 131	0.160 342	0.081 196	-0.022 246
1.2	-0.195 407	-0.152 186	0.017 674	-0.027 190	0.026 381	-0.094 125
1.4	-0.183 703	-0.073 165	-0.026 422	-0.041 854	-0.018 346	-0.094 170
1.5	-0.175 882	-0.043 612	-0.042 997	-0.037 947	-0.036 671	-0.087 016
1.6	-0.166 779	-0.018 002	-0.056 060	-0.029 951	-0.052 177	-0.076 716
1.8	-0.144 801	0.024 441	-0.072 019	-0.006 670	-0.074 373	-0.049 538
2.0	-0.117 903	0.057 434	-0.074 987	0.020 951	-0.084 251	-0.017 244
2.2	-0.086 194	0.081 738	-0.065 573	0.048 180	-0.081 178	0.016 267
2.4	-0.049 765	0.096 681	-0.044 321	0.070 841	-0.064 567	0.046 857
2.5	-0.029 803	0.100 026	-0.029 411	0.079 009	-0.051 010	0.059 461
2.6	-0.008 689	0.099 991	-0.011 721	0.084 050	-0.033 863	0.069 106
2.8	0.036 969	0.085 629	0.031 780	0.079 419	0.011 450	0.073 435
3.0	0.087 157	0.0	0.085 779	0.0	0.071 863	0.0
3.2	0.025 989	0.0	0.026 911	0.0	0.017 389	0.0
3.4	0.015 398	0.0	0.016 927	0.0	0.009 670	0.0
3.5	0.012 450	0.0	0.014 093	0.0	0.007 649	0.0
3.6	0.010 270	0.0	0.011 965	0.0	0.006 197	0.0
3.8	0.007 296	0.0	0.008 988	0.0	0.004 279	0.0
4.5	0.002 870	0.0	0.004 213	0.0	0.001 588	0.0

Table 5.2 (continued)

E/2V	Re{G(2,1,1)}	Im{G(2,1,1)}	Re{G(2,2,0)}	Im{G(2,2,0)}	Re{G(3,0,0)}	Im{G(3,0,0)}
0.0	0.0	0.185 788	0.0	-0.051 116	0.102 658	0.0
0.2	0.072 194	0.172 074	-0.035 756	-0.041 527	0.083 937	-0.059 520
0.4	0.136 244	0.131 113	-0.065 024	-0.012 416	0.031 616	-0.098 678
0.5	0.162 322	0.100 539	-0.074 501	0.010 095	-0.003 588	-0.104 662
0.6	0.182 491	0.063 357	-0.078 274	0.038 275	-0.041 447	-0.098 070
0.8	0.195 967	-0.030 276	-0.054 995	0.112 685	-0.106 839	-0.036 386
1.0	0.105 070	-0.148 591	0.169 476	0.214 911	-0.051 484	0.106 611
1.2	0.054 332	-0.122 370	0.131 214	-0.003 198	0.020 657	0.064 189
1.4	0.008 767	-0.117 578	0.088 374	-0.056 319	0.049 665	0.023 146
1.5	-0.011 436	-0.111 042	0.066 251	-0.070 590	0.051 810	0.002 450
1.6	-0.029 587	-0.101 889	0.044 256	-0.079 012	0.047 658	-0.015 996
1.8	-0.058 760	-0.076 806	0.002 587	-0.081 221	0.026 051	-0.042 000
2.0	-0.076 833	-0.044 847	-0.032 527	-0.067 062	-0.004 357	-0.049 882
2.2	-0.081 929	-0.009 132	-0.056 640	-0.040 003	-0.033 303	-0.038 901
2.4	-0.072 205	0.026 283	-0.064 998	-0.004 542	-0.051 000	-0.012 029
2.5	-0.061 218	0.042 154	-0.061 705	0.014 298	-0.052 691	0.005 294
2.6	-0.045 846	0.055 489	-0.052 563	0.032 255	-0.048 091	0.023 380
2.8	-0.001 059	0.067 703	-0.014 039	0.056 992	-0.015 603	0.052 086
3.0	0.063 931	0.0	0.056 110	0.0	0.055 090	0.0
3.2	0.012 685	0.0	0.009 106	0.0	0.009 231	0.0
3.4	0.006 405	0.0	0.004 293	0.0	0.004 594	0.0
3.5	0.004 862	0.0	0.003 180	0.0	0.003 499	0.0
3.6	0.003 792	0.0	0.002 428	0.0	0.002 749	0.0
3.8	0.002 443	0.0	0.001 513	0.0	0.001 809	0.0
4.5	0.000 743	0.0	0.000 428	0.0	0.000 611	0.0

E/2V	Re{G(3,1,0)}	Im{G(3,1,0)}	Re{G(4,0,0)}	Im{G(4,0,0)}
0.0	0.0	0.018 619	0.0	0.078 814
0.2	0.008 731	0.015 606	0.057 151	0.054 616
0.4	0.013 680	0.007 660	0.079 233	-0.006 106
0.5	0.013 690	0.002 605	0.069 084	-0.039 691
0.6	0.011 602	-0.002 458	0.043 454	-0.066 840
0.8	-0.000 094	-0.009 563	-0.039 781	-0.066 765
1.0	-0.054 831	-0.006 713	0.039 227	0.079 732
1.2	0.016 421	0.048 606	-0.033 782	-0.013 180
1.4	0.049 369	0.020 362	-0.031 992	0.025 213
1.5	0.054 050	0.002 144	-0.017 774	0.036 722
1.6	0.052 321	-0.015 436	-0.000 717	0.040 510
1.8	0.034 211	-0.042 968	0.028 959	0.027 341
2.0	0.004 549	-0.054 598	0.039 365	-0.002 085
2.2	-0.026 625	-0.047 353	0.025 540	-0.029 932
2.4	-0.048 773	-0.022 359	-0.005 390	-0.039 145
2.5	-0.053 043	-0.004 865	-0.021 874	-0.033 078
2.6	-0.050 898	0.014 238	-0.034 762	-0.019 427
2.8	-0.021 571	0.047 283	-0.032 872	0.023 128
3.0	0.051 046	0.0	0.040 578	0.0
3.2	0.007 147	0.0	0.003 578	0.0
3.4	0.003 231	0.0	0.001 389	0.0
3.5	0.002 360	0.0	0.000 962	0.0
3.6	0.001 782	0.0	0.000 695	0.0
3.8	0.001 091	0.0	0.000 396	0.0
4.5	0.000 299	0.0	0.000 094	0.0

6. Single Impurity Scattering

In this chapter we examine (using techniques developed in Chap.4) a model Hamiltonian describing the problem of a substitutional impurity in a perfect periodic lattice. We obtain explicit results for bound and scattering states. Certain important applications, such as gap levels in solids, Cooper pairs in superconductivity, resonance and bound states producing the Kondo effect, and impurity lattice vibrations, are presented.

6.1 Formalism

We consider here the case of a tight-binding Hamiltonian whose perfect periodicity has been destroyed at just one site (the $\underline{\ell}$ site); there the diagonal matrix element $\varepsilon_{\underline{\ell}}$ equals $\varepsilon_0 + \varepsilon$; at every other site it has the unperturbed value ε_0. This situation can be thought of physically as arising by substituting the host atom at the $\underline{\ell}$-site by a foreign atom having a level lying ε higher than the common level of the host atoms. Thus our TBH can be written as

$$H = H_0 + H_1 \tag{6.1}$$

where the unperturbed part H_0 is chosen as the periodic TBH examined in Chap.5, i.e.,

$$H_0 = \sum_{\underline{m}} |\underline{m}\rangle\varepsilon_0\langle\underline{m}| + V \sum_{\underline{n}\underline{m}}{}' |\underline{n}\rangle\langle\underline{m}| \tag{6.2}$$

and H_1 is the perturbation arising from the substitutional impurity

$$H_1 = |\underline{\ell}\rangle\varepsilon\langle\underline{\ell}| \quad . \tag{6.3}$$

The Green's function G_o corresponding to H_o was examined in detail for various lattices in Chap.5. In this chapter we would like to find the Green's function G corresponding to $H \equiv H_o + H_1$; having G (or equivalently the t-matrix T), one can extract all information about the eigenvalues and eigenfunctions of H.

Our starting point is (4.5), which expresses G in terms of G_o and H_1, i.e.,

$$G = G_o + G_o H_1 G_o + G_o H_1 G_o H_1 G_o + \dots \quad . \tag{6.4}$$

One could use the equivalent equation for T,

$$T = H_1 + H_1 G_o H_1 + H_1 G_o H_1 G_o H_1 + \dots \quad . \tag{6.5}$$

Substituting (6.3) in (6.5), we obtain

$$T = |\underline{\ell}>\varepsilon<\underline{\ell}| + |\underline{\ell}>\varepsilon<\underline{\ell}|G_o|\underline{\ell}>\varepsilon<\underline{\ell}| + |\underline{\ell}>\varepsilon<\underline{\ell}|G_o|\underline{\ell}>\varepsilon<\underline{\ell}|G_o|\underline{\ell}>\varepsilon<\underline{\ell}| + \dots$$

$$= |\underline{\ell}>\varepsilon\{1 + \varepsilon G_o(\underline{\ell},\underline{\ell}) + [\varepsilon G_o(\underline{\ell},\underline{\ell})]^2 + \dots\}<\underline{\ell}|$$

$$= |\underline{\ell}> \frac{\varepsilon}{1-\varepsilon G_o(\underline{\ell},\underline{\ell})} <\underline{\ell}| \tag{6.6}$$

where

$$G_o(\underline{\ell},\underline{\ell}) \equiv <\underline{\ell}|G_o|\underline{\ell}> \quad . \tag{6.7}$$

Having obtained a closed expression for T we have immediately G,

$$G = G_o + G_o T G_o = G_o + G_o|\underline{\ell}>\frac{\varepsilon}{1-\varepsilon G_o(\underline{\ell},\underline{\ell})} <\underline{\ell}|G_o \quad . \tag{6.8}$$

As was discussed in previous chapters, the poles of $G(E)$ [or $T(E)$] correspond to the discrete eigenvalues of H. In the present case the poles of G (or T) are given by

$$G_o(\underline{\ell},\underline{\ell};E_p) = 1/\varepsilon \tag{6.9}$$

as can be seen by inspection of (6.8) or (6.6). Note that the poles E_p must lie outside the band of H_o, because inside the band $G_o(\underline{\ell},\underline{\ell};E)$ has a nonzero imaginary part and consequently (6.9) cannot be satisfied. Using (6.8) we obtain for the residue of $G(\underline{n},\underline{m})$ at the pole E_p

$$\mathrm{Res}\{G(\underline{n},\underline{m};E_p)\} = -\frac{G_o(\underline{n},\underline{\ell};E_p)G_o(\underline{\ell},\underline{m};E_p)}{G_o'(\underline{\ell},\underline{\ell};E_p)} \tag{6.10}$$

where the prime denotes differentiation with respect to E. The degeneracy f_p of the level E_p can be found by using (3.6) and (6.10);

$$f_p = \mathrm{Tr}\{\mathrm{Res}\{\mathcal{G}(E_p)\}\} = \sum_{\underline{n}} \mathrm{Res}\{G(\underline{n},\underline{n};E_p)\}$$

$$= -\frac{1}{G_o'(\underline{\ell},\underline{\ell})}\sum_{\underline{n}} G_o(\underline{n},\underline{\ell})G_o(\underline{\ell},\underline{n}) = -\frac{\langle\underline{\ell}|(E_p-\mathcal{H}_o)^{-2}|\underline{\ell}\rangle}{G'(\underline{\ell},\underline{\ell})} = 1 \tag{6.11}$$

where the last step follows because of (1.31). Since the level E_p is nondegenerate, the discrete eigenfunction $|b)$ satisfies (3.7) which in the present case can be written as

$$\langle\underline{n}|b)(b|\underline{m}\rangle = \mathrm{Res}\{G(\underline{n},\underline{m};E_p)\} \quad . \tag{6.12}$$

Substituting (6.10) in (6.12) we obtain for $|b)$

$$|b) = \frac{\mathcal{G}_o(E_p)|\underline{\ell}\rangle}{\sqrt{-G_o'(\underline{\ell},\underline{\ell};E_p)}} \quad , \tag{6.13}$$

or more explicitly

$$|b) = \sum_{\underline{n}} b_{\underline{n}}|\underline{n}\rangle \tag{6.14}$$

where $b_{\underline{n}}$ is given by

$$b_{\underline{n}} = \frac{G_o(\underline{n},\underline{\ell};E_p)}{\sqrt{-G_o'(\underline{\ell},\underline{\ell};E_p)}} \quad . \tag{6.15}$$

We remind the reader that $G_o'(\underline{\ell},\underline{\ell};E)$ is always negative for E not belonging to the spectrum, see (1.31).

We have seen in Chap.5 that for E outside the band, $G_o(\underline{n},\underline{\ell};E)$ decays exponentially with the distance $R_{\underline{n}\underline{\ell}} \equiv |\underline{n} - \underline{\ell}|$, i.e.,

$$G_o(\underline{n},\underline{\ell};E) \xrightarrow[R_{\underline{n}\underline{\ell}}\to\infty]{} \text{const. } e^{-\alpha(E)R_{\underline{n}\underline{\ell}}} \tag{6.16}$$

with $\alpha(E)>0$. Hence, (6.14,15) imply that the eigenfunction $|b)$ is localized in the vicinity of the impurity site $\underline{\ell}$ and decays far away from it as $\exp[-\alpha(E_p)R_{\underline{n}\underline{\ell}}]$. The quantity $\alpha^{-1}(E_p)$ is a measure of the linear extent of the eigenfunction. In the 1-D case $\alpha(E)$ is given by

$$\alpha(E) = -\ell n\,[|E - \varepsilon_0|/B - \sqrt{(E - \varepsilon_0)^2/B^2 - 1}]/a \quad ; \quad |E - \varepsilon_0|>B. \qquad (6.17)$$

Let us now examine the effects of the perturbation on the continuous spectrum of H_0 (i.e., for E inside the band).

The partial DOS at the site $\underline{n}$ is given by $\rho(\underline{n};E) = -\,\mathrm{Im}\{<\underline{n}|G^+(E)|\underline{n}>\}/\pi$ which by using (6.8) becomes

$$\rho(\underline{n};E) = \rho_0(\underline{n};E) - \frac{\mathrm{Im}}{\pi}\left\{\frac{\varepsilon<\underline{n}|G_0^+(E)|\underline{\ell}><\underline{\ell}|G_0^+(E)|\underline{n}>}{1-\varepsilon G_0^+(\underline{\ell},\underline{\ell};E)}\right\} \quad . \qquad (6.18)$$

In particular the DOS at the impurity site $\underline{\ell}$ can be written after some simple algebraic manipulations as

$$\rho(\underline{\ell};E) = \frac{\rho_0(\underline{\ell};E)}{|1-\varepsilon G_0(\underline{\ell},\underline{\ell};E)|^2} \quad . \qquad (6.19)$$

Taking into account that $G_0(E) \xrightarrow[E\to\infty]{} 1/E$, we can easily see from (6.8) that $G(E) \xrightarrow[E\to\infty]{} 1/E$; hence $G(\underline{n},\underline{n};E) \equiv <\underline{n}|G(E)|\underline{n}> \xrightarrow[E\to\infty]{} 1/E$. This relation implies (by Cauchy's theorem) that

$$\int_{-\infty}^{\infty} \rho(\underline{n};E)dE = -\frac{\mathrm{Im}}{\pi}\left\{\int_{-\infty}^{\infty} G^+(\underline{n},\underline{n};E)dE\right\} = 1 \quad . \qquad (6.20)$$

The DOS $\rho(\underline{n};E)$ is a nonzero continuous function of E within the band (except possibly at isolated points); this means that the band of H coincides with the band of H_0. Unlike $\rho_0(\underline{n};E)$, the quantity $\rho(\underline{n};E)$ exhibits a δ-function singularity outside the band at the energy E_p of the bound state, where (6.18) can be recast as

$$\rho(\underline{n};E) = \frac{G_0(\underline{n},\underline{\ell};E_p)G_0(\underline{\ell},\underline{n};E_p)}{-G_0'(\underline{\ell},\underline{\ell};E_p)} \cdot \delta(E - E_p)$$

$$= |b_{\underline{n}}|^2\delta(E - E_p); \quad E \approx E_p \quad . \qquad (6.21)$$

To obtain the last result we have used (6.15). Equation (6.20) can be written then as

$$\int_{E_\ell}^{E_u} \rho(\underline{n};E)dE + \sum_p |b_{\underline{n}}|^2 = 1 \tag{6.22}$$

where the summation is over all the poles of G (if any), and E_ℓ and E_u are the lower and upper band edges, respectively. If we sum over all sites $\underline{n}$ and take into account that the bound eigenfunction(s) are normalized, we obtain

$$\int_{E_\ell}^{E_u} N(E)dE + P = N \tag{6.23}$$

where $P = \sum_p$ is the total number of poles (i.e., of discrete levels) and N(E) is the total DOS. Here N is the number of lattice sites which coincides with the total number of independent states in our Hilbert space. Equation (6.23) means that the discrete levels (if any) were formed at the expense of the continuous spectrum. Equation (6.22) means that this "sum rule" is valid for each site $\underline{n}$ and that the transfer of weight from the continuum at $\underline{n}$ to each discrete level equals the overlap $|b_{\underline{n}}|^2$ of this discrete level with the site $\underline{n}$.

The eigenstates in the band are of the scattering type and are given by (4.31) which in the present case can be written as

$$|\psi_E) = |\underline{k}) + \mathcal{G}_0^+(E)\mathcal{T}^+(E)|\underline{k}) \quad , \tag{6.24}$$

where $|\underline{k})$ is the Bloch wave given by (5.14); as was discussed in Chap.4, $\mathcal{G}_0^+\mathcal{T}^+$ selects the physically admissible outgoing solutions.

Substituting (6.6) in (6.24), we obtain (after some simple algebraic manipulations) the $\underline{n}$-site amplitude of $|\psi_E)$, $<\underline{n}|\psi_E)$

$$<\underline{n}|\psi_E) = <\underline{n}|\underline{k}) + \frac{<\underline{n}|\mathcal{G}_0^+(E)|\underline{\ell}>\varepsilon<\underline{\ell}|\underline{k})}{1-\varepsilon G_0^+(\underline{\ell},\underline{\ell};E)} \quad . \tag{6.25}$$

In particular the amplitude of $|\psi_E)$ at the site of the impurity is

$$<\underline{\ell}|\psi_E) = \frac{<\underline{\ell}|\underline{k})}{1-\varepsilon G_0^+(\underline{\ell},\underline{\ell};E)} \quad . \tag{6.26}$$

We have seen before in Chap.4 that the cross section $|f|^2$ for an incident wave $|\underline{k}_i)$ to be scattered to the final state $|\underline{k}_f)$ is proportional to $|(\underline{k}_f|T^+(E)|\underline{k}_i)|^2$ where T is the t-matrix. Using (6.6) and (5.14), we obtain for the scattering cross section

$$|f|^2 \propto \frac{\varepsilon^2}{|1-\varepsilon G_0^+(\underline{\ell},\underline{\ell};E)|^2} \quad . \tag{6.27}$$

We have pointed out before that $1-\varepsilon G_0^{\pm}(\underline{\ell},\underline{\ell};E)$ cannot become zero for E within the band. However, it is possible, under certain conditions, that the magnitude $|1-\varepsilon G_0^{\pm}(\underline{\ell},\underline{\ell};E)|^2$ becomes quite small for $E \approx E_r$. Then for $E \approx E_r$ the quantity $|<\underline{\ell}|\psi_E)|^2$ will become very large, as can be seen from (6.26). On the other hand, for $\underline{n}$ far away from $\underline{\ell}$, i.e., $R_{\underline{n}\underline{\ell}} \to \infty$, $|<\underline{n}|\psi_E)|^2 \to |<\underline{n}|\underline{k})|^2$, because $<\underline{n}|G_0(E)|\underline{\ell}>$ goes to zero as $R_{\underline{n}\underline{\ell}} \to \infty$. Thus the eigenfunction $|\psi_{E_r})$ reduces to the unperturbed propagating Bloch wave $|\underline{k})$ for $R_{\underline{n}\underline{\ell}} \to \infty$, while around the impurity site $\underline{\ell}$ it is considerably enhanced, as shown in Fig.6.1b.

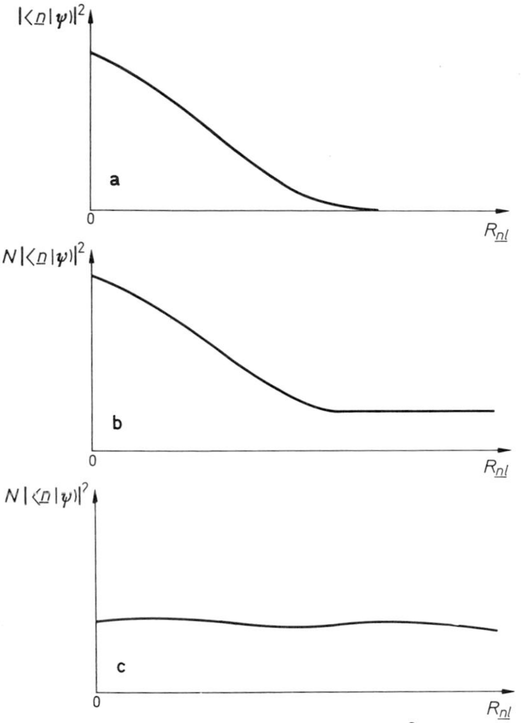

Fig.6.1. Schematic plot of $|<\underline{n}|\psi)|^2$ or $N|<\underline{n}|\psi)|^2$ vs. the distance $R_{\underline{n}\underline{\ell}}$ of the site $\underline{n}$ from the site $\underline{\ell}$; for an eigenstate $|\psi)$ localized around the site $\underline{\ell}$ (a); for a resonance eigenstate (b); and for an ordinary propagating (or extended) eigenstate (c)

For comparison we have also plotted schematically a bound eigenstate (Fig. 6.1a) and a regular (i.e., one for which $|1-\varepsilon G_0(\underline{\ell},\underline{\ell};E)|^2$ is not small) band state (Fig.6.1c). Eigenstates of the type shown in Fig.6.1b are called resonance eigenstates. They resemble, somehow, local (or bound) eigenstates in the sense that it is much more probable to find the particle around the site $\underline{\ell}$ than around any other site. On the other hand, they share with the other band states the property of being propagating and nonnormalizable.

Note that the resonance eigenenergies E_r will appear as peaks, both in the scattering cross section, see (6.27), and in the perturbed DOS $\rho(\underline{\ell};E)$, see (6.19). For the DOS $\rho(\underline{\ell};E)$, the peak may be shifted somehow if the unperturbed DOS $\rho_0(\underline{\ell};E)$ has a strong E dependence around E_r.

The quantity $|1-\varepsilon G_0^{\pm}(\underline{\ell},\underline{\ell};E)|^2$ can be written as $(1-\varepsilon \mathrm{Re}\{G_0\})^2 + \varepsilon^2(\mathrm{Im}\{G_0\})^2$. If $\mathrm{Im}\{G_0^{\pm}(\underline{\ell},\underline{\ell};E)\}$ is a slowly varying function of E (for E around E_r), then the resonance energy will be given approximately as a solution of

$$1-\varepsilon \mathrm{Re}\{G_0(\underline{\ell},\underline{\ell};E)\} \approx 0 \quad . \tag{6.28}$$

Obviously a sharp resonance requires that $\varepsilon^2(\mathrm{Im}\{G_0(\underline{\ell},\underline{\ell};E_r)\})^2$ must be much smaller than 1. If, furthermore, the derivative of $\mathrm{Re}\{G_0(\underline{\ell},\underline{\ell},;E)\}$ is a slowly varying function of E (for E around E_r), we can write

$$|1 - \varepsilon G_0(\underline{\ell},\underline{\ell};E)|^2 \approx A^{-2}[(E - E_r)^2 + \Gamma^2] \quad ; \quad E \simeq E_r \tag{6.29}$$

where $\Gamma = |\mathrm{Im}\{G_0(\underline{\ell},\underline{\ell};E_r)\}|/|\mathrm{Re}\{G_0'(\underline{\ell},\underline{\ell};E_r)\}|$ and $A^{-1} = |\varepsilon \mathrm{Re}\{G_0'(\underline{\ell},\underline{\ell},E_r)\}|$. Substituting the approximate expression (6.29) in (6.19), (6.26), and (6.27), we obtain

$$\rho(\underline{\ell};E) \approx \frac{A^2\rho_0(\underline{\ell};E)}{(E-E_r)^2+\Gamma^2} = \frac{A^2\rho_0(\underline{\ell};E)}{|E-z_r|^2} \quad , \tag{6.30}$$

$$|\langle\underline{\ell}|\psi_E\rangle|^2 \approx \frac{1}{N}\cdot\frac{A^2}{(E-E_r)^2+\Gamma^2} = \frac{A^2/N}{|E-z_r|^2} \quad , \tag{6.31}$$

$$|f|^2 \propto \frac{A^2\varepsilon^2}{(E-E_r)^2+\Gamma^2} = \frac{A^2\varepsilon^2}{|E-z_r|^2} \quad , \tag{6.32}$$

where

$$z_r \equiv E_r - i\Gamma \quad . \tag{6.33}$$

Thus, assuming the validity of (6.29), the resonance eigenstates appear as poles in the *lower* [*upper*] *half plane* of the *analytic continuation* of $G^+(\underline{\ell},\underline{\ell};E)$ $[(G^-(\underline{\ell},\underline{\ell};E)]$, across the branch cut; remember that $G^{\pm}(\underline{\ell},\underline{\ell};E)$ equals $G_0^{\pm}(\underline{\ell},\underline{\ell};E)/[1-\varepsilon G_0^{\pm}(\underline{\ell},\underline{\ell};E)]$. The inverse is also true: if the analytic continuation of $G^+(\underline{\ell},\underline{\ell};E)$ across the branch cut exhibits a pole near the real axis (with $\Gamma/B << 1$, B being the half bandwidth), then a resonance appears. Note though that the analytic continuation of $G^{\pm}(\underline{\ell},\underline{\ell};E)$ may possess a more complicated structure, in which case the quantities $\rho(\underline{\ell};E)$, $|\langle\underline{\ell}|\psi_E)|^2$ and $|f|^2$ do not have the simple form shown in (6.30-32). An example of this case is resonances appearing close to the band edges. We will see later that these resonances appear when the real pole E_p lies very close to the band edge.

Consider a particle, whose motion is described by H, placed initially ($t = 0$) at the site $\underline{\ell}$, i.e., $c_n(0) = \delta_{n\ell}$. Using (2.20,10,8) we obtain for the probability amplitude $c_{\underline{\ell}}(t)$ of still finding the particle at the same site later ($t>0$) the following expression:

$$c_{\underline{\ell}}(t) = \frac{i}{2\pi}\int_{-\infty}^{\infty} \exp(-iEt/\hbar)\, G^+(\underline{\ell},\underline{\ell};E)\, dE \quad . \tag{6.34}$$

The contribution to $c_{\underline{\ell}}(t)$ from a bound state [i.e., a pole of $G^+(\underline{\ell},\underline{\ell};E)$], $c_{\underline{\ell}b}(t)$, is

$$\begin{aligned} c_{\underline{\ell}b}(t) &= -2\pi i\,\frac{i}{2\pi}\,\mathrm{Res}\{G^+(\underline{\ell},\underline{\ell};E_p)\}\, e^{-iE_pt/\hbar} \\ &= |b_{\underline{\ell}}|^2\, e^{-iE_pt/\hbar} \end{aligned} \tag{6.35}$$

where (6.8) and (6.15) were taken into account. Equation (6.35) means physically that in the presence of the local eigenstate $|b)$ the particle will not diffuse away; even as $t \to \infty$ there will be a finite probability $|b_{\underline{\ell}}|^4$ of finding the particle at the site $\underline{\ell}$. The contribution to $c_{\underline{\ell}}(t)$ from ordinary band states (see Fig.6.1c), $c_{\underline{\ell}B}(t)$, is of the form

$$c_{\underline{\ell}B}(t) \sim e^{-|a|Bt/\hbar} \quad ; \quad t >> \hbar/B \quad , \tag{6.36}$$

where $|a|$ is of the order of unity. Thus, the ordinary band eigenstates help the particle to diffuse away at a characteristic time of the order $\hbar/B$ where B is the half bandwidth. Finally the contribution to $c_{\underline{\ell}}(t)$ from resonance eigenstates, $c_{\underline{\ell}r}(t)$, is, assuming a single pole representation of the resonance as in (6.30-32),

$$c_{\underline{\ell}r}(t) \sim e^{-\Gamma t/\hbar}\, e^{-iE_rt/\hbar} \quad . \tag{6.37}$$

Thus for $t \ll \hbar/\Gamma$ the resonance behaves similarly to the bound state as far as the quantity $c_{\underline{\ell}}(t)$ is concerned. The width Γ must be much smaller than B; otherwise the concept of a resonance eigenstate has no meaning at all.

To summarize this section: The discrete eigenvalues (if any) of H are found as solutions of (6.9). The corresponding eigenfunctions are local (or bound) and are given by (6.14) and (6.15). The DOS in the continuum is given by (6.18) and (6.19), and the scattering eigenfunctions by (6.25) and (6.26). Resonance eigenstates may appear in the continuum; these are propagating states which are considerably enhanced in the vicinity of the impurity site. The corresponding eigenenergies E_r appear as a sharp maximum in the DOS or in the scattering cross section. This maximum may come from a single pole z_r in the analytic continuation of $G^+(\underline{\ell},\underline{\ell};E)$ across the branch cut; then the real part of this pole gives the resonance eigenenergy and the inverse of the imaginary part determines a "lifetime" for the diffusion of a particle initially placed at the impurity site.

6.2 Explicit Results

6.2.1 Three-Dimensional Case

As was discussed in Chap.5, the typical 3-D DOS behaves as $C_\ell\sqrt{E-E_\ell}$ (or $C_u\sqrt{E_u-E}$) near the lower (or upper) band edge E_ℓ (or E_u). As a result of this, $\mathrm{Re}\{G_0(\underline{\ell},\underline{\ell};E_\ell)\} \equiv I_\ell$ and $\mathrm{Re}\{G_0(\underline{\ell},\underline{\ell};E_u)\} \equiv I_u$ are finite[1]; furthermore, because dG/dE is negative outside the band and $G(E)\to 0$ as $E \to\infty$, we have

$$I_\ell < G_0(\underline{\ell},\underline{\ell};E) < 0 \quad \text{for} \quad E < E_\ell \tag{6.38}$$

and

$$0 < G_0(\underline{\ell},\underline{\ell};E) \leq I_u \quad \text{for} \quad E_u \leq E \quad . \tag{6.39}$$

Hence, the solution of (6.9) requires us to distinguish several cases: 1) $\varepsilon \leq I_\ell^{-1}$; then (6.9) has one and only one solution E_p lying below E_ℓ. This is shown in Fig.6.2 for the case of a simple cubic lattice. 2) $I_\ell^{-1} < \varepsilon \leq 0$; then there is no solution of (6.9) because $1/\varepsilon$ lies below I_ℓ and consequently

[1] Exceptions appear in some cases, see Chap.5.

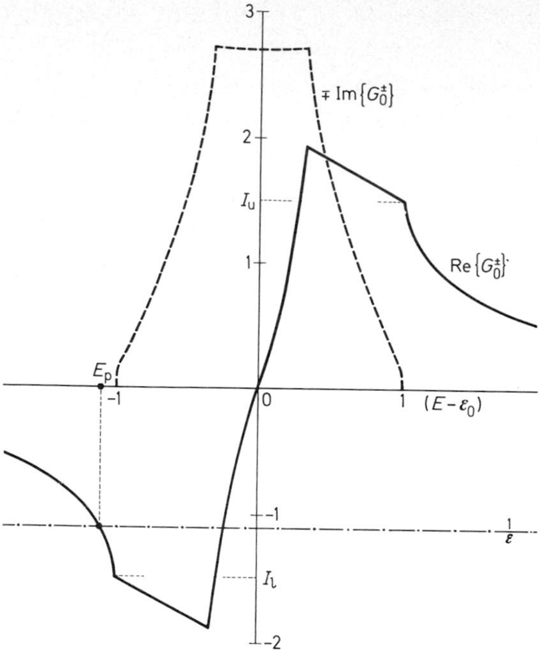

Fig.6.2. Graphical solution of the equation $\varepsilon G_0^{\pm}(\underline{\ell},\underline{\ell};E) = 1$ for the simple cubic lattice and for $\varepsilon < I_\ell^{-1} < 0$ (B = 1)

there is no intersection of G_0 with $1/\varepsilon$ (see Fig.6.2). The physical interpretation of this result is that an attractive ($\varepsilon<0$) impurity potential can create a bound eigenstate only when its strength $|\varepsilon|$ exceeds a critical value, $1/|I_\ell|$. 3) For positive ε, the behavior is similar: $|\varepsilon|$ needs to exceed a critical value, $1/I_u$, before one (and only one) bound level appears; this level lies above the upper band edge E_u. Thus a *repulsive* ($\varepsilon>0$) potential can trap a quantum particle in a level above the upper bound of the unperturbed continuum the same way that an attractive potential can trap a particle in a level below the lower bound of the continuum. This is a purely wave mechanical property of the same nature as the appearance of an *upper bound* to the continuum spectrum or the appearance of gaps. One can "understand" the connection better if one keeps in mind that in the vicinity of an upper bound of the spectrum the so-called effective mass is negative; hence the acceleration in a repulsive perturbing potential is attractive and a bound state is formed. Note that for $H_0 = p^2/2m$ (free particle) the continuous spectrum has a lower bound at E = 0 but no upper bound; in this case repulsive perturbing potentials cannot trap a particle.

We can conclude by pointing out that for $\varepsilon \leq -1/|I_\ell|$ there is one bound level below the continuum; for $-1/|I_\ell| < \varepsilon < 1/I_u$ there is no bound level; finally, for $I/I_u < \varepsilon$ there is one discrete (bound) level above the band.

It is interesting to see how the spectrum of H changes as $|\varepsilon|$ is gradually increasing from zero. To simplify the calculations, we use the Hubbard $G_0(\underline{\ell},\underline{\ell};E)$ (with $\varepsilon_0 = 0$), i.e., $G_0(\underline{\ell},\underline{\ell};E) = 2/(E \pm \sqrt{E^2 - B^2})$. For this simple case $|I_\ell|^{-1} = I_u^{-1} = B/2$, and $E_p = (B^2 + 4\varepsilon^2)/4\varepsilon$ $(|\varepsilon|>B/2)$. The DOS $\rho(\underline{\ell};E)$ is given by

$$\rho(\underline{\ell};E) = \frac{2}{\pi}\frac{\sqrt{B^2-E^2}}{B^2+4\varepsilon^2-4E\varepsilon} \ , \quad |E| < B \tag{6.40}$$

$\rho(\underline{\ell};E)$ exhibits a maximum at $E = E_r$, where

$$E_r = 4\varepsilon B^2/(B^2 + 4\varepsilon^2) \tag{6.41}$$

which can be considered as a resonance eigenenergy if the maximum is sharp enough. In Fig.6.3 we plot $\rho(\underline{\ell};E)$ for various negative values of ε. As $|\varepsilon|$ increases from zero, states are pushed towards the lower band edge; the maximum of $\rho(\underline{\ell},E)$ moves towards lower values and becomes sharper until a well-defined resonance near the lower band edge appears. This resonance coincides with the band edge for the critical value of $|\varepsilon| = B/2$. As $|\varepsilon|$ exceeds $B/2$, part of the resonance is split off the continuum to form a δ-function (which corresponds to the bound level) at $E = E_p$ of weight $|b_{\underline{\ell}}|^2 = \sqrt{E_p^2 - B^2}/|\varepsilon| = (4\varepsilon^2 - B^2)/4\varepsilon^2$. Further increase of $|\varepsilon|$ lowers the energy of the bound level, and increases its weight $|b_{\underline{\ell}}|^2$ at the expense of the continuum; at the same time the resonance recedes towards the center of the band, and becomes ill defined. This resonance is associated with the appearance of the real pole of G(E) at the vicinity of the band edge and not with a complex pole in the analytic continuation of $G^\pm$. Similar behavior is exhibited around the upper band edge E_u for repulsive $(\varepsilon>0)$ potentials.

In Fig.6.4 we summarize the above discussion by plotting: (a) the trajectories of the band edges E_ℓ and E_u, (b) the bound level E_p, and (c) the resonance [as defined from the peak of $\rho(\underline{\ell};E)$] as a function of the parameter ε, for the simple case of the Hubbard Green's function.

Note that in more complicated systems more than one bound level and resonance may appear. A typical case is shown schematically in Fig.6.5.

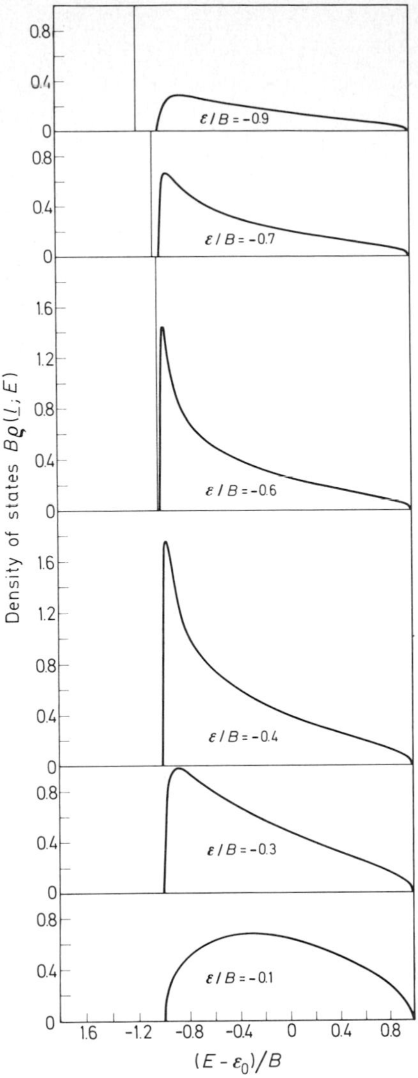

Fig.6.3. The perturbed DOS $\rho(\underline{\ell};E)$ vs. E for different values of the attractive local potential. As $|\varepsilon|$ increases from zero, states are pushed towards the lower band edge and at a critical value ($|\varepsilon|/B = 0.5$) a discrete level is split off the continuum. The unperturbed DOS $\rho_0(\underline{\ell};E)$ has been taken as in (5.55)

6.2.2 Two-Dimensional Case

As we discussed in Chap.5, the unperturbed DOS, $\rho_0(\underline{\ell};E)$, exhibits a discontinuity as E approaches the band edge, i.e.,

$$\rho_0(\underline{\ell};E) \xrightarrow[E \to E_\ell^+]{} \rho_d \quad . \tag{6.42}$$

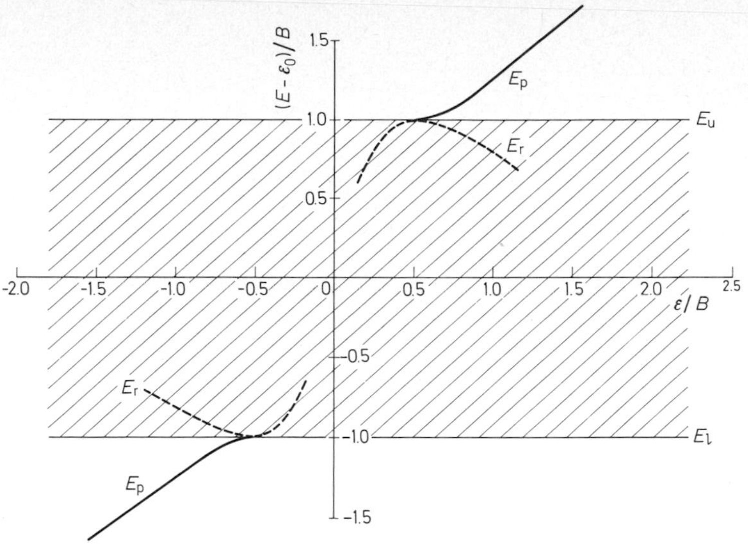

Fig.6.4. Discrete level (E_p), resonance level (E_r) and band edges (E_ℓ, E_u) vs. the local attractive ($\varepsilon<0$) or repulsive ($\varepsilon>0$) potential. The shaded area corresponds to the continuous spectrum. The unperturbed DOS $\rho_0(\underline{\ell};E)$ has been taken as in (5.55)

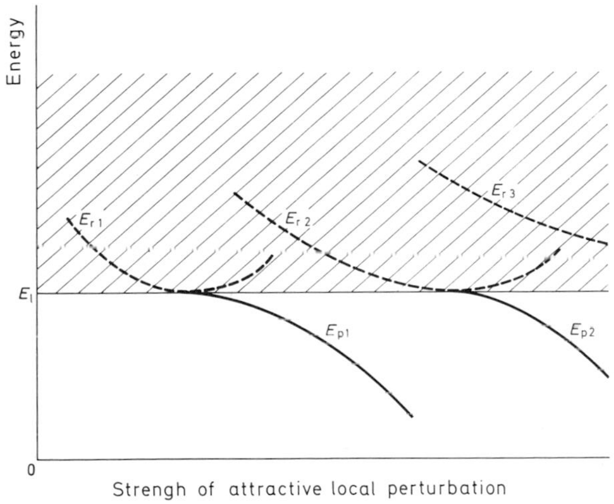

Fig.6.5. Schematic plot of various discrete (E_{pi}) and resonance (E_{ri}) levels vs. the magnitude of an attractive local perturbation. E_ℓ is the lower band edge

As a result

$$G_0(\underline{\ell},\underline{\ell};E) \xrightarrow[E \to E_\ell^-]{} \rho_d \, \ell n|(E - E_\ell)/C| \xrightarrow[E \to E_\ell^-]{} -\infty \tag{6.43}$$

where C is a positive constant. A similar behavior appears in the upper band edge. Thus, both I_ℓ and I_u are infinite, and consequently there is always one solution to (6.9) no matter how small $|\varepsilon|$ is. For very small $|\varepsilon|$, E_p lies very close to the band edge and one can use (6.43) for $G_0(\underline{\ell},\underline{\ell};E)$. Substituting in (6.9) we obtain

$$E_p - E_\ell \approx - C \exp\left(- \frac{1}{|\varepsilon|\rho_d}\right) \quad ; \quad \varepsilon \leq 0 \quad . \tag{6.44}$$

Thus the binding energy $|E_p - E_\ell|$ for weak perturbation depends exponentially on the strength of the perturbation. This dependence stems directly from the discontinuity of the unperturbed DOS at the band edge, which in turn is a characteristic feature of two-dimensional systems. For the case of the square lattice one obtains, by taking the limit $E \to \varepsilon_0 - B$ in (5.38),

$$G_0(\underline{\ell},\underline{\ell};E) \xrightarrow[E \to \varepsilon_0 - B]{} \frac{1}{\pi B} \ell n\left(\frac{|E-\varepsilon_0+B|}{8B}\right) ,$$

which means that $\rho_d = 1/\pi B$ and $C = 8B$. Thus for a square lattice

$$E_p - E_\ell \xrightarrow[\varepsilon \to 0^-]{} -8B \exp\left(- \frac{\pi B}{|\varepsilon|}\right) \quad . \tag{6.45}$$

It should be noted that (6.44) is applicable to the case of a particle of mass m moving in a 2-D potential well of depth V_0 and 2-D extent Ω_0; in this case $\rho_d = \rho_0\Omega_0$, where ρ_0 is the free particle DOS per unit area and $|\varepsilon| = V_0$. Substituting in (6.44) we obtain (4.76). The relevance of (6.44) to the theory of superconductivity will be discussed in the next section.

The interested reader may calculate explicitly various quantities of physical interest and construct a diagram of the type shown in Fig.6.4 by employing for the 2-D $G_0(\underline{\ell},\underline{\ell};z)$ the simple model (5.40).

6.2.3 One-Dimensional Case

In the 1-D case the unperturbed DOS near a band edge behaves as

$$\rho_0(\underline{\ell};E) \xrightarrow[E \to E_\ell^+]{} C(E - E_\ell)^{-1/2} \quad , \tag{6.46}$$

which implies that

$$G_0(\underline{\ell},\underline{\ell};E) \xrightarrow[E \to E_\ell^-]{} -\pi C(E_\ell - E)^{-1/2} \longrightarrow -\infty \quad . \tag{6.47}$$

Thus both I_ℓ and I_u are infinite, and consequently there is always a bound state no matter how small $|\varepsilon|$ is. For small negative ε the bound level is very close to the band edge and thus one can use expression (6.47) in (6.9). We then obtain for the binding energy E_b the following expression

$$E_b = |E_p - E_\ell| \xrightarrow[\varepsilon \to 0^-]{} \varepsilon^2\pi^2C^2 \quad . \tag{6.48}$$

For a particle of mass m moving in an ordinary potential well of depth V_0 and linear extent Ω_0, we have that $|\varepsilon| = V_0$ and $C = \Omega_0\hbar^{-1}\pi^{-1}(m/2)^{1/2}$, see (4.78). Then E_b is

$$E_b \xrightarrow[V_0\Omega_0 \to 0]{} \frac{mV_0^2\Omega_0^2}{2\hbar^2} \quad , \tag{6.49}$$

which agrees with (4.80).

The question of resonances in 1-D systems requires special attention. Firstly, instead of the scattering cross section, one defines the transmission and reflection coefficients $|t|^2$ and $|r|^2$, which in the present case are given by

$$|t|^2 = 1/|1 - \varepsilon G_0(\ell,\ell;E)|^2 \tag{6.50}$$

$$|r|^2 = |\varepsilon G_0(\ell,\ell;E)|^2/|1 - \varepsilon G_0(\ell,\ell;E)|^2 \quad . \tag{6.51}$$

Secondly, the quantity $G_0(\ell,\ell;E)$ is purely imaginary for E within the band so that $|r|^2+|t|^2 = 1$. The perturbed DOS $\rho(\ell;E)$ is given by combining (6.19) and (6.50); so

$$\rho(\ell;E) = \rho_0(\ell;E)|t|^2 \quad . \tag{6.52}$$

Similarly, the quantity $|\langle\ell|\psi)|^2$ is

$$|\langle\ell|\psi)|^2 = \frac{|t|^2}{N} \quad . \tag{6.53}$$

Since $|t|^2 \leq 1$ a resonance would appear as a sharp peak in $|t|^2$ versus E, where $|t|^2$ rises from values much smaller than one to values approaching one. As an example of such a behavior see [Ref.2.1, p.105].

For a 1-D lattice, $G_0(\ell,\ell;E)$ is (for $\varepsilon_0 = 0$)

$$G_0(\ell,\ell;E) = (E^2 - B^2)^{-1/2} \quad . \tag{6.54}$$

Thus, the bound level is

$$E_p = \bar{\varepsilon}(\varepsilon)\sqrt{B^2 + \varepsilon^2} \quad , \tag{6.55}$$

where the function $\bar{\varepsilon}(x)$ is defined by (2.37). The transmission coefficient is given by

$$|t|^2 = \frac{B^2-E^2}{B^2+\varepsilon^2-E^2} \quad , \tag{6.56}$$

which does not exhibit any resonance structure. Our model is too simple (no possibility for interference effects) for a resonance to appear.

We have seen that in 3-D, in contrast to 2-D or 1-D systems, the strength of H_1, $|\varepsilon|$, should exceed a critical value before a bound level appears. There is a simple physical interpretation of this basic result. Consider a potential well in a d-dimensional space; its depth is $-|\varepsilon|$ and its linear extent a. In order to trap a particle of mass m in a region of linear extent λ around the well the gain in the potential energy, ΔV, resulting from the particle being around the lower potential of the well, must overcome the corresponding increase in the kinetic energy ΔT. The quantity of ΔV is $\int V(\underline{r})|\psi|^2 d^d r \sim a^d|\varepsilon|/\lambda^d$, if $\lambda > a$ or $|\varepsilon|$ if $\lambda < a$. On the other hand, $\Delta T \approx p^2/2m \simeq h^2/2m\lambda^2$ where use of the uncertainty principle was made. For very small $|\varepsilon|$, which corresponds to large λ, $\Delta T \gg \Delta V$ when d>2, and $\Delta T \ll \Delta V$ when d>2. Thus, a bound state is formed for very small $|\varepsilon|$ if d<2, while no bound state can be formed if d>2. The d = 2 case is a borderline situation; we have seen that even for d = 2 a bound level is formed for very small $|\varepsilon|$, although the binding energy is extremely small.

6.3 Applications

6.3.1 Levels in the Gap

The Hamiltonian we considered in this chapter is an oversimplified model for describing what happens to the electronic properties of a crystalline solid in the presence of a substitutional impurity. The present model, in spite of its simplicity, retains the essential *qualitative* features of a real solid.
simplicity, retains the essential *qualitative* features of a real solid.
However, if quantitative features are of interest, one has to incorporate several complicating factors, e.g., the presence of more than one state for each lattice site $\underline{n}$, the possible nonorthogonality of the basis, etc. A detailed discussion can be found in the original papers by KOSTER and SLATER [6.1], in subsequent work by CALLAWAY and co-workers [6.2,3], in recent papers by PAPACONSTANTOPOULOS and ECONOMOU [5.45] and by BERNHOLC et al. [6.4], and in a book by LANNOO and BOURGOIN [6.5], where a more extensive list of references is given.

To perform realistic electronic calculations in the presence of isolated defects, one must determine the impurity potential. Even if this potential is known for a perfect solid consisting only of impurity atoms, still in the environment of the host material the electronic charge will be redistributed around the defect, thus modifying the local potential. To face this problem one has to carry out a self-consistent calculation [6.5,6]: start with a given impurity potential; calculate the Green's function as explained before; then determine the change in the electronic charge density [6.5]; using Poisson's equation obtain the local (impurity) potential; repeat the procedure until self-consistency is reached.

A further complication of the defect problem stems from the lattice relaxation (i.e., the displacement of the ions around the impurity to new equilibrium positions), which modifies directly as well as indirectly the electronic potential. The first principle determination of this relaxation is a very difficult problem which requires an accurate calculation of the total (electronic as well as ionic) energy for various plausible relaxations; the actual relaxation is the one which minimizes the total energy. Such a sophisticated calculation was carried out recently by BARAFF et al. [6.7] and by LIPARI et al. [6.8] for an isolated vacancy in Si.

6.3.2 The Cooper Pair and Superconductivity

Consider a system of fermions, each of mass m, interacting through an attractive pairwise potential: $V(r) = -V_0$ for $r \equiv |\underline{r}_i - \underline{r}_j| < a$ and zero otherwise. We will examine the motion of a pair of particles omitting the interactions of the pair with the rest of the particles. We shall take into account, however, that the Pauli principle forbids any member of the pair to be in an already occupied state.

In the absence of the interaction V, the eigenstate of the pair is $|\underline{k}_1, s_1, \underline{k}_2, s_2)$ where the momenta $\underline{k}_1$, $\underline{k}_2$ lie outside the Fermi sea defined by the rest of the particles. The spins s_1 and s_2 are arbitrary.

It is more convenient to transform to the center of mass and relative coordinates. Then the Hamiltonian of the pair, $\mathcal{H}_{12}$, can be written as

$$\mathcal{H}_{12} = \mathcal{H}_{CM} + \mathcal{H}_r = \frac{\hbar^2 K^2}{2M} - \frac{\hbar^2 \nabla_r^2}{2\mu} + V(\underline{r}) \quad , \tag{6.57}$$

where $M = 2m$, $\mu = m_1 m_2/(m_1 + m_2) = m/2$ and $\underline{r} \equiv \underline{r}_1 - \underline{r}_2$. The total momentum $\underline{K} = \underline{k}_1 + \underline{k}_2$ is a good quantum number of $\mathcal{H}_{12}$. Thus, the problem has been reduced to the examination of $\mathcal{H}_r$, which is the Hamiltonian of a spherical potential well, i.e., the problem examined in this chapter. All we need then is the unperturbed Green's function which in turn depends on the unperturbed DOS $\rho_0(E;\underline{K})$ per unit volume for the noninteracting pair of total momentum $\underline{K}$. The quantity $\rho_0(E;\underline{K})$ is given by the volume in $\underline{k}$-space subject to the restriction $E = \varepsilon(\underline{k}_1) + \varepsilon(\underline{k}_2)$, i.e.,

$$\rho_0(E;\underline{K}) = \frac{1}{(2\pi)^3} \int d\underline{k}\, \delta(E - \varepsilon(\underline{k}_1) - \varepsilon(k_2))\theta(\varepsilon(\underline{k}_1) - E_F)\theta(\varepsilon(\underline{k}_2) - E_F) \quad , \tag{6.58}$$

where $\underline{k} = (\underline{k}_1 - \underline{k}_2)/2$ is the momentum for the relative motion; $\underline{k}_1$, $\underline{k}_2$ in (6.58) must be replaced by $\underline{K}/2 + \underline{k}$ and $\underline{K}/2 - \underline{k}$, respectively. The two-step functions assure that neither electron of the pair can occupy levels below the Fermi energy E_F as required by Pauli's principle; for finite temperatures the product $\theta\theta$ has to be replaced by $[1 - f(\varepsilon(\underline{k}_1))][1 - f(\varepsilon(k_2))]$, where f is the Fermi distribution. Note that (6.58) remains valid even when the pair moves in an external periodic potential (such as in crystalline solids), in which case $\underline{k}_1$, $\underline{k}_2$, $\underline{K}$, $\underline{k}$ are crystal momenta and $\varepsilon(\underline{k})$ is in general more complicated than $\varepsilon(\underline{k}) = \hbar^2 k^2/2m$. It is clear from (6.58) that the presence of the fermion sea drastically modifies the unperturbed DOS reducing it to zero for $E < 2E_F$, i.e.

$$\rho_0(E;\underline{K}) = 0 \quad \text{for} \quad E < 2E_F \quad . \tag{6.59}$$

Furthermore for $\underline{K} = 0$, ρ_0 becomes

$$\rho_0(E;0) = \frac{1}{2}\rho(E/2)\theta(E/2 - E_F) \quad , \tag{6.60}$$

where $\rho(E)$ is the DOS per unit volume for a noninteracting particle of energy E and given spin orientation. Then it follows that

$$\rho_0(E;0) \xrightarrow[E \to 2E_F^+]{} \frac{1}{2}\rho_F \quad , \tag{6.61}$$

where $\rho_F = \rho(E_F)$. On the other hand

$$\rho_0(E;\underline{K}) \xrightarrow[E \to 2E_F^{\pm}]{} 0 \text{ for } \underline{K} \neq 0 \quad . \tag{6.62}$$

Thus, only for K = 0 does the unperturbed DOS drop discontinuously at $E = 2E_F$ from $\rho_F/2$ to 0. Under these conditions, as we have shown before, a bound state will be formed no matter how weak the attractive potential is. The binding energy of the K = 0 pair is given for weak attraction by (6.44), i.e.,

$$E_b(0) = C \exp\left(-\frac{2}{V_0\rho_F a^3}\right) \quad . \tag{6.63}$$

The basic result (6.63) was first obtained by COOPER [6.9]. For pairs with $K \neq 0$, the bound state, if it appears at all, will have weaker binding energy than $E_b(0)$. For spherically symmetric potentials $V(|\underline{r}_1 - \underline{r}_2|)$ (as the one we consider here) the bound pair, which is the ground state, is spherically symmetric and consequently remains invariant under the transformation $\underline{r} \to -\underline{r}$, i.e., under exchanging the two particles of the pair. Since the total wave function is odd under the exchange $1 \leftrightarrow 2$, the spin part must be odd, i.e., the pair must be in a singlet state. For spherically asymmetric potentials it is possible that the orbital ground state is odd under the exchange $1 \leftrightarrow 2$, and then the pair must be in a triplet spin state.

It should be stressed that the binding energy shown in (6.63) stemmed from the discontinuity in the unperturbed DOS for the relative motion of the pair. This discontinuity arose because of the presence of the other fermions which made the states below $2E_F$ unavailable to the pair. Thus, a system of fermions in the presence of an attractive interaction achieves its lowest energy by complete pair formation with each pair having a momentum $\underline{K} = 0$. This means that in the ground state (T = 0) all pairs undergo a quantum condensation. The existence of a condensate implies that independent pair excitations are suppressed. Thus the low-lying excitations are either pair-breaking individual fermions [each of which has a minimum energy $E_b(0)$ as in (6.63)], or collective density fluctuations (i.e., quantized sound waves for neutral fermions, or plasmons for charged fermions). Under these conditions, conservation of energy and momentum (see, e.g., LANDAU and LIFSHITZ [6.10]) implies that no elementary excitations can appear in a slow moving "condensed" system. Hence, such a system will be either superfluid (for neutral fermions as in He^3) or superconducting (for charged fermions as in various metallic superconductors).

The picture that emerged from our simple considerations is in qualitative agreement with the results of a sophisticated many-body theory of superconductivity [6.11-14]. However, there are quantitative discrepancies: the correct result for the binding energy of the pair is proportional to $\exp(-1/V_0\rho_F a^3)$ rather than $\exp(-2/V_0\rho_F a^3)$ as in (6.63). The resolution of this discrepancy is directly related with the answer to the following apparent paradox: Consider a system of noninteracting fermions moving in an external static weak potential unable to bound any particle. Clearly, it is enough to find the levels (none of which is bound) of a single fermion in this potential and then occupy these levels according to the Fermi distribution $f(E)$. However, if we consider a given fermion in the presence of all the others, then the available unperturbed DOS is $\rho(E)[1 - f(E)]$, which at $T = 0$ exhibits a discontinuity ρ_F; consequently, and in contrast to what was concluded before, there would be a bound level with binding energy proportional to $\exp(-1/|V|\rho_F)$

The origin of these difficulties is our omission of some processes, to be termed indirect, which take place as a result of the indistinguishability of the fermions. The DOS for the single fermion problem was found to be $\rho(E)[1 - f(E)]$, because it was assumed that all the other fermions of the Fermi sea play no role other than the passive one dictated by the Pauli principle. However, the fermions of the Fermi sea make possible additional processes (the indirect ones): A Fermi sea fermion can jump to the final state and the fermion under consideration fills up the created hole. Obviously the DOS for these indirect processes involving the occupied levels equals to $\rho(E)f(E)$. Adding this DOS to the direct processes DOS $\rho(E)[1 - f(E)]$ we obtain that the total DOS is $\rho(E)$, i.e., the same as in the absence of the Fermi sea which is the correct result.

There is an extensively studied case, where the cancellation of the $f(E)$-dependent term between the direct and the indirect processes is not complete: this case leads to the so-called Kondo effect, which will be summarized in the next subsection.

Returning now to the problem of the $\underline{K} = 0$ electronic pair we see that the omitted indirect processes involve two Fermi sea electrons jumping to the final states while the original electrons fill up the resulting holes. The corresponding DOS equals to $\rho(E/2)f(E/2)f(E/2)/2$. This DOS must be subtracted from the direct DOS $\rho(E)[1 - f(E)][1 - f(E)]/2$. The subtraction rather than addition of the two DOS has to do with the antisymmetry of the pair function under particle exchange. The total effective DOS is then

$$\rho_0(E;0) = \rho(E/2)[1 - 2f(E/2)]/2 \quad . \tag{6.64}$$

Equation (6.64) shows that the discontinuity at $T = 0$ and $E = 2E_F$ is now ρ_F instead of $\rho_F/2$ predicted by (6.60). This eliminates the spurious factor of 2 in the exponent of the rhs of (6.63).

Using the general expression

$$G_0(E;0) = \int dE' \frac{\rho_0(E';0)}{E - E'} \tag{6.65}$$

and (6.64) we can obtain $G_0(E;0)$. The integration limits in (6.65) are $2E_F - 2\hbar\omega_D$ and $2E_F + 2\hbar\omega_D$. The cutoff $2\hbar\omega_D$ is introduced because the attractive potential is nonzero only when the pair energy is in the range $[2E_F - 2\hbar\omega_D, 2E_F + 2\hbar\omega_D]$. For low temperatures, where $1 - 2f(E/2)$ varies rapidly at E_F, $|G_0(E;0)|$ exhibits a maximum at $E = 2E_F$. As one lowers the temperature this maximum becomes more pronounced until a critical temperature T_c is reached such that

$$|G_0(2E_F;0)|V_0 = 1 \quad . \tag{6.66}$$

For $T < T_c$, the equation $1 - V_0G_0(E;0) = 0$ will be satisfied and consequently a bound pair will be formed. Thus T_c is the critical temperature below which superconductivity appears. Substituting (6.64) in (6.65) and assuming that $\rho(E/2) \approx \rho_F$ and that $\hbar\omega_D/k_BT_c >> 1$, we obtain

$$G_0(2E_F;0) = -\rho_F \ln(2e^\gamma\hbar\omega_D/\pi k_BT) \quad . \tag{6.67}$$

Substituting (6.67) in (6.66) we obtain for T_c

$$T_c = \frac{2e^\gamma}{\pi}\frac{\hbar\omega_D}{k_B}\exp(-1/\lambda) \tag{6.68}$$

where $\lambda = \rho_FV_0a^3$ and $\gamma = 0.577...$ is Euler's constant. Equation (6.68) is exactly the BCS [6.11] result for T_c.

We conclude our brief summary of the theory of superconductivity by commenting on the origin of the attractive potential and on the so-called strong coupling modifications of (6.68).

The attraction required to overcome the Coulomb repulsion and bind the partners of each pair together can only be provided by the polarizable medium in which the electrons are embedded. The analogy of two persons on a mattress (which plays the role of the polarizable medium) suggests that the attractive interaction V between two electrons is proportional to the interaction of each electron with the polarizable medium V_{e-m}, and proportional to how easily the medium is polarizable which can be characterized by the inverse of a typical eigenfrequency ω_m of the medium. Thus V is given by

$$V \sim \frac{V_{e-m}^2}{h\omega_m} \quad . \tag{6.69}$$

In all well-understood cases the polarizable medium is the ionic lattice; then $V_{e-m} = V_{e-p}$ is the electron-phonon interaction and ω_m is a typical phonon frequency which can be taken equal to the Debye frequency ω_D. Many efforts have been made to find a V mediated by degrees of freedom other than the ionic ones but no positive results have been established. Taking into account that V_{e-p} depends on the momenta of each electron we see that actually V_{e-m}^2 in (6.69) is an appropriate Fermi surface average $\langle V_{e-p}^2\rangle$ of the electron-phonon interaction. Furthermore, there is not simply one eigenfrequency as was assumed in (6.69) but a continuum of eigenfrequencies characterized by a distribution $F(\omega)$. Thus (6.69) must be replaced by

$$V = \int d\omega F(\omega) \frac{\langle V_{e-p}^2\rangle}{\hbar\omega} \quad . \tag{6.70}$$

The quantity of interest is $\lambda = \rho_F V_0 a^3 = \rho_F \int V d^3r$ on which T_c depends exponentially. It is customary to express λ in terms of $\alpha^2(\omega) \equiv (\rho_F/2\hbar) \int d^3r \langle V_{e-p}^2\rangle$. Taking into account (6.70) one obtains

$$\lambda = 2 \int d\omega F(\omega) \frac{\alpha^2(\omega)}{\omega} \quad . \tag{6.71}$$

The phonon mediated electron-electron attraction V depends on the square of the electron-phonon interaction V_{e-p}^2. But the same quantity, V_{e-p}^2, determines the scattering probability of an electron by the lattice vibrations and consequently the phonon contribution to the electrical resistivity r. Thus one expects the materials with high lattice resistivity to have a high λ and hence to be high T_c superconductors. Such a correlation does indeed exist. As a matter of fact one can show that at high temperatures the derivative of the resistivity is given by

$$\frac{dr}{dT} = \frac{8\pi^2 k}{\hbar}\frac{1}{\omega_p^2}\lambda_{tr} \quad , \qquad (6.72)$$

where ω_p is the plasma frequency and λ_{tr} results from λ by substituting $\langle V^2_{e-p}\rangle$ by $\langle V^2_{e-p}(1 - \cos\theta)\rangle$, where θ is the angle between $\underline{k}_1$ and $\underline{k}_2$. In [6.15] λ_{tr}, as obtained by (6.72), is compared with λ, as obtained from T_c measurements, and/or tunneling experiments, and/or first principle calculations [6.16], for various metals. The overall agreement is impressive, suggesting thus that knowledge of ω_p, ω_D and dr/dT for a given metal provides a reasonable estimate of T_c through (6.72) and (6.68).

From (6.70) one can see that a large V can be obtained if there are many low-frequency phonons. This means that a soft, not so stable lattice may imply a high T_c. Indeed, most high T_c materials are not so stable structurally.

Let us examine finally how (6.68) can be improved. An obvious modification is the subtraction from V of a quantity V^* proportional to the screened Coulomb repulsion between the two electrons of the pair. This means that λ must be replaced by $\lambda - \mu^*$, where $\mu^* \equiv \rho_F \int d^3rV^*$. Usually μ^* is about 0.1. The origin of other modifications is more subtle and requires for its comprehension concepts which will be presented in Part III of this book. Nevertheless, a simplified exposition of these so-called strong coupling modifications will be attempted now.

In arriving at (6.68) we have implicitly assumed that the electron-phonon interaction V_{e-p} and the Coulomb repulsion V_c are so weak that they have no effect on the propagation of the individual electrons from which the pairs are made up. Actually both V_{e-p} and V_c modify the properties of each electron; it is these modified electrons (which are called dressed or quasi electrons) which combine to make up the pairs. One modification of importance is that the discontinuity at the Fermi level is reduced by a factor $w_F \equiv (1 - \partial\Sigma/\hbar\partial\omega)^{-1}$ for each electron of the pair (see Part III, Sects.9.3 and 11.1), where $\Sigma(\omega,\underline{k})$ is the so-called self-energy. One way to understand this reduction is by taking into account that only a fraction w_F of each electron propagates as a quasi electron, while the rest, $1 - w_F$, has no well-defined energy-momentum relation and thus does not produce any discontinuity at E_F. The net result is to multiply the DOS given by (6.64) by a factor w_F^2.

Another important effect of V_{e-p} and V_c is that they change the electron velocity at the Fermi level from its unperturbed value $\underline{v}_F$ to $\underline{v}'_F$ =

$(\underline{v}_F + \partial\Sigma/\hbar\partial\underline{k})w_F$. Since the single particle DOS is inversely proportional to the magnitude of the velocity, it follows that the DOS ρ_F must be replaced by $(v_F/v'_F)\rho_F$.

The result of the above two effects together is to multiply the quantity $\lambda - \mu^*$ (6.68) by a factor equal to $w_F^2(v_F/v'_F)$. To calculate this factor explicitly one needs to obtain the self-energy $\Sigma(\omega,\underline{k})$ which depends both on V_{e-p} and V_c. It is usually assumed that the Coulomb interaction V_c has no significant effects on $\partial\Sigma/\partial\omega$; furthermore, calculations employing the Hubbard dielectric function give that the effect of V_c on $\partial\Sigma/\partial\underline{k}$ is negligible for the usual electronic densities ($r_s \approx 2.5$) [6.14]. Thus in calculating Σ we usually keep only V_{e-p} and we employ second-order perturbation theory to obtain [6.12, 14,17]

$$\Sigma(\omega,\vec{p}) = -\lambda\hbar\omega \quad , \tag{6.73}$$

from which we get $w_F = (1 + \lambda)^{-1}$, $v_F/v'_F = (1 + \lambda)$ and $w^2_F(v_F/v'_F) = (1 + \lambda)^{-1}$. Hence the expression for T_c becomes

$$T_c = p \exp\left(-\frac{1 + \lambda}{\lambda - \mu^*}\right) \quad , \tag{6.74}$$

where the prefactor p is not equal to that of (6.68) because of contributions due to the non-quasiparticle smooth background of each electron propagator.

A more rigorous analysis based on the Eliashberg gap equations [6.17] gives the following expression for T_c [6.17,18]

$$T_c = p \cdot \exp\left(-\frac{1.04(1 + \lambda)}{\lambda - \mu^*(1 + 0.62\lambda)}\right) \tag{6.75}$$

which is remarkably close to our simplified result. According to McMILLAN, the prefactor in (6.75) is

$$p = \hbar\omega_D/1.45\ k_B \quad . \tag{6.76}$$

A more accurate value of p is given in [6.18].

6.3.3 The Kondo Problem

In most metals the electrical resistivity decreases when the temperature is lowered, as a result of the decreasing amplitude of the thermal ionic vibrations. In some metals containing magnetic impurities the resistivity *rises* as the temperature is lowered. In 1964 KONDO [6.19] examined a system of non-interacting electrons undergoing spin flip scatterings by external local moments. He was able to show that second-order contributions to the resistivity were increasing logarithmically with T as T was lowered. By keeping the most divergent contributions to all orders, ABRIKOSOV [6.20] concluded that the resistivity blows up at a characteristic temperature T_K. Many attempts were made (see, e.g., [6.21,22]) to remove the unphysical singularity and to understand the low-temperature ($T \ll T_K$) state of the system. KONDO [6.23] reviewed in detail the early literature. ANDERSON and co-workers [6.24] introduced the idea of scaling and they noticed the analogy of the Kondo problem with other problems in statistical physics; they managed thus to predict correctly the nature of the low-temperature state: one electron is bound to each local moment in a singlet state and each such combination is inaccessible to the rest of the electrons. The difficult task is how to follow continuously the solution from the high-temperature ($T \gg T_K$) weak scattering region to the low-temperature ($T \ll T_K$) bound state regime through the intermediate cross-over ($T \approx T_K$) region. Some contributions [6.25] are reviewed in [6.26]. Finally WILSON [6.27], by employing nonperturbative numerical treatment of the cross-over regime, succeeded in connecting the high T region to the low T regime and thus he obtained explicit numerical results for the T = 0 behavior. A clear presentation of the basic physical ideas is given in a review by NOZIERES [6.28]. Recently ANDREI [6.29] and, independently, WIEGMANN [6.30] managed to diagonalize exactly the Kondo Hamiltonian and to obtain quantities like the zero-temperature magnetic susceptibility in closed form.

Here we give a brief simplified explanation of why the bound state appears for $T \ll T_K$ and why the resistivity exhibits the logarithmic increase for $T \gg T_K$ by considering a tight-binding version of the Kondo Hamiltonian. Assume that one local moment is located at site $\underline{\ell}$. The scattering potential H_{1K} [instead of (6.3)] is given now by

$$H_{1K} = -J(|\underline{\ell}\downarrow\rangle S_+\langle\underline{\ell}\uparrow| + |\underline{\ell}\uparrow\rangle S_-\langle\underline{\ell}\downarrow| + |\underline{\ell}\uparrow\rangle S_z\langle\underline{\ell}\uparrow| - |\underline{\ell}\downarrow\rangle S_z\langle\underline{\ell}\downarrow|) \quad , \quad (6.77)$$

where the arrows indicate the direction of the electronic spin and $S_+ \equiv S_x + i\,S_y$, $S_- \equiv S_x - i\,S_y$, S_z are the spin components of the local moment. By considering the expansions (6.4) or (6.5) one sees immediately that the direct process DOS (for a given electronic spin), $\rho(1 - f)$, is between a product of S's, let us say S_-S_+. The corresponding indirect DOS, ρf, is sandwiched between S_+ and S_-; the reversal of the ordering of the S's reflect the reversal of the time sequence of the two consecutive spin flip events. Thus the total DOS (omitting terms which do not depend on f) is $(S_-S_+ - S_+S_-)\rho(1 - f) = -S_z\,2\rho(1 - f)$. Hence the problem is reduced to that of a local interaction J and an unperturbed DOS equal to $2\rho(1 - f)$. At T = 0, the latter has a discontinuity equal to $2\rho_F$ which for negative J (antiferromagnetic coupling) will be responsible for a bound state of energy E_b given by

$$E_b = C\,\exp(-1/2|J|\rho_F) \quad . \tag{6.78}$$

Furthermore, by the same method as for the Cooper pair, one obtains that the bound state appears only for $T < T_K$, where T_K is

$$T_K = \frac{2e^\gamma}{\pi}\frac{D}{k_B}\exp(-1/2|J|\rho_F) \tag{6.79}$$

with D being the cutoff. Finally for $T >> T_K$, the t-matrix is proportional to $J/(1 - JG_o)$, where G_o is the Green's function associated with the unperturbed DOS $2\rho(1 - f)$. Calculating G_o at $E = E_F$ [see (6.67)] and taking into account that the resistivity r is proportional to the square of the t-matrix, we obtain that

$$r \sim \frac{|J|^2}{[1 - 2|J|\rho_F \ln (T_o/T)]^2} \quad , \tag{6.80}$$

where T_o is the prefactor in the rhs of (6.79). Equation (6.80) exhibits the observed logarithmic increase with decreasing T (for $T >> T_K$).

We conclude this subsection by summarizing the physical origin of the Kondo effect: because of the S operators in (6.77), the f-dependent terms in the direct and indirect DOS do not cancel out as in the simple static case. As a result the effective total DOS is proportional to $(1 - f)$. This factor creates a discontinuity at $T = 0$ and hence a bound state. The same factor accounts for a decrease in the binding energy with increasing T, until a critical temperature T_K is reached beyond which no bound state exists. For $T > T_K$ there is, however, a resonance state which becomes less and less pronounced as T is further increased. This behavior of the resonance state accounts for the anomalous temperature dependence of the resistivity.

6.3.4 Lattice Vibrations in Crystals Containing "Isotope" Impurities

In this subsection we will examine the problem of small oscillations of a system of atoms of mass m placed at the sites {i} of a periodic lattice and experiencing an harmonic interaction of the form

$$V = -\frac{1}{2}\sum_{ij}\kappa_{ij}\,u_i\,u_j \quad , \tag{6.81}$$

where u_i is the displacement from equilibrium of the atom placed at the i site. We assume that $\kappa_{ij} = \kappa$ for i,j nearest neighbors, and $\kappa_{ij} = 0$ for i,j further away. Then one can show that $\kappa_{ii} = -Z\kappa$, where Z is the number of nearest neighbors. For simplicity we consider here a 1-D lattice, where $Z = 2$. Assuming a time dependence of the form $\exp(-i\omega t)$, we obtain the following equation of motion

$$-\omega^2 m_i u_i = F_i = -\frac{\partial V}{\partial u_i} = \sum_n \kappa_{in} u_n \tag{6.82}$$

or

$$(Z\kappa - \omega^2 m_i)u_i = \kappa \sum_n{}' u_n \quad , \tag{6.83}$$

where the sum extends over nearest neighbors only. Taking into account (5.10), we can rewrite (5.12) as follows

$$(E - \varepsilon_i)c_i = V \sum_n{}' c_n \tag{6.84}$$

where the sum extends over nearest neighbors. A simple inspection of (6.83) and (6.84) shows that the problem of lattice vibrations is mathematically equivalent to the TBM.

We consider now the so-called single isotope impurity case, where

$$m_i = m_o \quad \text{for} \quad i \neq \ell$$

$$m_i = m_o + \Delta m \quad \text{for} \quad i = \ell$$

corresponding to a single impurity in the TBM. Below we put in columns the corresponding quantities in the two equivalent problems

electrons	ZV	$E - \varepsilon_o$	ε	
lattice vibrations	$Z\kappa$	$Z\kappa - \omega^2 m_o$	$\Delta m \omega^2$.	(6.85)

The solution of the lattice problem at $\omega = 0$ is equivalent to the solution of the electronic problem at $E - \varepsilon_o = ZV$ for $\varepsilon = 0$, i.e., at the upper band edge without any perturbation. As ω^2 increases from zero the equivalent parameters $E - \varepsilon_o$ and ε in the TBM change: $\varepsilon = \omega^2 \Delta m$ and $E - \varepsilon_o = Z\kappa - \omega^2 m_o$; by eliminating ω^2 and substituting $Z\kappa$ by ZV we obtain

$$E - \varepsilon_o = ZV - \frac{m_o}{\Delta m}\varepsilon \quad . \tag{6.86}$$

As ω^2 varies, the parameters $E - \varepsilon_0$ and ε of the equivalent TBM move along the straight line defined by (6.86). In Fig.6.6 we plot the $E - \varepsilon_0$ versus ε diagram for the TBM; we also draw two straight lines according to (6.86) (O'AB corresponding to $\Delta m>0$ and O'CD corresponding to $\Delta m<0$). We see immediately from Fig.6.6 that for $\Delta m>0$ (i.e., the impurity mass heavier than the host mass) there is no discrete eigenfrequency; the spectrum is continuous and extends from $0'(\omega^2 = 0)$ to $B(\omega^2 = 2Z\kappa/m_0)$. The frequency corresponding to any point A is

$$\omega_A^2 = (2Z\kappa/m_0)\,\frac{(0'A)}{(0'B)} \quad . \tag{6.87}$$

Note that resonance eigenstates may appear at low frequencies if Δm becomes very large, so that the line O'B would rotate upwards until it crosses the dashed resonance line (if any) around the O' point.

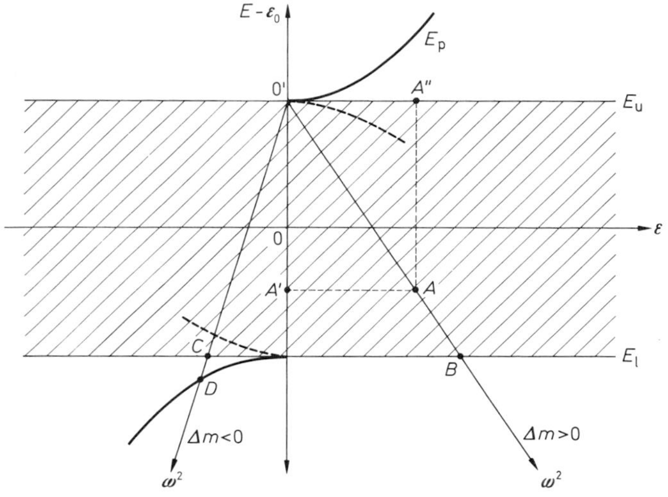

Fig.6.6. The electronic spectrum vs. ε diagram can be used to obtain the spectrum of the lattice vibration (see text)

For $\Delta m<0$ (i.e., impurity mass lighter than the host mass) there is a discrete eigenfrequency ω_D^2 because the ω^2 line crosses the E_p line (for typical 3-D cases $|\Delta m|$ must exceed a critical value for a discrete eigenfrequency to appear). The discrete eigenfrequency lies above the upper bound of the continuous spectrum and corresponds to a local oscillation, i.e., one confined around the impurity atom and decaying exponentially as one moves away from it. Using the analogies (6.85) and known results for the TBM one can obtain all information about the lattice vibration problems.

We can conclude this subsection by pointing out that a lighter impurity mass tends to push the eigenfrequencies up and to split off a discrete eigenfrequency above the upper bound(s) of the continuum(a). On the other hand, a heavier impurity mass tends to push the eigenfrequencies down; it cannot, of course, split a discrete eigenfrequency below $\omega^2 = 0$. However, if a lower band edge exists at a finite frequency (as in the case of an optic branch), then a heavier impurity mass can create a discrete level below the corresponding band.

6.4 Summary

In this chapter we examined (using the techniques developed in Chap.4) a model Hamiltonian describing the problem of a substitutional impurity in a perfect periodic lattice. The Hamiltonian is given by $H = H_0 + H_1$, where H_0 is a periodic TBH of the type studied in Chap.5 and H_1 describes the change of the potential as a result of the impurity and is given by

$$H_1 = |\underline{\ell}\rangle\varepsilon\langle\underline{\ell}| \quad . \tag{6.3}$$

As before each state $|\underline{n}\rangle$ is centered around the corresponding lattice site $\underline{n}$.

The Green's function corresponding to H can be evaluated exactly in terms of G_0 and ε. The result is

$$G(\underline{m},\underline{n};z) = G_0(\underline{m},\underline{n},z) + \varepsilon \frac{G_0(\underline{m},\underline{\ell};z)G_0(\underline{\ell},\underline{n};z)}{1-\varepsilon G_0(\underline{\ell},\underline{\ell};z)} \quad . \tag{6.8}$$

The discrete levels E_p, if any, are given by the poles of $G(z)$, i.e., by

$$G_0(\underline{\ell},\underline{\ell};E_p) = \frac{1}{\varepsilon} \quad . \tag{6.9}$$

In typical 3-D cases the strength $|\varepsilon|$ of the perturbing potential must exceed a critical value for a discrete level to appear. On the other hand, in 2-D systems a discrete level appears no matter how small $|\varepsilon|$ is. For small $|\varepsilon|$ the binding energy E_b is given by

$$E_b = \text{const} \cdot \exp[-1/(|\varepsilon|\rho_d)] \quad ; \tag{6.44}$$

where ρ_d is the discontinuity of the unperturbed DOS at the band edge. Equation (6.44) follows from (6.9) and the logarithmic behavior of $G_0(\underline{\ell},\underline{\ell};E)$ near the band edge. As was pointed out before, this logarithmic behavior is a direct consequence of the discontinuity of $\rho_0(\underline{\ell};E)$. In 1-D systems $G_0(\underline{\ell},\underline{\ell};E)$ has a square root infinity at the band edge. As a result, (6.9) has always a solution E_p which for small values of $|\varepsilon|$ behaves as

$$E_b = |E_p - E_B| = \text{const}.\varepsilon^2 \tag{6.48}$$

where E_B is the band edge.

The eigenstate $|b)$ associated with the nondegenerate discrete level E_p can be obtained as follows

$$|b) = \sum_{\underline{n}} b_{\underline{n}} |\underline{n}> \tag{6.14}$$

$$b_{\underline{n}} = \frac{G_0(\underline{n},\underline{\ell};E_p)}{\sqrt{-G_0'(\underline{\ell},\underline{\ell};E_p)}} \tag{6.15}$$

where the prime denotes differentiation with respect to E.

The eigenstates $|\psi_E)$ associated with the continuous part of the spectrum (scattering eigenstates) are given by

$$|\psi_E) = |\underline{k}) + \frac{G_0^+(E)|\underline{\ell}>\varepsilon<\underline{\ell}|\underline{k})}{1-\varepsilon G_0^+(\underline{\ell},\underline{\ell};E)} \tag{6.25}$$

where $|\underline{k})$ are the unperturbed Bloch functions.

The DOS at each site $\underline{n}$ can be obtained by taking the imaginary part of (6.8) with $\underline{m} = \underline{n}$. The DOS has a continuous part for E within the band and a δ-function contribution at the discrete level E_p, if the latter exists. As it was mentioned already, infinities of G outside the band correspond to discrete levels associated with eigenstates localized around the impurity. Sharp peaks of G (or equivalently of the t-matrix) within the band correspond to resonance levels associated with extended (or propagating) eigenstates which are considerably enhanced around the impurity site. The question of appearance of resonance eigenstates with increasing $|\varepsilon|$ is examined.

Among the various applications examined, two (superconductivity and Kondo effect) deserve special mention, because the present formalism provides a very simple derivation of their basic features.

Indeed, if one considers a system of identical fermions, it is a simple matter to show that the DOS for the relative motion of a pair of momentum $\underline{K} = 0$ is given by $\rho(1 - 2f)/2$, where f is the Fermi distribution. This DOS at T = 0 exhibits a discontinuity equal to ρ_F, and, hence, in the presence of an attraction between the partners of the pair, will be responsible for the creation of a bound state with binding energy of the form (6.44). Furthermore, for T > 0 this DOS exhibits no discontinuity at E_F but a maximum finite slope proportional to 1/T; as a result the Green's function $|G_0|$ does not blow up at E_F but exhibits a maximum of the form

$$|G_0|_{max} = \rho_F \ln (T_0/T) \quad , \tag{6.67}$$

where $T_0 \equiv 2e^{\gamma}\hbar\omega_D/\pi k_B$ is proportional to the cutoff $\hbar\omega_D$. Combining (6.67) with (6.9) one sees immediately that the bound state (and hence superconductivity) exists only for $T < T_c$, where T_c is obtained by equating $|G_0|_{max}$ to $1/|\varepsilon|$, i.e.

$$T_c = T_0 \exp(- 1/|\varepsilon|\rho_F) \quad . \tag{6.68}$$

Equation (6.68) is exactly the BCS result for T_c.

Similarly, for the Kondo problem (where a system of noninteracting fermions is scattered with strength J by a local moment undergoing spin flips) the DOS instead of being independent of f (as in the case of a static scatterer) includes a term of the form $2\rho(1 - f)$. Then, by the same simple reasoning as for the pair case above, one concludes that a bound electron-local moment state will be formed for $T < T_K$ [where T_K is given by (6.79)] and that the binding energy at T = 0 is as in (6.78). Furthermore, for $T_K < T$, the resistivity, which is proportional to the square of the t-matrix, will contain a factor $|1 - |J||G_0|_{max}|^{-2} = |1 - 2|J|\rho_F \ln (T_0/T)|^{-2}$; this factor accounts for the logarithmic increase of the resistivity with decreasing temperature.

7. Two or More Impurities; Disordered Systems

In this chapter we examine first a system consisting of two "impurities" embedded in a periodic tight-binding Hamiltonian (TBH). This prepares the way for the approximate treatment of a disordered system containing a nonzero concentration of impurities. We examine in particular the average DOS, which depends on the average Green's function $\langle G \rangle$, and transport coefficients like the electrical conductivity, which depends on an average of the form $\langle GAG \rangle$, where A is a known operator. Finally we discuss the important question of disorder-induced localization.

7.1 Two Impurities

We consider here the case where two substitutional impurities have been introduced at two different sites of the lattice, $\underline{\ell}$, $\underline{m}$. The Hamiltonian for such a system is assumed to be

$$H = H_0 + H_{\underline{\ell}} + H_{\underline{m}} \quad , \tag{7.1}$$

where H_0 is the periodic tight-binding Hamiltonian studied in Chap.5, and

$$H_{\underline{\ell}} = |\underline{\ell}\rangle\varepsilon\langle\underline{\ell}| \quad , \tag{7.2}$$

$$H_{\underline{m}} = |\underline{m}\rangle\varepsilon'\langle\underline{m}| \quad ; \tag{7.3}$$

we define further

$$H_{0\underline{\ell}} = H_0 + H_{\underline{\ell}} \quad , \tag{7.4}$$

$$H_{0\underline{m}} = H_0 + H_{\underline{m}} \quad . \tag{7.5}$$

Hence,

$$H = H_{o\underline{\ell}} + H_{\underline{m}} = H_{o\underline{m}} + H_{\underline{\ell}} \quad . \tag{7.6}$$

We denote by G_o, $G_{o\underline{\ell}}$, $G_{o\underline{m}}$ and G the Green's functions corresponding to H_o, $H_{o\underline{\ell}}$, $H_{o\underline{m}}$ and H, respectively.

We have seen in Chap.6 that

$$G_{o\underline{\ell}} = G_o + G_o T_{\underline{\ell}} G_o \quad , \tag{7.7}$$

where the t-matrix $T_{\underline{\ell}}$ associated with H_o and $H_{\underline{\ell}}$ is given by

$$T_{\underline{\ell}} = |\underline{\ell}>t_{\underline{\ell}}<\underline{\ell}| \; ; \quad t_{\underline{\ell}} = \frac{\varepsilon}{1-\varepsilon G_o(\underline{\ell},\underline{\ell})} \quad . \tag{7.8}$$

By considering $H_{o\underline{\ell}}$ as the unperturbed part and $H_{\underline{m}}$ as the perturbation, we obtain by applying the general formula (4.5)

$$G = G_{o\underline{\ell}} + G_{o\underline{\ell}} H_{\underline{m}} G_{o\underline{\ell}} + G_{o\underline{\ell}} H_{\underline{m}} G_{o\underline{\ell}} H_{\underline{m}} G_{o\underline{\ell}} + \dots \quad . \tag{7.9}$$

As before, the summation in (7.9) can be performed exactly, because of the simple form of $H_{\underline{m}}$. We obtain

$$G = G_{o\underline{\ell}} + G_{o\underline{\ell}} |\underline{m}> \frac{\varepsilon'}{1-\varepsilon' G_{o\underline{\ell}}(\underline{m},\underline{m})} <\underline{m}| G_{o\underline{\ell}} \quad . \tag{7.10}$$

Substituting (7.7,8) into (7.10) we have after some lengthy but straightforward algebra

$$G = G_o + G_o T G_o \quad , \tag{7.11}$$

where the t-matrix T associated with the unperturbed part H_o and the perturbation $H_{\underline{\ell}} + H_{\underline{m}}$ is given by

$$\begin{aligned} T &= f_{\underline{m}\underline{\ell}}(T_{\underline{\ell}} + T_{\underline{m}} + T_{\underline{\ell}} G_o T_{\underline{m}} + T_{\underline{m}} G_o T_{\underline{\ell}}) \\ &= f_{\underline{m}\underline{\ell}}(|\underline{\ell}>t_{\underline{\ell}}<\underline{\ell}| + |\underline{m}>t_{\underline{m}}<\underline{m}| \\ &\quad + |\underline{\ell}>t_{\underline{\ell}} G_o(\underline{\ell},\underline{m}) t_{\underline{m}}<\underline{m}| + |\underline{m}>t_{\underline{m}} G_o(\underline{m},\underline{\ell}) t_{\underline{\ell}}<\underline{\ell}|) \quad . \end{aligned} \tag{7.12}$$

The quantity $f_{\underline{m}\underline{\ell}}$ is

$$f_{\underline{m}\underline{\ell}} = \frac{1}{1-t_{\underline{m}}t_{\underline{\ell}}G_0(\underline{m},\underline{\ell})G_0(\underline{\ell},\underline{m})} \quad , \tag{7.13}$$

and $t_{\underline{m}}$ equals to $\varepsilon'/[1 - \varepsilon' G_0(\underline{m},\underline{m})]$.

The Green's function $G(\underline{j},\underline{i})$ is obtained by combining (7.11) and (7.12). The resulting expression can be represented in a diagrammatic way as shown in Fig.7.1, where we have drawn all continuous paths starting from site $\underline{i}$ and ending at site $\underline{j}$; the intermediate sites (if any) are the scattering centers $\underline{\ell}$ and $\underline{m}$. With each path (diagram) we associate a product according to the rules given in the caption of Fig.7.1. Thus the total contribution of the diagrams of subgroup (a) is

$$G_0(\underline{j},\underline{\ell})t_{\underline{\ell}}G_0(\underline{\ell},i) + G_0(\underline{j},\underline{\ell})t_{\underline{\ell}}G_0(\underline{\ell},\underline{m})t_{\underline{m}}G_0(\underline{m},\underline{\ell})t_{\underline{\ell}}G_0(\underline{\ell},i) + \dots$$

$$= G_0(\underline{j},\underline{\ell})t_{\underline{\ell}}G_0(\underline{\ell},i)\,\{1 + t_{\underline{\ell}}G_0(\underline{\ell},\underline{m})t_{\underline{m}}G_0(\underline{m},\underline{\ell}) + [t_{\underline{\ell}}G_0(\underline{\ell},\underline{m})t_{\underline{m}}G_0(\underline{m},\underline{\ell})]^2 + \dots\}$$

$$= G_0(\underline{j},\underline{\ell})t_{\underline{\ell}}G_0(\underline{\ell},i)f_{\underline{m}\underline{\ell}} \quad , \tag{7.14}$$

where $f_{\underline{m}\underline{\ell}}$ is given by (7.13). Similarly the total contribution of subgroups (b), (c) and (d) in Fig.7.1 are $G_0(\underline{j},\underline{m})t_{\underline{m}}G_0(\underline{m},i)f_{\underline{m}\underline{\ell}}$, $G_0(\underline{j},\underline{m})t_{\underline{m}}G_0(\underline{m},\underline{\ell})t_{\underline{\ell}}$ $G_0(\underline{\ell},\underline{i})f_{\underline{m}\underline{\ell}}$ and $G_0(\underline{j},\underline{\ell})t_{\underline{\ell}}G_0(\underline{\ell},\underline{m})t_{\underline{m}}G_0(\underline{m},\underline{i})f_{\underline{m}\underline{\ell}}$, respectively. Hence the summation of all diagrams of the type shown in Fig.7.1 is

$$G(\underline{j},\underline{i}) = G_0(\underline{j},\underline{i}) + \langle \underline{j}|G_0(T_{\underline{\ell}} + T_{\underline{m}} + T_{\underline{\ell}}G_0T_{\underline{m}} + T_{\underline{m}}G_0T_{\underline{\ell}})G_0|\underline{i}\rangle f_{\underline{m}\underline{\ell}} \tag{7.15}$$

which agrees with (7.11,12).

The diagrams shown in Fig.7.1 allow a physical interpretation of the various terms contributing to $G(\underline{j},\underline{i})$. Thus, if the thick line from $\underline{i}$ to $\underline{j}$ [i.e., $G(\underline{j},\underline{i})$] represents the propagation of the particle from site $\underline{i}$ to site $\underline{j}$, the various terms can be interpreted as follows. The first diagrams represent propagation without any scattering event [$G_0(\underline{j},\underline{i})$, zero-order contribution]; the first diagrams in subgroup (a) represents propagation from site $\underline{i}$ to site $\underline{\ell}$ [i.e., $G_0(\underline{\ell},\underline{i})$], scattering at the impurity site $\underline{\ell}(t_{\underline{\ell}})$ and then propagation from $\underline{\ell}$ to $\underline{j}$ i.e., $G_0(\underline{j},\underline{\ell})$; similar interpretation can be given to any other diagram of the type shown in Fig.7.1. Because $G(\underline{j},\underline{i})$ can be interpreted as representing the propagation from site $\underline{i}$ to site $\underline{j}$,

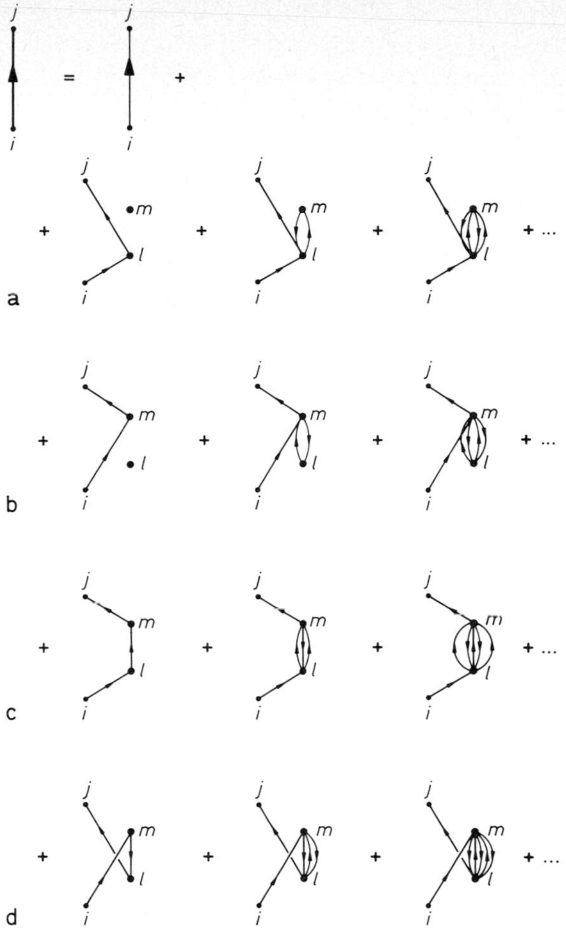

Fig.7.1. Diagrammatic representation of the various terms contributing to the Green's function G($\underline{j}$,$\underline{i}$) (thick line) associated with two scatterers at sites $\underline{\ell}$ and $\underline{m}$. Each diagram corresponds to a product of factors according to the following rules: a thin line starting from the arbitrary lattice site $\underline{i}$ and ending at the arbitrary lattice site $\underline{j}$ without visiting any other site corresponds to the factor $G_0(\underline{j},\underline{i})$; each time the scattering site $\underline{\ell}$($\underline{m}$) is visited by the diagram a factor $t_{\underline{\ell}}$($t_{\underline{m}}$) is introduced. The diagrams have been classified in four subgroups

it is customary to call the Green's function G($\underline{j}$,$\underline{i}$) a propagator. Some authors use the name locator for the diagonal matrix elements G($\underline{i}$,$\underline{i}$) for obvious reasons. Others use the name propagator for all matrix elements of G or for G itself so that the term propagator becomes synonymous to the term Green's function.

The diagrammatic representation has the advantage of allowing us to give a physical meaning to each term. This feature facilitates the development of various approximations, which in most cases are necessary. The vivid physical picture which emerges from the diagrams helps also in the memorization of the corresponding formula. It should be stressed that many types of diagrams appear in various branches of physics depending on the unperturbed part H_0, the perturbation H_1, the complete set of basic functions $|\underline{m}\rangle$, and the quantity calculated. The reader may verify this statement by comparing the diagrams of Fig.7.1 with those introduced in Appendix B.

We would like to point out that the t-matrix T associated with two scattering centers at $\underline{\ell}$ and $\underline{m}$ is not simply the sum of the t-matrices associated with the impurity at $\underline{\ell}$ and the impurity at $\underline{m}$. The difference is due to all multiple scattering terms shown in Fig.7.1. However, to the first order in $t_{\underline{\ell}}$ or $t_{\underline{m}}$, one keeps the zero-order diagram and the first diagrams in subgroups (a) and (b), and thus

$$T = T_{\underline{\ell}} + T_{\underline{m}} + O(T^2_{\underline{\ell}}, T^2_{\underline{m}}, T_{\underline{\ell}} T_{\underline{m}}) \quad . \tag{7.16}$$

Because $G_0(\underline{\ell},\underline{m})$ decays to zero as $|\underline{\ell} - \underline{m}|$ approaches infinity (except in the 1-D case and for E within the band), it follows that the multiple scattering terms which involves at least a factor $G_0(\underline{\ell},\underline{m})$ approach zero as $|\underline{\ell} - \underline{m}| \to \infty$; thus, in the limit $|\underline{\ell} - \underline{m}| \to \infty$, $T \to T_{\underline{\ell}} + T_{\underline{m}}$. The only exception is the 1-D case for E within the band, because in this case the reflected or transmitted wave from one scattering center propagates with constant amplitude throughout the linear chain, and thus it can be scattered again by the other impurity no matter how far apart the two impurities are.

It is interesting to examine the question of discrete level(s) in the present case of two impurity atoms. The most convenient way to find these levels is by finding the poles of G as given by (7.10). These poles are solutions of

$$1 - \varepsilon' G_{0\underline{\ell}}(\underline{m},\underline{m}) = 0 \quad . \tag{7.17}$$

The quantity $G_{0\underline{\ell}}(\underline{m},\underline{m})$ has been examined in detail in Chap.6. In Fig.7.2 we plot schematically the quantity $\mathrm{Re}\{G_{0\underline{\ell}}(\underline{m},\underline{m};E)\}$ versus E for the 1-D chain and for the special case where $\underline{\ell}$ and $\underline{m}$ are nearest neighbors; then we use this plot to obtain the qualitative features of the roots of (7.17). We examine first the case where both ε and ε' are negative. When $|\varepsilon|$ is small

(Fig.7.2.a) there is only one intersection and, hence, only one discrete level no matter how large $|\varepsilon'|$ is. This discrete level lies below the discrete level of the Hamiltonian $H_{0\ell}$. Thus, the additional attractive potential ε' simply pushes the discrete level further down. When $|\varepsilon|>B/2$, then $G_{0\ell}(E_\ell)<0$, and, consequently, there is the possibility of two intersections (Fig.7.2.b). Thus, when $|\varepsilon|>B/2$, the additional attractive potential not only pushes the first discrete level down but also creates a second level when $\varepsilon'<\varepsilon_c'$. The quantity ε_c' can be found as a solution of the equation $G_{0\ell}(E_\ell) = 1/\varepsilon'$ which gives

$$\frac{B}{\varepsilon_c'} + \frac{B}{\varepsilon} = -2 \quad . \tag{7.18}$$

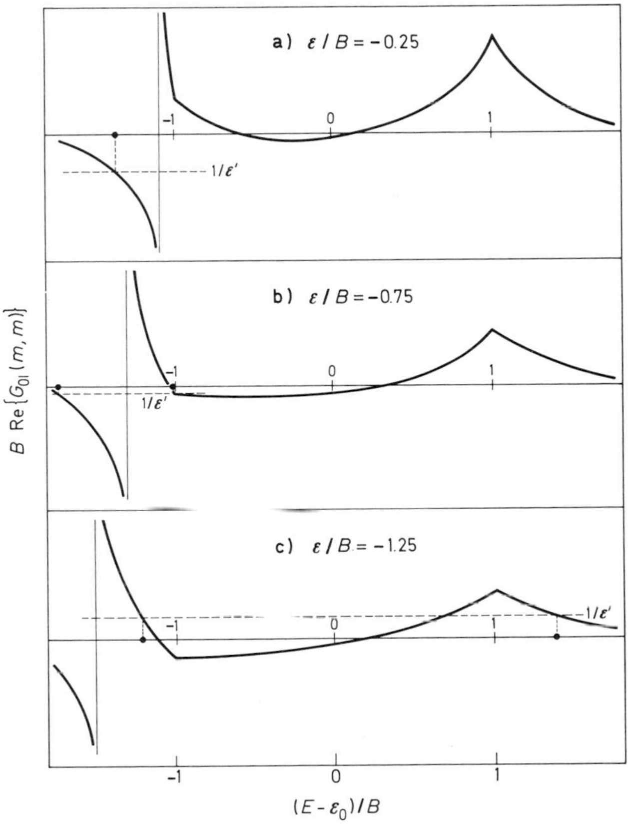

Fig.7.2. Re $G_{0\ell}(m,m)$ vs. E for three different values of ε in the 1-D case. The intersections of Re $G_{0\ell}(m,m;E)$ with $1/\varepsilon'$ are also shown. The sites ℓ,m are nearest neighbors

Let us now examine the case $\varepsilon<0$ but $\varepsilon'>0$ (Fig.7.2c). When ε' is very small there is only one intersection below the band but above the discrete level of $H_{0\ell}$. Thus, an additional weak repulsion pushes the discrete level up, and if $|\varepsilon|$ is small ($<B/2$) and ε' large ($\varepsilon'>\varepsilon_c'$), this level is pushed all the way inside the continuum. On the other hand, when ε' exceeds another critical value, ε_c'', a discrete level, appears above the band (see Fig.7.2c). The quantity ε_c'' is given as solution of the equation $G_{0\ell}(E_u) = 1/\varepsilon'$, which with the help of (5.29) and (6.8) becomes

$$\frac{B}{\varepsilon_c''} + \frac{B}{\varepsilon} = 2 \quad . \tag{7.19}$$

Similar behavior is exhibited in the case $\varepsilon>0$. The hyperbolae (7.18) and (7.19) separate the ε - ε' plane into regions such that one or two discrete levels appear below and/or above the band. This is shown in Fig.7.3a. It should be noted that the results depend on the separation of sites ℓ and m. As the distance $|\ell - m|$ increases the hyperbolae shown in Fig.7.3a approach the ε - ε' axes. In Fig.7.3b we plot the results of a similar study in a Bethe lattice with K = 4 and for the sites ℓ, m being nearest neighbors. The boundaries are the hyperbolae

$$\left(\frac{2\varepsilon}{B} \pm 1\right)\left(\frac{2\varepsilon'}{B} \pm 1\right) = \frac{1}{4} \quad . \tag{7.20}$$

The main qualitative difference from the 1-D case is the existence around the origin of the ε - ε' plane of a region where no discrete levels appear at all; in the 1-D case this region has collapsed to a single point, namely the origin.

We would like to remind the reader that intersections of $\mathrm{Re}\{G_{0\ell}(\underline{m},\underline{m};E)\}$ with $1/\varepsilon'$ within the band are not solutions of (7.17) since for E within the band $\mathrm{Im}\{G_{0\ell}(\underline{m},\underline{m};E)\}$ is not zero. However, such intersections may produce resonances if they occur at energies such that $\mathrm{Im}\{G_{0\ell}(\underline{m},\underline{m};E)\}$ is small. In the 1-D case this only occurs for E near the band edges. We leave it to the reader to find the positions of resonances (if any) as a function of the quantities ε and ε'.

We have derived the basic result (7.12) by repeating twice the method developed in Chap.6. Equation (7.12) could be obtained in one step by considering $H_1 \equiv H_{\underline{\ell}} + H_{\underline{m}}$ as the perturbation and applying (4.13). We have then

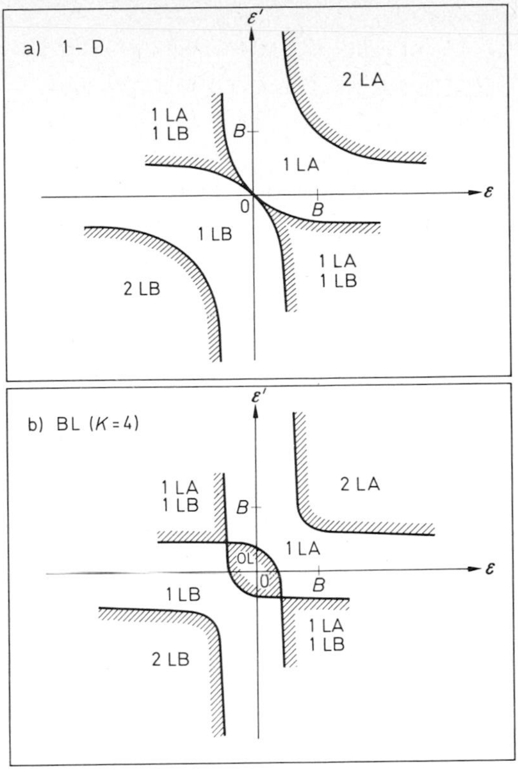

Fig.7.3. The ε - ε' plane is divided into regions where: no discrete levels (0L); one discrete level above the band (1LA); one discrete level below the band (1LB); two discrete levels one above and the other below (1LA, 1LB); two discrete levels above the band (2LA); and two discrete levels below the band (2LB) appear. In the one-dimensional case (a) the region 0L has collapsed to a single point (the origin). For a typical 3-D case such as a Bethe Lattice with K = 4 (b) the region 0L centered around the origin is of finite extent. The sites ℓ, m are nearest neighbors

$$T = H_1 + H_1 \sum_{\underline{n}} |\underline{n}\rangle\langle\underline{n}| G_0 \sum_{\underline{i}} |\underline{i}\rangle\langle\underline{i}| H_1 + \dots \quad ; \tag{7.21}$$

from the summation over all states we need to keep only two terms, namely $|\underline{m}\rangle\langle\underline{m}| + |\underline{\ell}\rangle\langle\underline{\ell}|$, because all other terms give zero as a result of the form of $H_1 = |\underline{\ell}\rangle\varepsilon\langle\underline{\ell}| + |\underline{m}\rangle\varepsilon'\langle\underline{m}|$. Now we can write

$$|\underline{\ell}\rangle\langle\underline{\ell}| + |\underline{m}\rangle\langle\underline{m}| = [\,|\underline{\ell}\rangle\,|\underline{m}\rangle\,] \begin{bmatrix} \langle\underline{\ell}| \\ \langle\underline{m}| \end{bmatrix} \quad ; \tag{7.22}$$

we denote by $|\alpha\rangle$ the matrix $[|\underline{\ell}\rangle |\underline{m}\rangle]$ and by $\langle\alpha|$ the adjoint matrix

$$\begin{bmatrix} \langle\underline{\ell}| \\ \langle\underline{m}| \end{bmatrix} .$$

With this notation (7.21) can be rewritten as

$$T = H_1 + H_1|\alpha\rangle\langle\alpha|G_0|\alpha\rangle\langle\alpha|H_1$$

$$+ H_1|\alpha\rangle\langle\alpha|G_0|\alpha\rangle\langle\alpha|H_1|\alpha\rangle\langle\alpha|G_0|\alpha\rangle\langle\alpha|H_1 + \dots \quad . \qquad (7.23)$$

The quantity $\langle\alpha|G_0|\alpha\rangle$ is a 2×2 matrix

$$\langle\alpha|G_0|\alpha\rangle = \begin{bmatrix} G_0(\underline{\ell},\underline{\ell}) & G_0(\underline{\ell},\underline{m}) \\ G_0(\underline{m},\underline{\ell}) & G_0(\underline{m},\underline{m}) \end{bmatrix} \quad ; \qquad (7.24)$$

similarly

$$\langle\alpha|H_1|\alpha\rangle = \begin{bmatrix} \varepsilon & 0 \\ 0 & \varepsilon' \end{bmatrix} , \qquad (7.25)$$

and

$$\langle\alpha|T|\alpha\rangle = \begin{bmatrix} \langle\underline{\ell}|T|\underline{\ell}\rangle & \langle\underline{\ell}|T|\underline{m}\rangle \\ \langle\underline{m}|T|\underline{\ell}\rangle & \langle\underline{m}|T|\underline{\ell}\rangle \end{bmatrix} \quad . \qquad (7.26)$$

We obtain then from (7.23)

$$\langle\alpha|T|\alpha\rangle = \langle\alpha|H_1|\alpha\rangle\,(1 + \langle\alpha|G_0|\alpha\rangle\langle\alpha|H_1|\alpha\rangle + \dots)$$

$$= \langle\alpha|H_1|\alpha\rangle\,(1 - \langle\alpha|G_0|\alpha\rangle\langle\alpha|H_1|\alpha\rangle)^{-1} \quad . \qquad (7.27)$$

By calculating the inverse of the 2×2 matrix $1 - \langle\alpha|G_0|\alpha\rangle\langle\alpha|H_1|\alpha\rangle$, one can easily show that (7.27) is equivalent to (7.12). This matrix approach is very convenient for generalization to three, four, etc., impurities. If the number of impurities becomes very large then the calculation becomes tedious because of the need to invert a large matrix. In this case it is sometimes useful to consider a third method of calculating the total t-matrix; this method is the following.

Let

$$H_1 = \sum_{\underline{m}} H_{\underline{m}} \tag{7.28}$$

be the perturbative part of the Hamiltonian; the quantity $H_{\underline{m}}$ has the form

$$H_{\underline{m}} = |\underline{m}>\varepsilon_{\underline{m}}'<\underline{m}| \quad , \tag{7.29}$$

and the summation runs over all sites $\underline{m}$; the sites occupied by the host atoms have $\varepsilon_{\underline{m}}' = 0$. We denote again by $T_{\underline{m}} = |\underline{m}>t_{\underline{m}}<\underline{m}|$ the t-matrix associated with the unperturbed part H_o and the perturbation $H_{\underline{m}}$; obviously, $t_{\underline{m}} = \varepsilon_{\underline{m}}'/[1 - \varepsilon_{\underline{m}}' G_o(\underline{m},\underline{m})]$.

If T is the total t-matrix associated with the unperturbed part H_o and the perturbation H_1, we have

$$\begin{aligned} T &= H_1 + H_1 G_o T \\ &= H_1(1 + G_o T) \\ &= \sum_{\underline{m}} H_{\underline{m}}(1 + G_o T) = \sum_{\underline{m}} Q_{\underline{m}} \quad , \end{aligned} \tag{7.30}$$

where $Q_{\underline{m}}$ is given by

$$Q_{\underline{m}} = H_{\underline{m}}(1 + G_o T) \quad . \tag{7.31}$$

Substituting (7.30) in (7.31) we have

$$Q_{\underline{m}} = H_{\underline{m}}\left(1 + G_o \sum_{\underline{n}} Q_{\underline{n}}\right)$$

or

$$(1 - H_{\underline{m}} G_o) Q_{\underline{m}} = H_{\underline{m}}\left(1 + G_o \sum_{\underline{n} \neq \underline{m}} Q_{\underline{n}}\right) \tag{7.32}$$

or

$$Q_{\underline{m}} = (1 - H_{\underline{m}} G_o)^{-1} H_{\underline{m}}\left(1 + G_o \sum_{\underline{n} \neq \underline{m}} Q_{\underline{n}}\right) \quad . \tag{7.33}$$

However, the quantity $(1 - H_{\underline{m}}G_0)^{-1}H_{\underline{m}}$ equals $T_{\underline{m}}$; thus, (7.33) becomes

$$Q_{\underline{m}} = T_{\underline{m}}\left(1 + G_0 \sum_{\underline{n}\neq\underline{m}} Q_{\underline{n}}\right) \quad . \tag{7.34}$$

If one solves (7.34) for all $Q_{\underline{m}}$ one has managed to express $T = \sum Q_{\underline{m}}$ in terms of $T_{\underline{m}}$'s and G_0. The reader is urged to solve (7.33) for the case of two impurities and recapture (7.12) in this way.

7.2 Infinite Number of Impurities

In this section we consider the case where a nonzero percentage of the lattice sites are occupied by impurity atoms in a random way. An example of such a system is the binary alloy A_xB_{1-x} where a fraction x of the lattice sites are occupied by atoms A with site energy $\varepsilon_0 + \varepsilon_A$, and the rest are occupied by atoms B with site energy $\varepsilon_0 + \varepsilon_B$. It should be mentioned that in reality we do not know the specific sites occupied by A and the specific sites occupied by B. What we may know is the probability for each particular configuration of A's and B's. Thus, we are led to the concept of a *random* or *disordered* system the Hamiltonian of which is not known; what we know is the probability distribution of the matrix elements of the Hamiltonian. In the specific example of the binary alloy, if no correlations among the random variables $\{\varepsilon'_{\underline{n}}\}$ are present, their probability distribution is given by

$$P(\{\varepsilon'_{\underline{n}}\}) = \prod_{\underline{n}} p(\varepsilon'_{\underline{n}}) \quad , \tag{7.35}$$

with

$$p(\varepsilon'_{\underline{n}}) = x\delta(\varepsilon'_{\underline{n}} - \varepsilon_A) + (1 - x)\delta(\varepsilon'_{\underline{n}} - \varepsilon_B) \quad . \tag{7.36}$$

In a random system we are interested in the average (over all configurations) of the physical quantities. More specifically, we would like to calculate the average of the Green's function $\langle G\rangle$ defined as

$$\langle G\rangle = \int d\{\varepsilon'_{\underline{n}}\}P(\{\varepsilon'_{\underline{n}}\})G \quad . \tag{7.37}$$

In general, $\langle G\rangle$ cannot be calculated exactly; a notable exception, examined by LLOYD [7.1], is the case where $p(\varepsilon'_{\underline{n}})$ is a Lorentzian

$$p(\varepsilon'_{\underline{n}}) = \frac{1}{\pi} \frac{\Gamma}{\varepsilon'^2_n + \Gamma^2} \quad .$$

Then, because $p(\varepsilon'_{\underline{n}})$ has two poles $\varepsilon'_{\underline{n}} = \pm i\Gamma$ in the complex $\varepsilon'_{\underline{n}}$ plane, each integration over $\varepsilon'_{\underline{n}}$ in (7.37) replaces $\varepsilon'_{\underline{n}}$ by $\pm i\Gamma$; thus finally $\langle G^{\pm}(E)\rangle = G_0(E \pm i\Gamma)$.

We mention that besides the binary alloy and the Lorentzian, the rectangular distribution

$$\begin{aligned} p(\varepsilon'_{\underline{n}}) &= \frac{1}{W} \quad \text{for} \quad |\varepsilon'_{\underline{n}}| < W/2 \\ &= 0 \quad \text{otherwise} \end{aligned}$$

has been used widely in the literature.

It is worthwhile to point out that our Hamiltonian is the simplest one to model a random system. Even within the framework of a TBH there are physically important aspects which we ignored: the off-diagonal matrix elements V_{ij} may actually be random variables as well (off-diagonal disorder); the quantity $\varepsilon'_{\underline{n}}$, and hence its probability distribution, may depend on the nature of other sites in the vicinity of $\underline{n}$, as, e.g., in the case of coupled pendulums examined in Chap.5 (environmental disorder); the random variables $\{\varepsilon'_{\underline{n}}\}$ may be statistically correlated (short-range order), etc. Furthermore, in real systems such as amorphous semiconductors there is disorder resulting not from small displacements from the crystal sites but associated with new bonding configurations (topological disorder). Such disorder is much more difficult to treat. For more details concerning the various types of disorder the reader is referred to ZIMAN's book [7.2].

In spite of the drastic simplifications in our model Hamiltonian, one is still forced to develop approximate schemes for obtaining $\langle G\rangle$. Here we review briefly three of the most widely used methods.

7.2.1 Virtual Crystal Approximation (VCA)

Our Hamiltonian is $H = H_0 + H_1$, where H_0 is periodic tight-binding and H_1 is given by (7.28,29). By averaging the basic equation (4.6) we obtain

$$\langle G\rangle = G_0 + G_0\langle H_1 G\rangle \quad ; \tag{7.38}$$

G_0 does not depend on the random variables $\{\varepsilon'_{\underline{n}}\}$, and hence $\langle G_0\rangle = G_0$. Now, we make the approximation

$$\langle H_1 G\rangle \simeq \langle H_1\rangle\langle G\rangle \quad , \tag{7.39}$$

and take into account that

$$\langle H_1 \rangle = \sum_{\underline{m}} \langle H_{\underline{m}} \rangle = \sum_{\underline{m}} |\underline{m}\rangle\langle\varepsilon'_{\underline{m}}\rangle\langle\underline{m}| = \langle\varepsilon'_{\underline{m}}\rangle \quad . \tag{7.40}$$

Because we are dealing with systems which are macroscopically homogeneous the average $\langle\varepsilon'_{\underline{m}}\rangle$ is independent of the site $\underline{m}$ and will be denoted by ε;

$$\varepsilon \equiv \langle\varepsilon'_{\underline{m}}\rangle \quad . \tag{7.41}$$

Substituting in (7.38) from (7.39-41) we obtain

$$\langle G \rangle = G_0 + G_0 \varepsilon \langle G \rangle$$

which can be rewritten as

$$\langle G(E) \rangle = \frac{1}{E - H_0 - \varepsilon} = \frac{1}{E - \langle H \rangle} \quad . \tag{7.42}$$

Equation (7.42) is known as the VCA for $\langle G \rangle$; this approximation consists in calculating $\langle G(H) \rangle$ as $G(\langle H \rangle)$ and simply shifts the energy levels by $\varepsilon = \langle\varepsilon'_{\underline{m}}\rangle$. This shift can be made equal to zero (which means that the VCA $\langle G \rangle$ equals G_0) by choosing $H_0 \equiv \langle H \rangle$ and $H_1 \equiv H - \langle H \rangle$. From now on we adopt this notation.

The VCA fails completely for large values of $\langle\varepsilon'^2_{\underline{m}}\rangle$. For more comments about this very simple method see the review article by ELLIOTT et al. [7.3] and references therein.

7.2.2 Average t-Matrix Approximation (ATA)

If T is the t-matrix associated with H_0 and H_1 we have

$$\langle G \rangle = G_0 + G_0 \langle T \rangle G_0 \quad . \tag{7.43}$$

Here T is a complicated function of the individual $t_{\underline{m}}$'s

$$T = f(\{t_{\underline{m}}\}) \quad , \tag{7.44}$$

where

$$t_{\underline{m}} = \varepsilon'_{\underline{m}}/(1 - \varepsilon'_{\underline{m}} G_0(\underline{m},\underline{m})) \quad . \tag{7.45}$$

The basic approximation in ATA is to put

$$\langle T\rangle \approx f(\{\langle t_{\underline{m}}\rangle\}) \quad . \tag{7.46}$$

Because the rhs of (7.46) has the same form as (7.44) with an $\underline{m}$-independent $\langle t_{\underline{m}}\rangle$, it follows that $\langle T\rangle$ is the same as the t-matrix T_e, associated with H_0 and a periodic H_{1e} of the form

$$H_{1e} = \sum_{\underline{m}} |\underline{m}\rangle \, \Sigma \, \langle\underline{m}| = \Sigma \tag{7.47}$$

provided [as can be seen from (7.45)] that Σ is chosen to satisfy the relation

$$\langle t_{\underline{m}}\rangle = \Sigma \, [1 - \Sigma \, G_0(\underline{m},\underline{m})]^{-1} \quad . \tag{7.48}$$

Substituting T_e for $\langle T\rangle$ in (7.43) we obtain immediately

$$\langle G(E - H)\rangle = (E - H_0 - H_{1e})^{-1} = G_0(E - \Sigma) \quad . \tag{7.49}$$

Thus the ATA calculates $\langle G\rangle$ from the periodic effective Hamiltonian $H_e \equiv H_0 + H_{1e}$, which shifts the energy by the complex, energy-dependent quantity Σ; within the ATA the so-called self-energy Σ is obtained from (7.48), which can be recast as

$$\Sigma = \frac{\langle t_{\underline{m}}\rangle}{1 + \langle t_{\underline{m}}\rangle G_0(\underline{m},\underline{m})} \quad . \tag{7.50}$$

We mention that the above derivation can be translated into a diagrammatic language [7.3].

In the limit of weak scattering $\varepsilon'_{\underline{m}} \to 0$, we obtain, in lowest order, from (7.50,45) that

$$\Sigma \to \langle \varepsilon'^2_{\underline{m}}\rangle G_0(\underline{m},\underline{m}) \quad . \tag{7.51}$$

This is the correct limit. The ATA is also successful in the very dilute limit, where the concentration of impurities is $\ll 1$. The ATA has been used by EHRENREICH [7.4], SCHWARTZ [7.5], BANSIL [7.6-9] and co-workers to calculate the electronic structure of real random alloys, such as Cu_xNi_{1-x}.

7.2.3 Coherent Potential Approximation (CPA)

As in the ATA case, the CPA calculates $\langle G\rangle$ through an effective Hamiltonian H_e, which in the simplest case is characterized by a single, complex energy-dependent self-energy Σ. In contrast to the ATA, the CPA is based upon expanding G in terms of $G_e \equiv (E - H_e)^{-1}$ and $H'_1 \equiv H - H_e$, i.e.,

$$G = G_e + G_e T' G_e \quad , \qquad (7.52)$$

where the t-matrix T' is

$$T' = f(\{t'_{\underline{m}}\}) \quad , \qquad (7.53)$$

with

$$t'_{\underline{m}} = (\varepsilon'_{\underline{m}} - \Sigma)[1 - (\varepsilon'_{\underline{m}} - \Sigma)G_e(\underline{m},\underline{m})]^{-1} \quad .$$

The central approximation is again similar to that of the ATA

$$\langle T' \rangle \approx f(\{\langle t'_{\underline{m}} \rangle\}) \quad . \qquad (7.54)$$

Taking the average of (7.52) and recalling that by definition $\langle G \rangle = G_e$, we obtain $\langle T' \rangle = 0$,

which in view of (7.54) leads to

$$\langle t'_{\underline{m}} \rangle = 0 \quad . \qquad (7.55)$$

Equation (7.55) is an implicit equation for the self energy Σ, which can be recast as

$$\Sigma = \left\langle \frac{\varepsilon'_{\underline{m}}}{1 - (\varepsilon'_{\underline{m}} - \Sigma)G_e(\underline{m},\underline{m})} \right\rangle \qquad (7.56)$$

or as

$$\left\langle \frac{1}{1 - (\varepsilon'_{\underline{m}} - \Sigma)G_e(\underline{m},\underline{m})} \right\rangle = 1 \quad . \qquad (7.57)$$

In (7.53,55-57) the quantity $G_e(\underline{m},\underline{m})$ depends on Σ through (7.49)

$$G_e(\underline{m},\underline{m};E) = G_0(\underline{m},\underline{m};\ E - \Sigma) \quad . \qquad (7.58)$$

The CPA has been proven impressively successful. It behaves correctly in the weak scattering limit [where it reduces to (7.51) as it can be seen from (7.56)], in the strong scattering (or atomic) limit (i.e., when $\langle \varepsilon'^2_{\underline{m}} \rangle \gg V^2$), and in the dilute limit; and it interpolates properly between these limits.

Having mentioned the success of the CPA let us discuss now its limitations and failures. The only approximation we have employed is that $\langle t'_{\underline{m}} \rangle = 0$ implies $\langle T' \rangle = 0$. The physical meaning of this approximation is revealed if we express T' in terms of the $T'_{\underline{m}}$'s by iterating (7.34) and substituting in (7.30).

$$T' = \sum_{\underline{m}} T'_{\underline{m}} + \sum_{\underline{n}\neq\underline{m}} T'_{\underline{n}} G'_o T_{\underline{m}} + \sum_{\underline{n}\neq\underline{m}\neq\underline{r}} T'_{\underline{n}} G'_o T'_{\underline{m}} G'_o T'_{\underline{r}} + \dots \quad . \tag{7.59}$$

In the present case where $\{\varepsilon'_{\underline{m}}\}$ are independent random variables the quantities $\{t'_{\underline{m}}\}$ are independent random variables. By averaging (7.59) we can easily see that the average of the first three terms on the right-hand side of (7.59) are proportional to $<t'_{\underline{m}}>$; the fourth term is not, in general, proportional to $<t'_{\underline{m}}>$ because of contributions of the type $\sum_{\underline{n}\neq\underline{m}} <T'_{\underline{n}} G'_o T'_{\underline{m}} G'_o T'_{\underline{n}} G'_o T'_{\underline{m}}> = \sum_{\underline{n}\neq\underline{m}} |\underline{n}><t'^2_{\underline{n}}><t'^2_{\underline{m}}> G_o(\underline{n},\underline{m}) G_o(\underline{m},\underline{n}) G_o(\underline{n},\underline{m}) <\underline{m}|$, which are proportional to $<t'^2_{\underline{n}}>$. Similarly higher-order terms in the expansion (7.59) corresponding to multiple scattering events from clusters of a fixed number of sites will introduce terms which are proportional to $<t'^n_{\underline{m}}>$ where n is any integer. Since, in general, $<t'_{\underline{m}}> = 0$ does not necessarily imply $<t'^n_{\underline{m}}> = 0$, we can conclude that the CPA incorrectly treats multiple scattering terms associated with clusters of a fixed number of sites. These multiple scattering events are of great importance for eigenstates which are significantly enhanced in the vicinity of a certain cluster of sites, because this enhancement is produced through the multiple-scattering terms which trap the particle around the cluster for long time. Hence, at energies where the density of states is dominated by contributions from resonance or localized (around a cluster of sites) eigenstates, the CPA is expected to fail. As we will discuss below, the tails of the band are due entirely to such cluster-localized eigenstates and consequently the CPA is a poor approximation there; actually, the CPA predicts no tail at all. Cluster-localized eigenstates may, under certain circumstances, be responsible for the appearance of considerable structure in the DOS by creating peaks at certain energies or depleting states from other energies. Again the CPA tends to eliminate this structure. It should be noted that when the probability of occurrence of a cluster capable of trapping a particle in it is small, the role of cluster-localized eigenstates is small. If short-range order is absent, the probability of occurrence of special clusters is small and decreases with increasing dimensionality. Thus, the CPA works better in 3-dimensional systems than in 1-dimensional ones. A sketch in the preface of a book edited by THORPE [5.56] summarizes the CPA situation best.

The CPA combines two basic ideas: one is to calculate the average of a given quantity associated with a random medium by introducing a periodic effective medium; the second is to determine this effective medium by a self-consistency requirement, i.e, by demanding that the fluctuations of the given quantity due to *local* fluctuations around the effective medium average out to zero. As was pointed out by SEN [7.10], these ideas can be traced back to MAXWELL [7.11]. HUBBARD [7.12] was probably the first to use the CPA in an electronic structure calculation. The CPA was brought to its present form by TAYLOR [7.13] and SOVEN [7.14], who are usually credited for the invention of the method. The basic equation (7.55), which determines the effective medium, has been rederived by various techniques. Among them one must mention diagrammatic methods based upon the expansion of G in powers of H_1 [7.15-17]; the development of these methods started by EDWARDS [7.18], LANGER [7.19], KLAUDER [7.20] and by MATSUBARA and TOYOZAWA [7.21] with important contributions by LEATH and GOODMAN [7.22] and by YONEZAWA and MATSUBARA [7.23]. Other diagrammatic methods used to derive the CPA equations are based upon the expansion of G in powers of the off-diagonal matrix elements of the Hamiltonian. This so-called locator expansion, which is similar in spirit to the RPE of Appendix B, was used by LEATH [7.24] and by MATSUBARA and KANEYOSHI [7.25]. Contributions to our understanding of the CPA were made among others by VELICKY et al. [7.26], ONODERA and TOYOZAWA [7.27], YONEZAWA [7.28], BUTLER [7.29] and by BROUERS et al. [7.30]. In addition to [7.3,4] we mention also the review article by YONEZAWA and MORIGAKI [7.31] and the book edited by THORPE [5.56].

7.2.4 Direct Extensions of the CPA

It must be stressed that in our derivation of the ATA and the CPA it was necessary to have the random part of the Hamiltonian H_1 as a sum of local terms. The simple ATA and CPA we presented can be easily generalized in the case where each of the additive parts in H_1 involves not just one local orbital but a finite number of them, e.g., if with each site $\underline{m}$ a finite number of orbitals $|\underline{m},\nu\rangle$, $\nu = 1,2,\ldots,j$ is associated, the random part of the Hamiltonian may have the form

$$H_1 = \sum_{\underline{m}} \sum_{\nu\nu'} |\underline{m},\nu\rangle \varepsilon'_{\underline{m},\nu\nu'} \langle\nu',\underline{m}| \quad . \tag{7.60}$$

In this case the effective part of the Hamiltonian will have the form

$$H_{1e} = \sum_{\underline{m}} \sum_{\nu\nu'} |\underline{m},\nu\rangle \sum_{\nu\nu'} \langle\nu',\underline{m}| \quad ; \tag{7.61}$$

the $j \times j$ matrix Σ is determined by a matrix equation of the form (7.50) for the ATA or of the form (7.55) for the CPA. Care must be exercised regarding the ordering of the matrices involved in $t_{\underline{m}}$. Another example is the special case of off-diagonal disorder where

$$V_{\underline{nm}} = V_{\underline{n}} + V_{\underline{m}} \quad ; \tag{7.62}$$

then H_1 is the sum of local terms each of which has the form

$$|\underline{m}> \sum_{\underline{n}} (\varepsilon'_{\underline{m}}\delta_{\underline{mn}} + V_{\underline{m}}) <\underline{n}| \quad ,$$

where the summation is over the nearest neighbors of $\underline{m}$. In the special binary alloy case, (7.62) implies that $V_{AB} = (V_{AA} + V_{BB})/2$. A third, less obvious example is the case where

$$\varepsilon'_{\underline{m}} = -\sum_{\underline{n}} V_{\underline{mn}} \quad ; \tag{7.63}$$

this is realized in the case of pendulums with random spring constants but fixed uncoupled eigenfrequencies and in the case of lattice vibrations, etc. When (7.63) is satisfied, H_1 can be decomposed in *bond* contributions of the form

$$|\underline{n}>V_{\underline{nm}}<\underline{m}| + |\underline{m}>V_{\underline{mn}}<\underline{n}| - |\underline{m}>V_{\underline{mn}}<\underline{m}| - |\underline{n}>V_{\underline{nm}}<\underline{n}| \ .$$

In this connection it is worthwhile to mention the so-called homomorphic CPA developed by YONEZAWA and ODAGAKI [7.32], where *bond* additivity is forced by writing H_1 as a sum of terms of the form

$$|\underline{n}>V_{\underline{nm}}<\underline{m}| + |\underline{m}>V_{\underline{mn}}<\underline{n}| + |\underline{m}>\frac{\varepsilon'_{\underline{m}}}{Z}<\underline{m}| + |\underline{n}>\frac{\varepsilon'_{\underline{n}}}{Z}<\underline{n}| \quad .$$

Here Z is the number of nearest neighbors; if $\{\varepsilon'_{\underline{m}}\}$ are not random this decomposition is reasonable. However, if diagonal disorder is present, these terms are not in general statistically independent and hence the above decomposition may lead to erroneous results.

These extensions of the simple CPA have been employed to study the electronic structure, lattice vibrations, and magnetic excitations in real disordered systems. Thus the case (7.60) was examined by FAULKNER [7.33], FAULKNER and STOCKS [7.34], PAPACONSTANTOPOULOS et al. [7.35], and PAPACONSTANTOPOULOS and ECONOMOU [7.36] to treat electronic excitations. A variation of this case, developed in the framework of real space representation (instead of the TBM), is reviewed in [7.5,6,37]; this so-called muffin tin CPA avoids the difficulties associated with the off-diagonal randomness in TB CPA. In addition, we mention contributions by STOCKS et al. [7.38], KORRINGA and MILLS [7.39], ROTH [7.40], and BALANOVSKI [7.41]. The case (7.62) was examined by SCHWARTZ et al. [7.42] for the electronic problem, by KAPLAN and MOSTOLLER [7.43] for lattice vibrations, and by HARRIS et al.

[7.44] for the magnetic excitations. The case (7.63) has been studied by TAHIR-KHELI and co-workers [7.45]; a generalization of this method incorporating environmental disorder was treated in [7.3 and 46].

According to what was said above the CPA can treat off-diagonal disorder in the special case (7.62). SHIBA [7.47] noted that the special case $V_{AB} = (V_{AA}V_{BB})^{\frac{1}{2}}$ can also be reduced to the simple CPA; this becomes apparent if one uses the locator expansion to derive the CPA. Among the efforts to treat off-diagonal disorder we mention work by FOO et al. [7.48] and by BROUERS et al. [7.30]. BLACKMAN et al. [7.49] obtained a solution to the problem for the case of binary distribution by introducing a 2×2 matrix version of the simple CPA, where the diagonal matrix elements refer to the AA, or BB configurations and the off-diagonal to the AB configuration; their method can be generalized to the n-component alloy distribution by working with $n \times n$ matrices. Recently WHITELAW [7.50] succeeded to incorporate environmental disorder in addition to the off-diagonal disorder; his method is briefly reviewed by LEATH [7.51].

Up to now we dealt with direct extensions of the CPA which map the problem back into the simple CPA. As a result, these extensions share with the simple CPA its desirable features: correct reproduction of the limiting cases, successful interpolation in between, and proper analytic behavior [7.52,53], which assures the nonnegativeness of the DOS. Some of these extensions incorporated cluster scattering as well. However, it must be realized that this was achieved only because of the special form of the random Hamiltonian. The question which arises is whether a systematic way to include cluster scattering in the simple CPA can be devised no matter what the form of the Hamiltonian is. In the next subsection we review briefly the attempts to answer this question.

7.2.5 Cluster Generalizations of the CPA

It was already pointed out that the limitations of the simple CPA stem from its omission of multiple scattering from clusters of neighboring sites. Hence, it is obviously worthwhile to devise a general scheme for the incorporation of cluster scattering features in the simple CPA. A conceptuallly straightforward way to achieve this purpose is the following. Replace the random part of the Hamiltonian H_1 by an effective Hamiltonian H_{1e} which is characterized by a number of yet undetermined parameters $\Sigma_1, \Sigma_2, \dots, \Sigma_j$. Obviously H_{1e} must possess the same periodicity as $\langle H_1 \rangle + H_0$. Then determine the self-energies $\Sigma_1, \dots, \Sigma_j$ by demanding that $\langle t_c \rangle = 0$, where t_c is the t-matrix associated with the fluctuations around the effective Hamiltonian within a chosen cluster c. BUTLER and NICKEL [7.54] implemented such a scheme for a cluster of two neighboring sites. They found that serious nonanalyticities appear in the solutions when the degree of disorder is large [7.55]. Furthermore, DUCASTELLE [7.53] demonstrated that this cluster generalization fails to reproduce the strong scattering limit. The nonanalyticities and the incorrect limiting behavior appeared routinely in the early schemes for cluster generalizations. LEATH [7.56] attributed these failures to the fact that a particular site, as a result of the average periodicity, does not belong to a single fixed cluster but it can be assigned to a number of overlapping, on the average equivalent, clusters. According to this argument one must break the average periodicity in order to obtain properly analytic solutions. An artificial way to do this is by introducing an effective Hamiltonian H_{1e} which possesses a superlattice periodicity; the primitive cell of this superlattice equals the chosen cluster. This superperiodic H_{1e} is characterized by self-energies Σ_i's connecting sites belonging to the same primitive cell; no self-energy refers to sites belonging to different primitive cells. Under these assumptions the problem becomes equivalent to that defined by (7.60,61), with the orbitals within the $\underline{m}^{th}$ supercell corresponding to the orbitals of the $\underline{m}^{th}$ site in (7.60). Thus the problem has been mapped to that of the simple CPA, and consequently the solution is properly analytic. This has been achieved at the cost of introducing the incorrect superperiodicity on H_{1e}, which produces small errors to quantities like $\langle G(\underline{m},\underline{n})\rangle$ when both $\underline{m}$, $\underline{n}$ are well inside the supercell; on the other hand, when $\underline{m}$, $\underline{n}$ belong to different supercells the error is expected to be substantial.

A better way to break the average periodicity is by embedding a cluster in an effective medium. This means that the cluster is allowed to take all its possible configurations while the rest of the system is kept at its effective medium values. Again this method is appropriate for calculating $\langle G(\underline{m},\underline{n})\rangle$ only when both $\underline{m}$ and $\underline{n}$ belong to the cluster. The scheme can be subdivided into three categories depending on how the effective medium is determined. The simplest method is to determine it, let us say, from the simple CPA without attempting to make it consistent with the final result [7.57,58]. The method is easy to implement and has been very useful in treating real systems. In this connection one must mention the recursion method recently reviewed by HAYDOCK [7.59] and the continued fraction method reviewed by CYROT-LACKMANN and KHANNA [7.60]. In the second category there are many effective media, one for each particular configuration of the cluster [7.61,62]; their determination is achieved by a slightly modified version of the simple CPA. Finally in the third category one classifies all cases where an attempt is made to determine the effective medium so as to be consistent with the final result [7.30,40,54, 63-70]. There is no guarantee that the self-consistency requirement will not produce nonanalyticities; there is also no established optimal way for implementing the self-consistency.

In Fig.7.4 we plot the average density of states per site versus E for an 1-D disordered system where the site energies $\{\varepsilon'_m\}$ have the binary distribution shown in (7.35,36) with $x = 0.5$, $|\varepsilon_A - \varepsilon_B|/B = 2$, $(B = 2V)$ and $\varepsilon_A + \varepsilon_B = \varepsilon_0 = 0$. It should be stressed that the 1-D case with a binary alloy distribution is the most severe test of any approximation. The reason is that the 1-dimensionality and the binary distribution strongly enhance the probability of occurrence of certain special clusters capable of trapping electrons within them. Such clusters are responsible for the sharp peaks shown in Fig.7.4e which is based upon the numerical solution of the exact equation (obtained by SCHMIDT [7.71]) for $\int^E \langle G(E')\rangle dE'$. Figure 7.4 shows that the ATA is a definite improvement over the VCA and that the CPA is a better approximation than the ATA. However, all these approximations fail to produce the fine structure of the exact results because, as was discussed before, even the CPA does not treat correctly the multiple scattering events responsible for the fine structure. On the other hand, the approximation termed CSSA [7.62] treats exactly clusters of up to three nearest-neighbor sites and as a result produces most of the fine structure. We would like to point out once again that in higher dimensionality and/or for a smoother probability distribution there is no fine structure, and the CPA is a good approximation indeed; as a matter of fact in the case where the probability distribution has a lorentzian form both the ATA and the CPA reproduce the exact result, because they replace each $\varepsilon'_{\underline{m}}$ by $\pm i\Gamma$ (for $E \pm is$) (see Sect.7.2).

In spite of the impressive success that embedded cluster CPA has in reproducing fine structure in the DOS (compare Figs.7.4d and e), it cannot be considered as a complete solution to the problem of incorporating cluster effects because of its inherent inapplicability to the calculation of $\langle G(\underline{n},\underline{m})\rangle$ for large $|\underline{n} - \underline{m}|$.

Recently there is much progress towards a general solution of the problem. One important advance is the work by MILLS and RATANAVARARAKSA [7.72], who developed a systematic diagrammatic way of assuring that a partial summation of a perturbation expansion is going to produce properly analytic results.

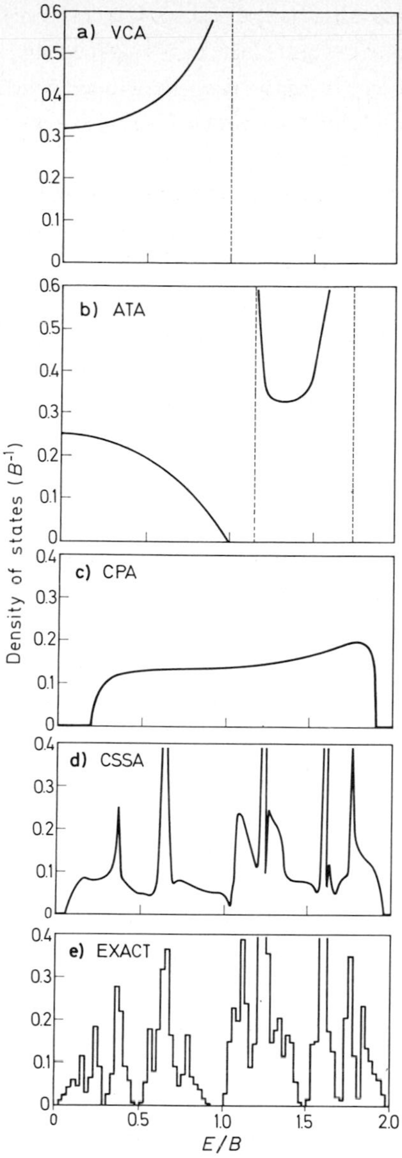

Fig.7.4. Average density of states versus E for a 1-D disordered system where the site energies have a binary distribution of the type shown in (7.35,36) with x = 0.5 and $[|\varepsilon_A - \varepsilon_B|/B] = 2$. Panel d (CSSA) is based on a generalization of the CPA, see [7.62]. In the present case $\rho(E) = \rho(-E)$

Then they proceeded to sum up all diagrams associated with pair scattering, and thus they managed to produce the first properly analytic, properly periodic, fully self-consistent theory incorporating pair scattering.

Another powerful idea is to augment the Hilbert space through a new parameter which determines the component of the binary alloy occupying any site (this can be immediately generalized to multicomponent distributions). Then the process of configurational averaging corresponds to taking the "ground-state quantum" average in this augmented space. This augmented space was first introduced by MOOKERJEE [7.73]. KAPLAN et al. [7.74] combined the augmented space idea and the pair-scattering method of [7.72] to produce a theory which is applicable not only to diagonal disorder but to off-diagonal and environmental disorder as well. Even short-range order can be treated within the framework of this theory [7.75]. These important recent developments are reviewed by LEATH [7.51] and by KAPLAN and GRAY [7.76].

7.3 Electrical Conductivity

Disorder has a much more pronounced effect on transport properties than on the DOS. As a matter of fact, the DC electrical conductivity is finite and not infinite (at T = 0), because of the presence of disorder no matter how weak. As the disorder increases further, it may produce a metal-insulator transition, i.e., it may prevent the propagation of the carriers altogether, making the conductivity equal to zero. In recent years there have been impressive developments in our understanding of these phenomena, and Green's functions have played a central role as a theoretical tool. In this section we present the formalism pertinent to electrical conductivity. This material is important in its own right as well as in connection with the recent developments which will be reviewed in Sect.7.4.

7.3.1 Definition and Some Basic Results

The presence of an electric field $\underline{\mathscr{E}}$ in a system induces a current density $\underline{j}$. The conductivity is defined as the coefficient of the linear (in $\underline{\mathscr{E}}$) part of $\underline{j}$:

$$j_\alpha(\underline{r},t) = \int_0^\infty d\tau \int d\underline{r}'\sigma_{\alpha\beta}(\underline{r},\underline{r}';\tau)\mathscr{E}_\beta(\underline{r}', t - \tau) \quad , \qquad (7.64)$$

where the subscripts α,β denote cartesian coordinates and a summation is implied over the repeated index β. In what follows we assume that both $\underline{\mathscr{E}}$ and $\underline{j}$ are along the x-axis, so that we need to consider only σ_{xx}; for simplicity we drop the subscripts. At the end it is easy to deduce from σ_{xx} the form of the other components of σ. Usually $\mathscr{E}$ and j vary slowly over distances of the order of ℓ_0, where ℓ_0 is determined by the condition $\sigma \approx 0$ for $|\underline{r} - \underline{r}'| \gg \ell_0$. In this case one can perform the integration over $\underline{r}'$ and the average over $\underline{r}$ to obtain

$$j(t) = \int_0^\infty d\tau\ \sigma(\tau)\mathscr{E}(t - \tau) \quad , \tag{7.65}$$

where obviously

$$\sigma(\tau) = \frac{1}{\Omega} \int d\underline{r}\ d\underline{r}'\sigma(\underline{r},\underline{r}';\tau) \quad . \tag{7.66}$$

From now on we shall consider $\sigma(\tau)$ or its Fourier transform

$$\sigma(\omega) = \int_0^\infty d\tau\ \sigma(\tau)e^{i\omega\tau} \quad . \tag{7.67}$$

At the end we shall indicate how one can obtain $\sigma(\underline{r},\underline{r}';\omega)$ from $\sigma(\omega)$. If $\mathscr{E}(t)$ is given by

$$\mathscr{E}(t) = Fe^{-i\omega t} + F^*e^{i\omega t} \quad , \tag{7.68}$$

we have from (7.65) and (7.67) that

$$j(t) = \sigma(\omega)Fe^{-i\omega t} + \sigma(-\omega)F^*e^{i\omega t} \quad . \tag{7.69}$$

The reality of j requires that $\sigma(-\omega) = \sigma^*(\omega)$, from which it follows that the real (imaginary) part $\sigma_1(\sigma_2)$ of σ is an even (odd) function of ω.

Note that as a result of causality, $\sigma(\tau)$ is nonzero only for $\tau \geq 0$; consequently, as it can be seen from (7.67), $\sigma(\omega)$ is analytic for $\mathrm{Im}\{\omega\} \geq 0$ with the possible exception of $\omega = 0$. Taking into account this analyticity and that $\sigma(\omega) \to 0$ for $\omega \to \infty$ (as we show later), we can see that the integral $\int_{-\infty}^{\infty} d\omega'\sigma(\omega')/(\omega' + is - \omega)$ is zero. By taking the limit $s \to 0^+$, and using (1.20), we obtain for the real (σ_1) and the imaginary (σ_2) parts of σ

$$\sigma_1(\omega) = \frac{1}{\pi} P \int \frac{d\omega'\sigma_2(\omega')}{\omega' - \omega} \quad , \tag{7.70}$$

$$\sigma_2(\omega) = -\frac{1}{\pi} P \int \frac{d\omega'\sigma_1(\omega')}{\omega' - \omega} + \frac{A}{\omega} \quad , \tag{7.71}$$

where iA is the residue (if any) of $\sigma(\omega)$ at $\omega = 0$. Thus knowing $\sigma_1(\omega)$ for $\omega \neq 0$ one can calculate $\sigma_2(\omega)$ and consequently $\sigma(\omega)$; the constant A can be obtained from the behavior of $\sigma(\omega)$ at infinity.

The simplest and crudest way to obtain approximately $\sigma(\omega)$ is by employing Newton's equation for the electronic drift velocity υ in the presence of the field F exp($-i\omega t$) and of a friction force $-m\upsilon/\tau_{tr}$ (τ_{tr} is the transport relaxation time):

$$-i\omega m\upsilon = -m\upsilon/\tau_{tr} + eF \quad . \tag{7.72}$$

Since the current is given by $j = nev$, where n is the electronic density, we obtain for $\sigma(\omega)$

$$\sigma(\omega) \approx \frac{ne^2\tau_{tr}}{m(1 - i\omega\tau_{tr})} \quad . \tag{7.73}$$

For $\omega \to \infty$ the electronic motion is classical, all scattering is negligible, and only the electronic inertia (as measured by its mass m) matters. Hence, (7.73) (without the friction term) becomes exact:

$$\sigma(\omega) \xrightarrow[\omega \to \infty]{} i \frac{ne^2}{m\omega} \quad . \tag{7.74}$$

A slight generalization of this approach (capable of treating nonspherical Fermi surfaces) consists in considering the whole equilibrium distribution transposed rigidly in $\underline{k}$-space by an amount $\delta k = mv/\hbar$, where v is again given by (7.72) [5.3]. A more sophisticated approach (but still semiclassical) is the one based on Boltzmann's equation, which determines the electron distribution in $\underline{k}$-space in the presence of an applied electric field and scattering mechanism, which tends to restore equilibrium [7.77]. The advantage of this approach is that the phenomenological parameter τ_{tr} is related to the scattering potential $V(\underline{r})$ by

$$\frac{1}{\tau_{tr}} = \frac{2\pi}{\hbar} \rho_F n_{imp} \frac{1}{4\pi} \int d\Omega V^2(\underline{q})(1 - \cos\theta) \quad , \tag{7.75}$$

where ρ_F is the DOS per volume per spin direction, n_{imp} is the number of scatterers per unit volume, the integration is over all directions of $\underline{q}$ ($E(\underline{q}) = E_F$) and $V(\underline{q})$ is the Fourier transform of $V(\underline{r})$ [see (4.23)]. Equation (7.75) follows immediately from Fermi's golden rule (4.48) for the transition rate from a given state on the Fermi surface to all others. The extra factor $(1 - \cos\theta)$ accounts for the fact that what matters for transport is the momentum change along the direction of initial propagation. If $V^2(\underline{q})$ is isotropic (or more generally, if it does not contain a p-spherical harmonic) then the $\cos\theta$ term averages out to zero, and $\tau_{tr} = \tau$, where τ is given by (7.75) without the $\cos\theta$ term.

7.3.2 A General Formula for Conductivity

We proceed now to obtain a general quantum mechanical expression for $\sigma_1(\omega)$, $\omega \neq 0$ [7.78]. Then $\sigma(\omega)$ will be calculated by using (7.71) and (7.74). An alternative way to obtain $\sigma(\omega)$ based on linear response theory is given in [6.12] and [7.79]. Using (7.68,69) we obtain for the time average power $P = \Omega\overline{\mathcal{E}j}$ consumed by the system the expression

$$P = 2\Omega|F|^2\sigma_1(\omega) \quad , \quad \omega \neq 0 \quad . \tag{7.76}$$

The average power P can also be calculated by multiplying the energy $\varepsilon_{\beta\alpha} \equiv \hbar\omega_{\beta\alpha} = E_\beta - E_\alpha$ (absorbed by the system during the field-induced transition $|\alpha\rangle \to |\beta\rangle$) by the transition rate $p_{\alpha\beta}$ and by summing over all possibilities ($|\alpha\rangle \neq |\beta\rangle$)

$$P = \sum_{\alpha\beta} \varepsilon_{\beta\alpha}p_{\alpha\beta} = \frac{1}{2}\sum_{\alpha\beta} \varepsilon_{\beta\alpha}(p_{\alpha\beta} - p_{\beta\alpha}) \quad . \tag{7.76'}$$

The summation includes a spin degeneracy factor equal to 2; the probability per unit time $p_{\alpha\beta}$ is given by

$$p_{\alpha\beta} = f_\alpha(1 - f_\beta)W_{\alpha\beta} \quad , \tag{7.77}$$

where $f_\alpha \equiv f(E_\alpha)$ is the Fermi distribution, and $W_{\alpha\beta}$ can be obtained (like Fermi's golden rule) by substituting in (4.44) the perturbation; in the present case the perturbation is

$$H_1(t) = ex\mathcal{E}(t) \quad , \tag{7.78}$$

where the field is given by (7.68). The result is

$$W_{\alpha\beta} = \frac{2\pi}{\hbar}e^2|F|^2|\langle\alpha|x|\beta\rangle|^2[\delta(\hbar\omega - \varepsilon_{\beta\alpha}) + \delta(\hbar\omega + \varepsilon_{\beta\alpha})] \quad . \tag{7.79}$$

Combining (7.76',77,79) and comparing with (7.76) we obtain

$$\sigma_1(\omega) = \frac{\pi e^2}{\Omega}\sum_{\alpha\beta}|\langle\alpha|x|\beta\rangle|^2\omega_{\beta\alpha}(f_\alpha - f_\beta)\delta(\hbar\omega - \hbar\omega_{\beta\alpha}) \quad . \tag{7.80}$$

We can recast this expression in terms of the momentum p_x matrix elements. Indeed, since $p_x = m\,dx/dt = im[Hx - xH]/\hbar$, we have

$$\langle\alpha|p_x|\beta\rangle = im\omega_{\alpha\beta}\langle\alpha|x|\beta\rangle \quad . \tag{7.81}$$

Substituting (7.81) in (7.80) we obtain

$$\sigma_1(\omega) = \frac{\pi e^2}{\Omega m^2}\sum_{\alpha\beta}|\langle\alpha|p_x|\beta\rangle|^2\,\frac{f_\alpha - f_\beta}{\omega_{\beta\alpha}}\,\delta(\hbar\omega - \hbar\omega_{\beta\alpha}) \quad . \tag{7.82}$$

Because of the δ function in (7.82) it is trivial to perform the integration in (7.71); the constant A can be obtained by comparing with (7.74), and thus $\sigma_2(\omega)$ is calculated. Combining the expressions for $\sigma_1(\omega)$ and $\sigma_2(\omega)$ we have finally:

$$\sigma(\omega + is) = \left(\frac{e^2 n}{m} - \frac{e^2}{\Omega m^2} \sum_{\alpha\beta} |\langle\alpha|p_x|\beta\rangle|^2 \frac{f_\alpha - f_\beta}{\hbar\omega_{\beta\alpha}}\right) \frac{i}{\omega + is}$$

$$- i \frac{e^2}{\Omega m^2} \sum_{\alpha\beta} |\langle\alpha|p_x|\beta\rangle|^2 \frac{f_\alpha - f_\beta}{\hbar\omega_{\beta\alpha}} \frac{1}{\omega_{\beta\alpha} - \omega - is} \quad . \tag{7.83}$$

The term $ie^2 n/m\omega \equiv \sigma_d(\omega)$ is the so-called diamagnetic contribution to $\sigma(\omega)$, and the rest, $\sigma_p(\omega)$, is the paramagnetic one, which can be recast as

$$\sigma_p(\omega) = -i \frac{e^2}{\Omega m^2 \hbar\omega} \sum_{\alpha\beta} |\langle\alpha|p_x|\beta\rangle|^2 \frac{f_\alpha - f_\beta}{\omega_{\beta\alpha} - \omega - is} \quad . \tag{7.84}$$

The local conductivity $\sigma(\underline{r},\underline{r}';\omega)$ can be obtained from (7.83) by replacing the velocity, p_x/m, matrix elements by the current operator matrix elements, i.e., $|\langle\alpha|p_x|\beta\rangle|^2/\Omega \rightarrow \lambda_{\alpha\beta}(\underline{r})\lambda^*_{\alpha\beta}(\underline{r}')$, where

$$\lambda_{\alpha\beta}(\underline{r}) = \frac{-i\hbar}{2}\left[\psi^*_\alpha(\underline{r})\frac{\partial\psi_\beta(\underline{r})}{\partial x} - \psi_\beta(\underline{r})\frac{\partial\psi^*_\alpha(\underline{r})}{\partial x}\right] \quad ; \tag{7.85}$$

$\psi_\alpha(\underline{r}) = \langle\underline{r}|\alpha\rangle$ and $\psi_\beta(\underline{r}) = \langle\underline{r}|\beta\rangle$. To obtain $\sigma_{\mu\nu}$ one replaces $|\langle\alpha|p_x|\beta\rangle|^2$ by $\langle\alpha|p_\mu|\beta\rangle\langle\beta|p_\nu|\alpha\rangle$ in (7.83).

For fixed boundary conditions the term in parentheses in (7.83) vanishes identically. This can be proved by noticing that (7.81) implies the following equality:

$$|\langle\alpha|p_x|\beta\rangle|^2/\omega_{\beta\alpha} = -im\langle\alpha|x|\beta\rangle\langle\beta|p_x|\alpha\rangle \quad .$$

It is worthwhile to point out that for a *perfect* conductor with *periodic boundary* conditions the term in parentheses in (7.83) is not zero, because then (7.81) fails; in this case the only term which survives is the diamagnetic one.

MOTT and DAVIS [7.78] gave an instructive way to obtain (7.73) (for $\omega = 0$) starting from (7.82). Their main assumption is that for weakly disordered systems the eigenfunctions have essentially constant amplitude (as in perfectly ordered systems) but phase coherence is maintained only over a distance of the order of the transport mean free path $\ell_{tr} \equiv \upsilon_F\tau_{tr}$ (for ordered systems $\ell_{tr} = \infty$). If the total volume Ω is divided into cells, each of volume $\upsilon_i \approx \ell_{tr}^d$, then

$$<\alpha|p_x|\beta> = \sum_{\upsilon_i} \int_{\upsilon_i} d\underline{r}\ \psi_\alpha^* p_x \psi_\beta \quad . \tag{7.86}$$

Multiplying (7.86) by its complex conjugate and cancelling the cross terms due to their random phases, one obtains

$$|<\alpha|p_x|\beta>|^2 = \sum_i |\int_{\upsilon_i} d\underline{r}\ \psi_\alpha^* p_x \psi_\beta|^2 = \frac{\Omega}{\upsilon} |\int_\upsilon \psi_\alpha^* p_x \psi_\beta d\underline{r}|^2 \quad . \tag{7.87}$$

Within the volume υ, ψ_α, ψ_β are assumed to be plane waves so that the integral can be performed explicitly. Substituting the result in (7.82) one obtains $\sigma_1 \approx e^2 n\ell_{tr}/2m\upsilon_F$, which coincides with (7.73) apart from a factor of 2 (for $\omega = 0$) since $\ell_{tr} = \upsilon_F\tau_{tr}$.

THOULESS [7.80] has also obtained (7.73) starting from (7.82) by assuming that ψ_α (and ψ_β) is a linear combination of plane waves with coefficients which are uncorrelated Gaussian random variables whose variance is $(\pi/\ell_{tr}k^2\Omega)[(k - k_\alpha)^2 + \ell_{tr}^2/4]$.

The general formulae we derived for σ or σ_1, are usually referred to in the literature as the KUBO [7.81] - GREENWOOD [7.82] formulae.

7.3.3 The Conductivity in Terms of Green's Functions

Now we recast the general expression for $\sigma_1(\omega)$ or $\sigma(\omega)$ in an invariant form involving Green's functions. Among the various slightly different versions, we discuss here two. In the first used by ABRIKOSOV et al. [7.83] and by DONIACH and SONDHEIMER [7.84], $\sigma(\omega)$ is epxressed in terms of the causal Green's function $g(E)$, which is of central importance in many-body theory (Part III). For the present purposes (where electrons are not interacting with each other), $g(E)$ can be defined as $G^-(E)$ for $E < E_F$ and $G^+(E)$ for $E > E_F$, i.e.,

$$g(E) \equiv [E + is\bar{\varepsilon}(E - E_F) - H]^{-1} \tag{7.88}$$

from which it follows that $<\alpha|g(E)|\beta> = \delta_{\alpha\beta}[E + is\bar{\varepsilon}(E - E_F) - E_\alpha]^{-1}$. At T = 0 one has

$$\frac{f_\alpha - f_\beta}{\varepsilon_{\beta\alpha} - \hbar\omega - is} = \frac{-1}{2\pi i}\int_{-\infty}^{\infty} dE \, \langle\alpha|g(E)|\alpha\rangle\langle\beta|g(E + \hbar\omega)|\beta\rangle \quad . \tag{7.89}$$

To prove (7.89) close the integration path in its rhs by a semicircle either in the upper or in the lower half-plane and use residue theory.

Substituting (7.89) in (7.84) and performing the summation over intermediate states by employing (1.4'), one obtains

$$\sigma(\omega) = \frac{ie^2 n}{m\omega} + \frac{e^2}{\Omega m^2 \omega}\int \frac{dE}{2\pi} \mathrm{Tr}\{p_x g(E) p_x g(E + \hbar\omega)\} \quad , \tag{7.90}$$

where the Tr operation includes a factor of 2 due to spin summation. By using standard many-body techniques (see, e.g., Part III of this book), one can generalize (7.90) to nonzero temperatures. ABRIKOSOV et al. [7.83] as well as DONIACH and SONDHEIMER [7.84], starting from (7.90), expand the two g's in a perturbation series and after a lengthy calculation they obtain the ensemble average of $\sigma(\omega)$ (over all configuraitons of the scatterers), which agrees with (7.73) and (7.75).

It is probably simpler and more convenient to express $\sigma_1(\omega)$ in terms of $\tilde{G} = G^+ - G^-$. For that purpose one notices first that

$$\frac{f_\alpha - f_\beta}{\omega_{\beta\alpha}} \delta(\hbar\omega - \varepsilon_{\beta\alpha}) = \int dE \delta(E - E_\alpha)\delta(E - E_\beta + \hbar\omega)\frac{f(E) - f(E + \hbar\omega)}{\omega} \, . \tag{7.91}$$

Substituting (7.91) in (7.82) we have

$$\begin{aligned} \sigma_1(\omega) &= \frac{\pi e^2 \hbar}{\Omega m^2}\int_{-\infty}^{\infty} dE \, \frac{f(E) - f(E + \hbar\omega)}{\hbar\omega} \mathrm{Tr}\{p_x \delta(E + \hbar\omega - H) p_x \delta(E - H)\} \\ &= \frac{e^2 \hbar}{\pi\Omega m^2}\int_{-\infty}^{\infty} dE \, \frac{f(E) - f(E + \hbar\omega)}{\hbar\omega} \mathrm{Tr}\{p_x \mathrm{Im} G^+(E + \hbar\omega) p_x \, \mathrm{Im} G^+(E)\} \, , \end{aligned} \tag{7.92}$$

where the Tr operation includes a factor of 2 due to spin summation. One can express $\sigma_1(\omega)$ in terms of $G^\pm$ by noticing that $\mathrm{Im}G^+ = i(G^- - G^+)/2$, and substituting it in (7.92).

It is important to remember that in all expressions (7.81-84,90,92) for $\sigma_1(\omega)$ or $\sigma(\omega)$ an appropriate ensemble average over all possible configurations of the scatterers has to be taken. One usually takes the arithmetic mean. Recently, [7.85-88] explicit results in one-dimensional systems demonstrated that it is the probability distribution of $\ln\sigma(0)$, and not that of $\sigma(0)$ itself which is sharply peaked; thus the geometric mean of $\sigma(0)$, and not the arithmetic, is representative of the ensemble. This anomalous behavior seems to be related with the exponential decrease of $\sigma(0)$ as one increases the linear dimension of 1-D systems.

To calculate the arithmetic mean of $\sigma_1(\omega)$ one has to average the product of two G's, which in general is different from the product of the averages. We shall return to this important question later, but for the time being let us replace each G in (7.92) by its average value $\langle G\rangle$, which can be calculated with the use of the CPA. Working in the $\underline{k}$-representation where both p and $\langle G\rangle$ are diagonal we obtain from (7.92) in the limit $\omega \to 0$

$$\sigma^{(0)}(0) = \frac{2e^2\hbar}{\pi\Omega} \int dE \left(-\frac{\partial f}{\partial E}\right) \sum_{\underline{k}} |\langle\underline{k}|(p_x/m)|\underline{k}\rangle|^2 |\mathrm{Im}\{G^+(E,\underline{k})\}|^2 \quad . \qquad (7.93)$$

Taking into account that $\langle\underline{k}|(p_x/m)|\underline{k}\rangle = \partial E(\underline{k})/\hbar\partial k_x \equiv \upsilon_x(\underline{k})$, that $G^+(E,\underline{k}) = [E - \Sigma(E) - E(\underline{k})]^{-1}$, where $\Sigma = \Sigma_1 - i\Sigma_2$ is the self-energy, and that $(-\partial f/\partial E) \to \delta(E - E_F)$ as $T \to 0$ we obtain for the zero temperature conductivity

$$\sigma^{(0)}(0) = \frac{2e^2\hbar}{\pi\Omega} \sum_{\underline{k}} \upsilon_x^2(\underline{k}) \frac{\Sigma_2^2}{\{[E_F - \Sigma_1 - E(\underline{k})]^2 + \Sigma_2^2\}^2} \quad . \qquad (7.93')$$

The summation over $\underline{k}$ is facilitated if one introduces an integration over $\delta(E' - E(\underline{k}))dE'$, so that

$$\sigma^{(0)}(0) = \frac{2e^2\hbar}{\pi} \int dE' \left[\frac{1}{\Omega} \sum_{\underline{k}} \upsilon_x^2(\underline{k})\delta(E' - E(\underline{k}))\right] \frac{\Sigma_2^2}{[(E_F - \Sigma_1 - E')^2 + \Sigma_2^2]^2} \quad . \qquad (7.94)$$

The quantity in the brackets depends only on the form of $E(\underline{k})$. For lattices of cubic symmetry υ_x^2 can be replaced by $|\underline{\upsilon}|^2/d$, where d is the dimensionality. Furthermore the $\underline{k}$ summation can be replaced by an integration over the surface of constant energy E', so that

$$\frac{1}{d\Omega} \sum_{\underline{k}} |\underline{\upsilon}(\underline{k})|^2 \delta(E' - E(\underline{k})) = \frac{1}{(2\pi)^d d} \int dS_k \frac{1}{\hbar|\underline{\upsilon}(\underline{k})|} |\underline{\upsilon}(k)|^2 = \frac{1}{(2\pi)^d d\hbar} S(E')\upsilon(E')$$

where S is the area of the surface of energy E' and $\upsilon(E')$ is the average of the magnitude of the velocity over this surface. Equation (7.94) can be simplified considerably in the weak scattering limit, where Σ_2 is small, and $S\upsilon$ can be taken as a constant equal to its value at $E_F - \Sigma_1 \approx E_F$. Then the integration over E' can be performed explicitly giving for σ

$$\sigma^{(0)}(0) = \frac{e^2 \upsilon_F}{|\Sigma_2|} \frac{S_F}{(2\pi)^d d} \quad . \tag{7.95}$$

For a spherical surface of constant energy $S_F/d(2\pi)^d$ equals to $n/2k_F$, where n is the electronic density. Thus

$$\sigma^{(0)}(0) = e^2 n \frac{\upsilon_F}{k_F} \frac{1}{2|\Sigma_2|} \quad . \tag{7.95'}$$

In the simple case, where $E(k) = \hbar^2 k^2/2m$, $\upsilon_F/k_F = \hbar/m$, and (7.95') coincides with (7.73), (7.75) if we identify $2|\Sigma_2|/\hbar$ with $1/\tau_{tr}$. Comparing (7.75) with (7.51) we see that $2|\Sigma_2|/\hbar = 1/\tau$. Thus our calculation reproduces the standard result apart from the $\cos\theta$ term, which is not present in $|\Sigma_2|$. The reason for this omission is our neglect of the difference $p_x\langle Gp_xG\rangle - p_x\langle G\rangle p_x\langle G\rangle$ which is usually called vertex correction. In the next subsection we present a CPA-like scheme for obtaining the vertex corrections. In the weak scattering limit this scheme reduces to the standard theory for conductivity [7.83,84] and reproduces the $\cos\theta$ correction.

One can rewrite (7.95) (with the $\cos\theta$ correction) in terms of τ_{tr} or $\ell_{tr} = \upsilon_F\tau_{tr}$ as follows

$$\sigma^{(0)}(0) = \frac{2}{(2\pi)^d d} \frac{e^2}{\hbar} \upsilon_F\tau_{tr} S_F = \frac{2}{(2\pi)^d d} \frac{e^2}{\hbar} \ell_{tr} S_F \quad . \tag{7.95''}$$

(For $d = 2$, S_F is the length of the Fermi line, and for $d = 1$, $S_F = 2$.)

To summarize: Starting from the general equation (7.92) and utilizing the CPA to obtain both $\langle G\rangle$ and the vertex corrections produces a very satisfactory approximate result for $\sigma_1(\omega)$. In the weak scattering limit this result reduces to the standard expression for $\sigma_1(\omega)$ [7.18,83,84,89]. Furthermore, the vertex corrections in the weak scattering limit are responsible for the appearance of the $\cos\theta$ term in $1/\tau_{tr}$.

It is worthwhile to mention that in the simple TBM the scattering is isotopic in $\underline{k}$ space (as a matter of fact it is $\underline{k}$-independent) and as a result the $\cos\theta$ term gives zero contribution to $1/\tau_{tr}$. This means that the CPA vertex corrections for the *conductivity* and for the *simple TBM* are zero. In the next subsection we shall see that this is actually the case.

7.3.4 CPA for the Vertex Corrections

As was mentioned before, transport properties depend on combinations of the form

$$\langle AG(z)BG(z')\rangle \quad ,$$

where A and B are nonrandom operators. Following VELICKY [7.90] we define the operator $\tilde{\Gamma}$ by the relation

$$\langle G(z)\rangle\tilde{\Gamma}\langle G(z')\rangle = \langle G(z)BG(z')\rangle - \langle G(z)\rangle B\langle G(z')\rangle \tag{7.96}$$

Within the CPA $\langle G(z)\rangle = G_e(z)$. Furthermore, according to (7.52) we can express G in terms of G_e and the t-matrix T', which is given by (7.30), i.e.,

$$T' = \sum_{\underline{m}} Q'_{\underline{m}} = \sum_{\underline{m}} \tilde{Q}'_{\underline{m}} \quad . \tag{7.97}$$

The operators $Q'_{\underline{m}}$, $\tilde{Q}'_{\underline{m}}$ satisfy the following equations

$$Q'_{\underline{m}} = T'_{\underline{m}}(1 + G_e \sum_{\underline{n}\neq\underline{m}} Q'_{\underline{n}}) \quad , \tag{7.98}$$

$$\tilde{Q}'_{\underline{m}} = (1 + \sum_{\underline{n}\neq\underline{m}} \tilde{Q}'_{\underline{n}}G_e)T'_{\underline{m}} \quad , \tag{7.99}$$

where $T'_{\underline{m}} = |\underline{m}\rangle t'_{\underline{m}}\langle \underline{m}|$. Substituting the above relations in (7.96) we have

$$\tilde{\Gamma} = \sum_{\underline{m}\underline{n}} \tilde{\Gamma}_{\underline{m}\underline{n}} \tag{7.100}$$

$$\tilde{\Gamma}_{\underline{m}\underline{n}} = \langle T'_{\underline{m}}(1 + G_e \sum_{\underline{\ell}\neq\underline{m}} Q'_{\underline{\ell}})G_eBG_e(1 + \sum_{\underline{s}\neq\underline{n}} \tilde{Q}'_{\underline{s}}G_e)T'_{\underline{n}}\rangle \quad . \tag{7.100'}$$

The CPA for the vertex part $\tilde{\Gamma}_{\underline{m}\underline{n}}$ is again to replace the average of the product by a product of averages as follows

$$\tilde{\Gamma}_{\underline{m}\underline{n}} \approx \langle T'_{\underline{m}}\langle(1 + G_e \sum_{\underline{\ell}\neq\underline{m}} Q'_{\underline{\ell}})G_eBG_e(1 + \sum_{\underline{s}\neq\underline{n}} \tilde{Q}'_{\underline{s}}G_e)\rangle T'_{\underline{n}}\rangle \quad . \tag{7.101}$$

Taking into account that the $t'_{\underline{m}}$ are independent random variables with zero mean (due to the CPA condition) we see that the rhs of (7.101) is nonzero only when $\underline{m} = \underline{n}$, i.e.,

$$\tilde{\Gamma}_{\underline{m}\underline{n}} = \delta_{\underline{m}\underline{n}}\tilde{\Gamma}_{\underline{n}} = \delta_{\underline{m}\underline{n}}|\underline{n}\rangle\tilde{\gamma}_{\underline{n}}\langle\underline{n}| \quad . \tag{7.102}$$

Out of the four terms inside the inner average sign in (7.101) two are proportional to $\langle Q'_{\underline{\ell}}\rangle$ or $\langle Q'_{\underline{s}}\rangle$ and hence are zero within the CPA; the third is G_eBG_e, and the fourth is $\sum_{\underline{\ell},\underline{s}\neq\underline{n}} G_e\langle\tilde{Q}'_{\underline{\ell}}G_eBG_e\tilde{Q}'_{\underline{s}}\rangle G_e$, which in view of (7.98-101) and (7.102), equals to $G_e \sum_{\ell \neq n} \tilde{\Gamma}_\ell G_e$. Hence we can rewrite (7.101) as

$$\tilde{\Gamma}_n = \langle T_n' G_e (B + \sum_{\underline{\ell} \neq \underline{n}} \tilde{\Gamma}_{\underline{\ell}}) G_e T_n' \rangle \quad . \tag{7.103}$$

The system (7.103) for the unknown quantities Γ_n is a closed one, and consequently the problem has been solved formally. To be more explicit, we take matrix elements in the local basis; we also write $\sum_{\underline{\ell} \neq \underline{n}} \tilde{\Gamma}_{\underline{\ell}} = \sum_{\underline{\ell}} \tilde{\Gamma}_{\underline{\ell}} - \tilde{\Gamma}_n$. Then (7.103) becomes

$$\tilde{\gamma}_n [1 + \langle t_{\underline{n}}' t_{\underline{n}}' \rangle G_e(\underline{n}\underline{n}) G_e(\underline{n}\underline{n})] = \sum_{\underline{\ell}\underline{m}} \langle t_{\underline{n}}' t_{\underline{n}}' \rangle G_e(\underline{n},\underline{\ell}) [B_{\underline{\ell}\underline{m}} + \tilde{\gamma}_{\underline{\ell}} \delta_{\underline{\ell}\underline{m}}] G_e(\underline{m},\underline{n}) \quad . \tag{7.104}$$

To display the energy dependences explicitly and to make the expression more compact we introduce the quantity $u(z,z')$ defined by

$$u(z,z') \equiv \langle t_{\underline{n}}'(z) t_{\underline{n}}'(z') \rangle [1 + \langle t_{\underline{n}}'(z) t_{\underline{n}}'(z') \rangle G_e(\underline{n},\underline{n};z) G_e(\underline{n},\underline{n};z')]^{-1} . \tag{7.105}$$

Then $\tilde{\gamma}_n(z,z')$ becomes

$$\tilde{\gamma}_{\underline{n}}(z,z') = \sum_{\underline{\ell}\underline{m}} u(z,z') G_e(\underline{n},\underline{\ell};z) [B_{\underline{\ell}\underline{m}} + \tilde{\gamma}_{\underline{\ell}}(z,z') \delta_{\underline{\ell}\underline{m}}] G_e(\underline{m},\underline{n};z') \quad . \tag{7.106}$$

Then $\tilde{\gamma}_{\underline{n}}$ can be factorized as follows

$$\tilde{\gamma}_{\underline{n}} = \sum_{\underline{m}\underline{\ell}\underline{s}} \gamma_{\underline{n}\underline{m}} G_e(\underline{m}\underline{\ell}) B_{\underline{\ell}\underline{s}} G_e(\underline{s}\underline{m}) \quad . \tag{7.107}$$

Substituting (7.107) in (7.106) we find that $\gamma_{\underline{n}\underline{m}}$ obeys the B-independent equation

$$\gamma_{\underline{n}\underline{m}} = u \delta_{\underline{n}\underline{m}} + \sum_{\underline{\ell}} u G_e(\underline{n},\underline{\ell}) G_e(\underline{\ell},\underline{n}) \gamma_{\underline{\ell}\underline{m}} \quad , \tag{7.108}$$

where the energy dependences are not shown explicitly. The quantity $\gamma_{\underline{n}\underline{m}}$ corresponds more closely than $\tilde{\gamma}_{\underline{n}}$ to what is called vertex part in many-body theory (Part III). Equation (7.108) can be solved immediately in $\underline{k}$ space

$$\gamma(\underline{q};z,z') = u(z,z') [1 - u(z,z') A(\underline{q};z,z')]^{-1} \quad , \tag{7.109}$$

where

$$A(\underline{q};z,z') = \sum_{\underline{m}} e^{-i\underline{q}\underline{m}} G_e(\underline{0},\underline{m};z) G_e(\underline{m},\underline{0};z')$$

$$= \frac{1}{N} \sum_{\underline{k}} G_e(\underline{k};z) G_e(\underline{k} - \underline{q};z') \tag{7.110}$$

and

$$\gamma_{\underline{n}\underline{m}}(z,z') = \frac{1}{N}\sum_{\underline{q}} \exp[i\underline{q}(\underline{n}-\underline{m})]\gamma(\underline{q};z,z') \quad . \tag{7.111}$$

The basic equation (7.108), the series that results by iterating it, equation (7.107), and the vertex correction to $\mathrm{Tr}\{A\langle G(z)BG(z')\rangle\}$ are shown graphically in Fig.7.5 for the local basis representation. The same diagrams, with slightly different correspondence rules, can also be used for the $\underline{k}$-representation.

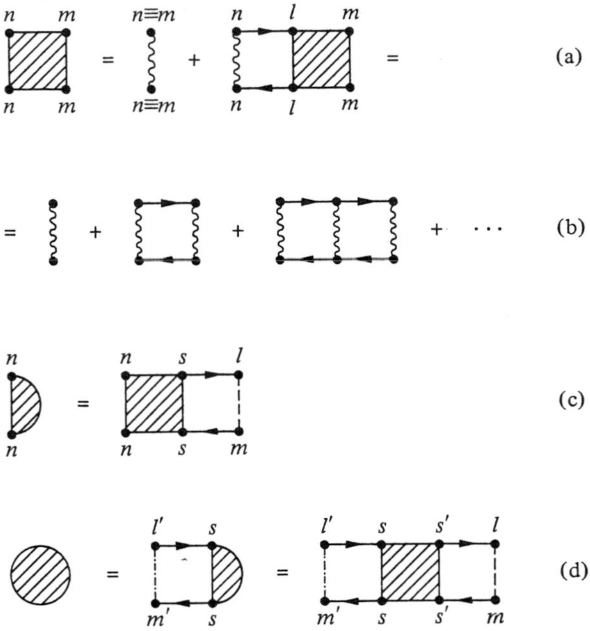

Fig.7.5a-d. Diagrammatic representation of the CPA for the vertex corrections in the local representation. Directed solid lines from $\underline{n}$ to $\underline{\ell}$ correspond to $G_e(\underline{\ell},\underline{n})$; the upper ones have an energy argument z', and the lower ones have z. The wavy lines, which always connect identical sites, correspond to u, see (7.105). The shaded square is $\gamma_{\underline{n}\underline{m}}$, the shaded semicircle is $\tilde{\gamma}_{\underline{n}}$, and the shaded circle is the vertex correction for the quantity $\mathrm{Tr}\{A\langle G(z)BG(z')\rangle\}$; the dashed lines correspond to $B_{\underline{m}\underline{\ell}}$ and the dashed-dot lines to $A_{\underline{\ell}'\underline{m}'}$. Summation over all internal points is implied. (a) is equivalent to (7.108) and (b) to its iteration; (c) is equivalent to (7.107)

Let us comment on the CPA results. In the weak scattering limit, as it follows from (7.105), $u \to \langle \varepsilon_{\underline{n}}'^2\rangle$. Then the terms summed in Fig.7.5b are exactly the same as those used in the standard approach for the conductivity [7.83, 84,89]. Hence, in the weak scattering limit the CPA for the vertex corrections reduces to the standard result.

If we are interested in the vertex corrections to the conductivity tensor, we must take $A = B = \underline{p}$. Now the momentum operator $\underline{p}$ in the k representation is given by $m\partial E(\underline{k})/\hbar\partial\underline{k}$ and is an odd function of $\underline{k}$ due to time reversal symmetry. This implies that $p_{\underline{\ell s}}$ is an odd function of $\underline{\ell} - \underline{s}$, which means that $\tilde{\gamma}_{\underline{n}}$ for this case is zero (7.107). As we mentioned before, the vanishing of the vertex corrections for σ is a consequence of the isotropy of the scattering potential in our model. The vertex corrections for other quantities do not vanish. We shall see in the next section that a quantity of importance is $\langle G(\underline{0},\underline{m};z)G(\underline{m},\underline{0};z')\rangle$ which corresponds to $A = |\underline{0}\rangle\langle\underline{0}|$ and $B = |\underline{m}\rangle\langle\underline{m}|$. Combining (7.109,107 and 96) we obtain

$$\langle G(\underline{m},\underline{0};z')G(\underline{0},\underline{m};z)\rangle = \frac{1}{N}\sum_{\underline{k}} e^{i\underline{k}\underline{m}} \frac{A(\underline{k};z,z')}{1 - u(z,z')A(\underline{k};z,z')} \quad . \tag{7.112}$$

The CPA does not of course keep all the terms which contribute to Γ. But the ones kept are very important at least in the case where $z = E + is$ and $z' = E - is$: they are positive definite so that no possibility of self-cancellations exists, in contrast to other terms, which involve unpaired $G_e(\underline{\ell},\underline{m})$'s with $|\underline{\ell} - \underline{m}|$ larger than the phase coherence length. As we shall see in the next section, the CPA terms can give a very large contribution, and last but not least, they can be summed explicitly to produce a closed expression [see (7.109)]. It is also worthwhile to point out that the CPA preserves an exact relation between the self-energy and the vertex part $\tilde{\Gamma}$ for $B = 1$. By observing that $G(z)G(z') = (z - z')^{-1}[G(z') - G(z)]$ and by taking into account (7.96), one obtains (for $B = 1$)

$$\tilde{\Gamma}(z,z') = -(z - z')^{-1}[\Sigma(z) - \Sigma(z')] \quad , \tag{7.113}$$

where the exact self-energy $\Sigma(z)$ satisfies by definition the relation $\langle G(z)\rangle = [z - \Sigma(z) - H_0]^{-1}$. The CPA $\tilde{\Gamma}$ and Σ's satisfy (7.113).

7.3.5 Vertex Corrections Beyond the CPA

We now discuss a subgroup of terms contributing to the vertex corrections but omitted within the CPA. These terms have a structure quite similar to the CPA terms and, as a result, may be equally important. They are shown diagrammatically in Fig.7.6, from which one has

$$\mathrm{Tr}\{A\langle G(z)BG(z')\rangle\} - \mathrm{Tr}\{A\langle G(z)\rangle B\langle G(z')\rangle\}$$

$$\approx \sum A_{\underline{\ell s}}\, G_e(\underline{s},\underline{m};z)\gamma'_{\underline{mn}}(z,z')\, G_e(\underline{n},\underline{i};z)\, B_{\underline{ij}}\, G_e(\underline{j},\underline{m};z')\, G_e(\underline{n},\underline{\ell};z') \quad . \tag{7.114}$$

The quantity $\gamma'_{\underline{mn}}(z,z')$ can be calculated explicitly by working in $\underline{k}$-space, and the result is similar to that of $\gamma_{\underline{nm}}(z,z')$, except that A is now replaced by A' where

$$A'(\underline{q}) = \sum_{\underline{m}} \exp(-i\underline{q}\underline{m})\, G_e(\underline{0},\underline{m})\, G_e(\underline{0},\underline{m}) = \frac{1}{N}\sum_{\underline{k}} G_e(\underline{k})\, G_e(\underline{q}-\underline{k}) \quad , \tag{7.115}$$

and the first term of the series is $u^2A'(\underline{q})$ and not u. Thus

$$\gamma'(\underline{q}) = u^2A'(\underline{q})[1-uA'(\underline{q})]^{-1} \tag{7.116}$$

$$\gamma'_{\underline{mn}} = \frac{1}{N}\sum_{\underline{q}} \exp[i\underline{q}(\underline{m}-\underline{n})]\gamma'(\underline{q}) \quad . \tag{7.117}$$

It must be pointed out that time reversal symmetry of the Hamiltonian implies that $G_e(\underline{0},\underline{m})$ equals to $G_e(\underline{m},\underline{0})$, which in turn means that $A(\underline{q}) = A'(\underline{q})$. We remind the reader that the presence of an external magnetic field destroys time reversal.

The post-CPA terms shown in Fig.7.6b (or c) produce results for the vertex part $\gamma'_{\underline{nm}}$, which are essentially equivalent to the CPA for $\gamma_{\underline{nm}}$. The main difference between the two cases is the way the vertex part is connected to the G's [compare Fig.7.5d with Fig.7.6a: in Fig.7.5d the two G's which connect to *A* (or *B*) start and end at the same site of the vertex part, while in Fig. 7.6a the sites are different]. As a result of this difference the vertex cor-

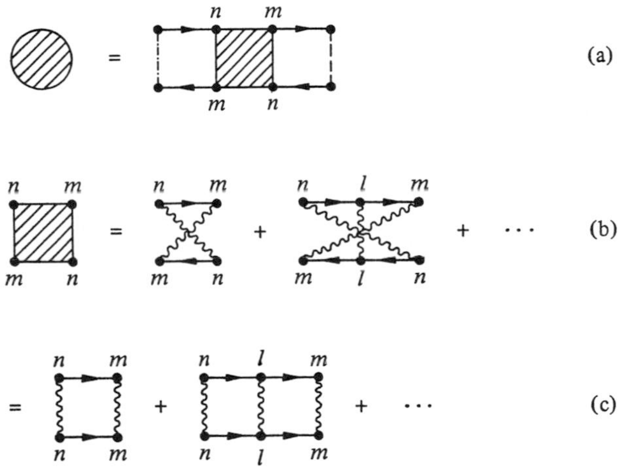

Fig.7.6a-c. (a) The shaded circle represents additional (beyond the CPA shown in Fig.7.5) vertex corrections for the quantity Tr{$A\langle G(z)BG(z')\rangle$}. In the present case the vertex part (shaded square) is given by the diagrams shown in (b); they can be redrawn (without changing their values) as in (c). The rules are as in Fig.7.5

rections to the conductivity, and for our simple model Hamiltonian, are identically zero for the CPA terms (because of the oddness of $P_{\ell m}$ with respect to the indices), while it is not zero for the post-CPA terms shown in Fig.7.6. As a matter of fact, these terms make an extremely important contribution to the conductivity, which will be discussed in some detail in the next section in connection with the problem of localization.

7.4 Disorder and Localization

In Sect.7.3.2 we mentioned that the basic result for the DC conductivity [see (7.73,75)] is a consequence of the phase incoherence created by the disorder for distances larger than the mean free path ℓ. On the other hand, it was implicitly assumed there that the amplitude of the eigenfunctions is essentially unaffected by the disorder. Transport properties calculations in the metallic regime are usually based on this important assumption: Disorder creates phase incoherence characterized by ℓ (or τ), while the amplitude remains unaffected. The validity of this assumption has been questioned over the last twenty five years. As a result a new field has emerged, which recently established a firm connection with experimental data. In this section we present the basic ideas and results as well as a brief review (far from complete) of this fast developing area.

In principle, one has reasons to expect that disorder, if strong enough, may affect not only the phase but the amplitude as well. For this purpose consider the elementary example of two coupled pendulums. The transfer of the motion from one to the other is facilitated by a strong mutual coupling and is opposed by a large *frequency mismatch*. If we have an array of pendulums of random individual eigenfrequencies and then we couple them, it is not difficult to imagine that a wave propagating in such a medium may find regions of large frequency mismatch; such regions will be almost inaccessible to it. As a matter of fact one may even think of the possibility of the wave surrounded by such regions so that it cannot escape to infinity; in other words the wave may be confined to a finite region of space decaying to zero (usually in an exponential way) as one moves away from this region. These decaying eigenstates are termed localized. The ordinary propagating eigenstates are usually called extended.

In previous chapters we examined localized eigenstates associated with one (or two) isolated impurities; these states belonged to the discrete part of the spectrum. The continuous part of the spectrum was associated with extended eigenstates. Now imagine that a disordered system is created by allowing the number of impurities to become infinite. Then one expects the discrete spectrum to be smeared out to form a continuum which usually joins the main band as a tail. Obviously the question arises whether or not the eigenstates (at least those in the tails where they started as localized) become extended or remain localized.

The answer to this question can be sought within the framework of the random TBM (7.28,29,35). This by now familiar model (also known as the Anderson model) is the simplest one which incorporates the essential competition between the transfer strength V and the energy (or frequency) mismatch, which is characterized by the width $\delta\varepsilon$ ($\delta\varepsilon = |\varepsilon_A - \varepsilon_B|$, Γ, W for the binary, Lorentzian, and rectangular distributions, respectively) of the probability distribution of the random variable $\varepsilon'_{\underline{n}}$. Thus the important parameter for localization is the dimensionless ratio

$$Q = V/\delta\varepsilon \quad . \tag{7.118}$$

As we shall see another equally important parameter is the dimensionality d. Other aspects, such as the shape of the probability distribution and the type of lattice are of secondary importance and are believed not to influence the alleged universal features of the problem.

Before we proceed with our presentation we shall state here the main results in the field. These results have not yet been rigorously proven but are accepted on the basis of considerable evidence in their support:

i) There is a critical dimensionality $d = 2$ for the localization problem. For $d \leq 2$ all eigenstates are localized, no matter how weak the disorder is. For $d > 2$, and for weak disorder, the tails of the band consist of localized eigenstates while the interior corresponds to extended states. These regions are separated by critical energies termed mobility edges E_c. As the disorder increases, the mobility edges move eventually towards the center of the band; they may merge together eliminating the region of extended states. This behavior is shown in Fig.7.7, where the mobility edge trajectories (as the disorder W varies) separate the E - W plane into regions of localized and extended states.

ii) For $d \leq 2$ the vertex corrections to the conductivity, for *weak* disorder, is given by

$$\delta\sigma_1 \sim -\int_{1/L_M}^{1/L_m} \frac{dq}{q^2} \quad , \tag{7.119}$$

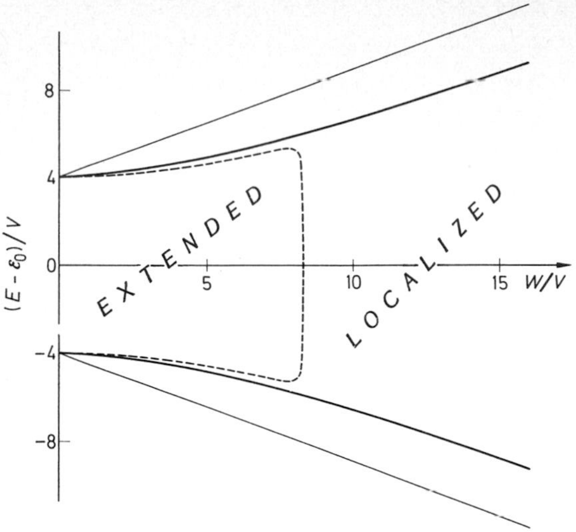

Fig.7.7. Mobility-edge (dashed line) and band-edge trajectories (heavy solid line is based on the CPA; light solid line is exact) for a diamond lattice with a rectangular distribution of site energies centered at ε_0 and of total width W. [7.91]

where L_M is dominated by the shortest of several upper cutoff lengths which may be present in the system, and L_m, the lower cutoff length, is believed to be very close to the mean free path ℓ_{tr}, i.e.,

$$L_m \sim (D\tau_{tr})^{\frac{1}{2}} = \ell_{tr}/\sqrt{d} \quad , \tag{7.120}$$

where d is the dimensionality, and D is the diffusion coefficient which is related to the conductivity by the Einstein relation

$$\sigma = 2e^2 \rho_F D \quad ; \tag{7.121}$$

ρ_F is the DOS at E_F per spin per unit volume (area, or length). Examples of upper cutoff lengths L_M, besides the geometrical dimension L, are: the diffusion length during the *inelastic* relaxation time τ_{in}

$$L_T = (D\tau_{in})^{\frac{1}{2}} \quad ; \tag{7.122}$$

the diffusion length during the time ω^{-1}, where ω is the frequency of an external AC field

$$L_\omega = (D/\omega)^{\frac{1}{2}} \quad . \tag{7.123}$$

The presence of an external magnetic field H introduces two characteristic lengths:

$$L'_H = (D/\omega_0)^{\frac{1}{2}} \tag{7.124}$$

where $\omega_0 = eH/mc$, and the cyclotron radius

$$L_H = \alpha(\hbar c/eH)^{\frac{1}{2}} \quad ; \tag{7.125}$$

the constant α must be taken as $\exp(\gamma/2) \approx 1.33$ in order to reproduce more detailed calculations in two-dimensional systems. For weak disorder $L'_H >> L_H$, and hence L_H is the relevant length. KAVEH [7.92] has pointed out that in the presence of more than one upper cutoff length a possible choice for L_M is

$$L_M^{-2} \approx L^{-2} + L_T^{-2} + L_\omega^{-2} + L_H^{-2} + \dots \quad . \tag{7.126}$$

For $d = 2$, (7.119) yields

$$\delta\sigma_1 = -\frac{e^2}{\pi^2\hbar} \ln\frac{L_M}{L_m} \quad , \tag{7.127}$$

while for $d = 1$ (where L_m can be taken as zero)

$$\delta\sigma_1 \sim - L_M \quad . \tag{7.128}$$

Equations (7.127,128) are very significant because they show that the corrections do not disappear as the upper cutoff length L_M becomes very large. On the contrary, they increase with L_M until L_M becomes so large that $\sigma \to 0$. Thus a truly metallic behavior is not possible for $d \leq 2$. This is, of course, consistent with the statement that all eigenstates are localized for $d \leq 2$. The nonmetallic nature of the conduction for $d \leq 2$ is usually masked by the fact that at high enough temperatures L_T (and hence L_M) becomes quite short. It is clear from (7.119) that when $L_M \approx L_m$, $\delta\sigma \approx 0$.

iii) As we shall see in Sect.7.4.6 the vertex correction (7.119) comes from the post-CPA terms shown in Fig.7.6. The CPA terms, although quite similar to the post-CPA ones, produce no correction to the conductivity because of exact mutual cancellation of the terms shown in Fig.7.5d. The presence of electron-electron interactions eliminates this cancellation. Indeed, if the electron between the points $\underline{\ell}'$ and $\underline{s}$ (Fig.7.5d) interacted with the electron between $\underline{s}'$ and $\underline{\ell}$, then the terms resulting from the exchange $\underline{\ell} \rightleftarrows \underline{m}$ (or $\underline{\ell}' \rightleftarrows \underline{m}'$) are no longer opposite to each other and hence the CPA vertex corrections, in the presence of electron-electron interaction, do not vanish. As a matter of fact they produce a result like (7.119) with a similar but not identical proportionality factor. The main difference is that the cutoff L_M depends now on the frequency ω or the temperature T, whichever is larger,

$$L_M^{-2} \approx k_B T/\hbar D + \omega/D \quad . \tag{7.129}$$

It must be pointed out that L_M in the present case, and in contrast to the previous case, does not depend significantly on the external magnetic field H. The reason is that the CPA terms involve $G^+_{nm}G^-_{mn}$, which remains unaffected by H, while the post-CPA terms of Fig.7.6 involve $G^+_{\underline{nm}}G^-_{\underline{nm}}$, which acquires an extra phase factor in the presence of H.

iv) The theoretical results presented in (ii) and (iii) are in impressive agreement with experimental data especially for the two-dimensional case.

v) There are proposals that (7.119) can be used even for three-dimensional systems as we approach the mobility edge from the metallic side. For $d = 3$, (7.119) gives

$$\delta\sigma_1 \sim 1/L_M \quad . \tag{7.130}$$

Equation (7.130) combined with (7.125) predicts a negative magnetoresistance proportional to $H^{\frac{1}{2}}$. Such a behavior has been observed in impurity bands at low temperatures ($\sim$ 1 K) and remained a puzzle until the advance of the present explanation.

7.4.1 One-Dimensional Case

Even during the early studies of one-dimensional disordered systems there were suggestions that the eigenstates may become localized [7.93]. MOTT and TWOSE [7.94] were the first to propose that all eigenstates in a 1-D random system are localized. One may visualize this proposal by assuming that each back-scattered wave at every elementary scattering event is lost as far as the propagation is concerned due to destructive interference; this picture suggests that the localization length λ, defined by the relation

$$\lambda^{-1} = -\lim_{m\to\infty} (\langle \ln|\psi_m|\rangle/m) \quad , \tag{7.131}$$

is about the same as the mean free path ℓ; ψ_m is the amplitude of the eigenfunction at m.

BORLAND [7.95] showed that for a random δ-function array the solutions of the differential equation (with fixed values of the function and its derivative at one end) grow exponentially on the average with the distance. This exponential growth was shown to be a direct result of the phase incoherence. The picture which emerged on the basis of BORLAND's proof is that at every energy there are two independent solutions of Schrödinger's equation, one increasing and the other decreasing exponentially (on the average) with increasing x. At a set of particular energies (the eigenenergies) the left growing solution matches at some point the right growing solution to form the eigenfunction which decays to zero as $|x|\to\infty$. According to this picture (which was not proven rigorously) the localization length is the same as the inverse of the average rate of growth of the solution of the differential equation. WEGNER [7.96] demonstrated the localization of eigenfunctions through a different method. Borland's method is discussed very clearly in a review article by HALPERIN [7.97]. Relevant material can be found in books by HORI [7.98], and by LIEB and MATTIS [7.99].

The exponential (on the average) growth of the solutions of the differential equation can be shown rigorously [7.100] by employing the transfer matrix technique. The transfer matrix is a 2×2 matrix, which connects, e.g., ψ, $d\psi/dx$ at x with ψ, $d\psi/dx$ at x'. The existence of such a 2×2 matrix is a direct consequence of the uniqueness of the path connecting x and x' in 1-D. The concept of the transfer matrix transforms the propagation of the wave in 1-D to a product of random 2×2 matrices. There are exact theorems [7.101] which state that under quite general conditions the product of random matrices grows exponentially.

The transfer matrix technique can be used for the calculation of the transmission ($|t|^2$) or reflection ($|r|^2$) coefficient associated with a disordered segment of total length L. Using FÜRSTENBERG's theorem [7.100,101] it is easy to prove that

$$\frac{1}{L}<\ell n|t|^2> = -\frac{2}{\lambda} \quad , \tag{7.132}$$

where $1/\lambda$ is the average rate of exponential growth of the solution. There is recently increased interest in $|t|^2$ because it is related with the conductance or the resistance R of the system. LANDAUER [7.102] considered a disordered segment of length L connected to a DC current source. Using semiclassical arguments, he was able to show that its resistance R is

$$R = \frac{\pi\hbar}{e^2}\frac{|r|^2}{|t|^2} = \frac{\pi\hbar}{e^2}\frac{1-|t|^2}{|t|^2} \quad . \tag{7.133}$$

ECONOMOU and SOUKOULIS [7.85] derived (7.133) from the Kubo-Greenwood formula. They also examined the case where the disordered segment is connected to a perfect wire whose length approaches infinity; they found that the current-field relation is *nonlocal*. As a result the DC resistance of a finite disordered one-dimensional system depends on what happens to the rest of the loop employed for its measurement, e.g., if the only voltage in the loop is across the disordered segment, its resistance is

$$R = \frac{\pi\hbar}{e^2}\frac{1}{|t|^2} \quad ; \tag{7.134}$$

however, such an arrangement produces charge fluctuations in the perfect wire. If an additional voltage is applied across the perfect wire to eliminate these charge fluctuations, one recovers for R equation (7.133) [7.103-105]. To avoid these uncertainties one may define R as L/σ, where the DC conductivity σ (for a finite length L) can be obtained by averaging the AC conductivity $\sigma_1(\omega)$

$$\sigma = \frac{1}{\omega_0}\int_0^{\omega_0} \sigma_1(\omega)d\omega \quad ; \tag{7.135}$$

ω_0 equals to the level spacing at E_F. It is worthwhile to mention that the so-defined R does not, in general, coincide with either (7.133) or (7.134). In the weak disorder limit, it is consistent with (7.134) (R is larger than $\pi\hbar/e^2$), while in the large disorder limit it grows with L more slowly than (7.133,134). The physical reason for this discrepancy is that the resistance defined through (7.135) depends on the overlap of two eigenstates of about equal energy, while (7.133) or (7.134) depend on the propagation of a wave at an energy E (which does not coincide in general with an eigenenergy) from the one end to the other end of the system [7.106].

In addition to the above uncertainties in defining properly the DC resistance of a finite one-dimensional system [7.107-110], it was found [7.85-88,111-113] that the probability distribution of $|t|^2$ possesses long tails which are responsible for a rather peculiar behavior of various averages. ABRIKOSOV [7.88] finds that

$$\langle \ln|t|^2 \rangle = -\,2L/\lambda \quad , \tag{7.136}$$

$$<|t|^{-2}> = \frac{1}{2}[1+\exp(4L/\lambda)] \quad , \qquad (7.137)$$

$$<|t|^{-4}> = \frac{1}{3} + \frac{1}{2}\exp(4L/\lambda) + \frac{1}{6}\exp(12L/\lambda) \quad , \qquad (7.138)$$

$$<|t|^{2n}> \underset{L\to\infty}{\longrightarrow} C_n(\lambda/2L)^{3/2}\exp(-L/2\lambda) \quad , \qquad (7.139)$$

where the constant C_n depends only on n. The standard deviation of $\ln|t|^2$ is proportional to $(L/\lambda)^{\frac{1}{2}}$ for large values of L/λ. Hence, in the strong disorder limit $\ln|t|^2$ becomes sharply distributed, with a probability distribution which seems to be Gaussian. The interested reader may find additional material in [7.114-117] and in review articles by ABRIKOSOV and RYZHKIN [7.118], and by ERDÖS and HERNDON [7.119]. It is worthwhile to point out that a special case [7.112,113,120,121] exists such that $1/\lambda = 0$. The vanishing of $1/\lambda$ does not necessarily imply extended states. Indeed in the case examined in [7.112,113], $\ln|t|^2 \sim -\sqrt{L}$ as $L\to\infty$; on the other hand, the special case examined by TONG [7.122] shows the existence of a nondecaying wave at a particular energy.

In the remainder of this subsection we shall show how the RPE (Appendix B) can be used to study the question of localization in a disordered TBM. In Sects.7.4.5,6 we shall discuss in a more systematic way how the Green's functions can be used in the localization problem.

The localization length can be obtained from the Green's function as follows:

$$\lambda^{-1} = -\frac{1}{|\ell - m|}<\ln|G(\ell,m)|> \qquad |\ell - m| \to \infty \quad . \qquad (7.140)$$

According to (B.6) we have

$$G(\ell,m) = G(\ell,\ell)VG(\ell+1,\ell+1\ [\ell])V \ldots G(m,m[m-1]) \quad . \qquad (7.141)$$

In writing (7.141) we have taken into account that there is only one self-avoiding path connecting ℓ to m in 1-D, and that $G(n,n\ [n-1])$ does not depend on ε_{n-2}, ε_{n-3}, ... From (B.12,13) we have in the present case (where $K=1$)

$$G(\ell+1,\ \ell+1\ [\ell]) = \frac{1}{E - \varepsilon'_{\ell+1} - V^2G(\ell+2,\ \ell+2\ [\ell+1])} \quad . \qquad (7.142)$$

Equation (7.142) allows us to express the probability distribution of $x \equiv G(\ell+1,\ \ell+1\ [\ell])$, $f_{\ell+1}(x;E)$, in terms of the probability distribution of $\varepsilon'_{\ell+1}$, $p(\varepsilon'_{\ell+1})$, and the probability distribution of $x' \equiv G(\ell+2, \ell+2\ [\ell+1])$, $f_{\ell+2}(x';E)$. The result is (in units where $V=1$)

$$f_{\ell+1}(x;E) = \frac{1}{x^2} \int p(E - x' - \frac{1}{x})\, f_{\ell+2}(x')\, dx' \quad . \qquad (7.143)$$

Because each site is equivalent on the average, $f_{\ell+1} = f_{\ell+2} = f$. Thus (7.143) is an integral equation for f. From (7.140,141) we have (taking the lattice spacing $a = 1$)

$$\lambda^{-1} = -<\ln|G(\ell+1, \ell+1\ [\ell])|> = -\int_{-\infty}^{\infty} f(x;E) \ln|x|\, dx \quad . \qquad (7.144)$$

Substituting in (7.144) from (7.143) and changing variables we can recast the expression for λ^{-1} in the following form

$$\lambda^{-1} = \frac{1}{2} \int_{-\infty}^{\infty} d\varepsilon \int_{-\infty}^{\infty} dx\ p(E+\varepsilon) f(x) \ln|1 + \frac{\varepsilon}{x}| \quad , \qquad (7.145)$$

which is more convenient in the weak disorder regime. An alternative, but equivalent, expression for λ^{-1} can be obtained by observing that the product in (7.141) can be expressed in terms of the eigenenergies of H and those of H with the sites ℓ, $\ell+1$, ... m removed, see (B.18). When the states are localized and $|\ell - m| \to \infty$ the eigenenergies of H can be separated into three groups: those associated with the semiinfinite segment to the left of ℓ, those associated with the segment to the right of m, and those associated with the segment ℓ, $\ell+1$, ..., m. Only the latter do not cancel in (B.18). Hence

$$\begin{aligned}
\lambda^{-1} &= \lim_{|\ell-m|\to\infty} \frac{1}{|\ell-m|} <\sum_j \ln|E - E_j|> \\
&= \lim_{|\ell-m|\to\infty} \int_{-\infty}^{\infty} dE' \left< \frac{\sum_j \delta(E' - E_j)}{|\ell - m|} \right> \ln|E - E'| \\
&= \int_{-\infty}^{\infty} <\rho(E')> \ln|E - E'|\, dE' \quad . \qquad (7.146)
\end{aligned}$$

Equation (7.146) was first obtained by THOULESS [7.123]. We point out that the average DOS per site $<\rho(E')> = -\mathrm{Im}\{<G^+(\ell,\ell;E')>/\pi\}$ can be expressed in terms of f(x) since $\Delta(\ell;E')$ in (B.10) is

$$\Delta(\ell;E') = V^2 G(\ell+1, \ell+1\ [\ell]) + V^2 G(\ell-1, \ell-1\ [\ell]) \quad . \qquad (7.147)$$

Hence (in units where $V=1$)

$$\langle\rho(E')\rangle = \langle\delta(E'-\varepsilon_\ell-\Delta(\ell;E'))\rangle$$

$$= \int d\varepsilon_\ell\, dx\, dx'\, p(\varepsilon_\ell)\, f(x;E')\, f(x';E)\, \delta(E'-\varepsilon_\ell-x-x')$$

$$= \int_{-\infty}^{\infty} dy\; f(y;E')\, f(1/y;E') \quad . \tag{7.148}$$

The last step follows by integrating over ε_ℓ, taking (7.143) into account, and changing the remaining integration variable from x to $y = 1/x$.

As an application of (7.145) we shall calculate λ^{-1} to order σ^2 (where $\sigma^2 = \langle\varepsilon^2\rangle$) at $E = 0$. To this order $f(x)$ in (7.145) can be replaced by $f_0(x)$, where $f_0(x)$ is the limit of $f(x)$ as $\sigma \to 0$. From (7.143) we find that

$$f_0(x) = \frac{1}{x^2} f_0(-1/x) \quad , \tag{7.149}$$

the solution of which is

$$f_0(x) = \frac{C}{\sqrt{1+x^4}} \quad ; \tag{7.150}$$

the normalization constant $C = [2\mathbb{K}(\sqrt{1/2})]^{-1} \approx 0.27$. Substituting (7.150) in (7.145) and performing the integration we find (to order σ^2)

$$\lambda^{-1} = \frac{2\mathbb{E}(\sqrt{1/2}) - \mathbb{K}(\sqrt{1/2})}{4\mathbb{K}(\sqrt{1/2})}\,\sigma^2 \approx 0.1142\,\sigma^2 \quad . \tag{7.151}$$

For the rectangular distribution, where $\sigma^2 = W^2/12$, (7.151) becomes $\lambda = 105.045/W^2$. This result was recently obtained through different methods by KAPPUS and WEGNER [7.124] and by SARKER [7.125].

ECONOMOU and co-workers [7.126] obtained with a similar technique an integral equation for the joint probability distribution of $x = \lim \mathrm{Re}\{G(\ell+1, \ell+1\,[\ell];\, E+is\}$ and $y = -\lim \mathrm{Im}\{G(\ell+1, \ell+1\,[\ell];\, E+is)/s\}$ as $s \to 0^+$. This allowed them to obtain averages of transport quantities like $sG^+(\ell,m;\, E+is) \cdot G^-(m,\ell;\, E-is)$, see (7.112); on the basis of their results they concluded that in the weak disorder limit the eigenfunctions are concentrated in a very small fraction of the length λ. To be more specific let us introduce $L_e \equiv (\sum_m|\psi_m|^4)^{-1}$, as a measure of the number of sites participating in the eigenfunction $|\psi\rangle = \sum_m \psi_m|m\rangle$. ECONOMOU and co-workers [7.126] found that while $\lambda \sim W^{-2}$ as $W \to 0$, the quantity L_e seems to behave like $W^{-\nu}$ with ν very close to 1.

Finally we comment on the connection between the localization length λ and the mean free path ℓ. The latter is defined as $\ell = \upsilon\tau = \hbar\upsilon/2|\Sigma_2|$ where υ

is the velocity at the energy E. The so-defined mean free path ℓ determines the exponential decay of the average of $G(m,n;E)$ or of $\psi_m\psi_n$, i.e.,

$$\langle G(m,n;E)\rangle \sim \langle \psi_m\psi_n \rangle \sim e^{-|m-n|/2\ell} \qquad |\ell - m| \to \infty \quad . \tag{7.152}$$

We remind the reader that the transport mean free path $\ell_{tr} = \upsilon\tau_{tr}$ is in general different from ℓ because of the $1 - \cos\theta$ term in (7.75). However, for the simple model we consider here $\ell_{tr} = \ell$. In 1-D, where θ takes only two values ($\theta = 0$, forward scattering and $\theta = \pi$, backward scattering), ℓ^{-1} can be decomposed into two terms ℓ_f^{-1} and ℓ_b^{-1} corresponding to forward and backward scattering:

$$\ell^{-1} = \ell_f^{-1} + \ell_b^{-1} \quad , \tag{7.153}$$

$$\ell_{tr}^{-1} = 2\ell_b^{-1} \quad . \tag{7.154}$$

For our simple model $\ell_f = \ell_b = 2\ell = 2\ell_{tr}$. In the weak scattering limit, using second-order perturbation expansion, see (7.51), and taking into account that $\hbar\upsilon/a = |G_0(m,m)|^{-1} = \sqrt{(2V)^2 - E^2}$, we find

$$\ell \approx \frac{a[(2V)^2 - E^2]}{2\sigma^2} \quad , \tag{7.155}$$

where a is the lattice spacing and $\sigma^2 = \langle\varepsilon_n'^2\rangle$. For a rectangular distribution of width W and at the center of the band ($E = 0$) we have (putting $a = 1$, $V = 1$)

$$\ell \approx 24/W^2 \quad . \tag{7.156}$$

Second-order perturbation theory gives $\langle\rho(E')\rangle = -\,\mathrm{Im}\{G_0^+(m,m;E-\Sigma)/\pi\}$ with $\Sigma = \sigma^2 G_0(m,m;E)$, see (7.51). Substituting in (7.146) and integrating by parts one obtains

$$\lambda \approx -\frac{2a}{\sigma^2 G_0^2(m,m;E)} = \frac{2a(4V^2 - E^2)}{\sigma^2} \quad , \tag{7.157}$$

which is four times the mean free path ℓ. In the rectangular case and at the center of the band one obtains $\lambda = 96/W^2$ [7.80] to be compared with $\lambda = 105.045/W^2$ which is the exact result to second order in W. An explanation of this discrepancy is given in [7.124].

It must be pointed out that the decay shown in (7.152) is partly due to the exponential decay of the amplitude (localization) and partly due to the phase incoherence. The latter is present even when there is no localization, and hence the exponential decay of $\langle G(m,n;E)\rangle$ as $|m-n| \to \infty$ cannot be used to deduce the existence of localization. To see to what extent the decay shown in (7.152) is due to localization, one may consider the average

$$\langle |G(m,n;E)| \rangle \sim \langle |\psi_m \psi_n| \rangle \sim e^{-|m-n|/\lambda'} \quad . \tag{7.158}$$

To calculate λ' one may approximately assume that $\ln|G(m,n;E)|$ has a Gaussian probability distribution with mean equal to $-|m-n|/\lambda$ and standard deviation equal to $(|m-n|/\lambda)^{\frac{1}{2}}$. Then $\lambda' = 2\lambda$, which to second order in W^2 and at the center of the band yields $\lambda' = 210.09/W^2$, to be compared with $\lambda' = 212.59/W^2$ obtained in [7.124].

7.4.2 Scaling Approach

In this subsection we present a work which played a decisive role in shaping our ideas in the field of localization. The basic idea in this work is the assumption that there is just one parameter Q (equal to the ratio of transfer strength V over an average energy mismatch $\delta\varepsilon$) which completely characterizes the answer to the localization question for each dimensionality d. The existence of just *one* parameter Q is obvious in the simple TBM we have examined in Chaps.5-7. What is not obvious is that a single parameter Q can still characterize more complicated models, where there are many energies per site and more than one transfer matrix element. Such complicated Hamiltonians may arise even from the simple TBM if, e.g., one enlarges the unit cell of the lattice by a factor x so that the new "unit" cell has linear dimension $L_1 = ax$ and contains $n_1 = x^d$ original sites (a is the lattice spacing). If each new "unit" cell is considered as an effective "site", there must be (according to our initial assumption) a quantity $Q(L_1)$ which characterizes the localization properties of the reformulated problem. By repeating the procedure of unit cell augmentation one produces a sequence $Q(L_2)$, $Q(L_3)$ ... $Q(L_m)$ ..., where the "unit" cell after the mth step has linear dimension L_m equal to ax^m and contains $n_m = x^{md}$ sites. ABRAHAMS et al. [7.127] realized that the derivative of $Q(L)$ with respect to L, or for that matter, the logarithmic derivative β, where

$$\beta = \frac{d \ln [Q(L)]}{d \ln L} \tag{7.159}$$

is an extremely important quantity for localization. Indeed, if β is larger than a positive number, no matter how small, the successive transformations will monotonically increase Q towards infinity as $L \to \infty$, which means that the initial problem is mapped into one, where the effective energy mismatch is zero and, hence, the eigenstates are extended. If β is less than a negative number, Q monotonically decreases towards zero; this means that the initial problem has been mapped into one where the effective transfer matrix element is zero, and consequently the eigenstates are localized. Hence the quantity β completely characterizes the localization problem. The assumption that there is just one parameter Q which determines localization forces the conclusion that β is a function of Q:

$$\beta = f(Q) \quad . \tag{7.160}$$

To proceed further with these ideas one needs to identify what Q is in the general case. THOULESS and co-workers [7.80,128] argued that in the general case, where the "unit" cell has linear dimension L, the role of V in (7.118) is played by ΔE (where ΔE is a measure of how much the eigenenergies of an isolated "unit" cell change upon changing the boundary conditions from periodic to antiperiodic) and the role of $\delta\varepsilon$ is played by the level spacing, which is just the inverse of the total DOS. In other words

$$Q = \rho L^d \Delta E \quad . \tag{7.161}$$

It was further argued [7.128] that $\Delta E = \hbar/\tau'$, where τ' is the time it takes a particle to diffuse from the center to the boundary of the "unit" cell: $\tau' = (L/2)^2/D$, where D is the diffusion coefficient. Taking into account (7.121) we obtain that $\Delta E = 4\hbar D/L^2$ equals to $2\hbar\sigma/e^2\rho L^2$. Substituting in (7.161) we obtain that $Q = 2\hbar\sigma L^{d-2}/e^2$. But σL^{d-2} is the DC conductance R^{-1} of the unit cell,

$$R^{-1} = \sigma L^{d-2} \quad . \tag{7.162}$$

Thus we reached the very important conclusion that the quantity Q is nothing else than the dimensionless DC conductance, i.e.,

$$Q = \frac{\hbar}{e^2} R^{-1} \quad . \tag{7.163}$$

[In (7.163) we redefined Q as to eliminate a factor of 2].

In view of the fact that higher values of both Q and β favor extended states, it is not unreasonable to assume, as ABRAHAMS et al. [7.127] did, that β is a monotonically increasing function of Q or of ℓnQ. We are now in a position to find the qualitative features of β versus ℓnQ. In the weak scattering limit, when the conductance Q is very large, one is almost in the metallic regime, where σ is independent of L, and $Q \sim L^{d-2}$. In this limit $\beta = d\,\ell n\,Q/d\,\ell n\,L$ approaches d - 2. In the other extreme of very strong disorder the localization length is much less than L, and the conductance decays exponentially with L. Hence $\beta = \ell n\,Q$ in this limit. This asymptotic behavior together with the assumption of monotonicity leads to a β versus $\ell n\,Q$ relation as shown in Fig.7.8. For d = 1,2, β is always negative, hence by increasing the length L we decrease $\ell n\,Q$ which in turn further decreases β. Thus, as we increase L we slide down the curves as shown in Fig.7.8 and at sufficiently large lengths L we approach the regime of exponential localization, independently of what our starting point was. Hence all eigenstates are localized and a truly metallic behavior is not possible. Only a quasimetallic behavior can be observed when L is much larger than the total mean free path and much smaller than the localization length. In 1-D systems in the presence of only elastic scattering the mean free path ℓ is comparable to the localization length and consequently even a quasimetallic behavior is not possible. However, inelastic scattering changes this conclusion drastically, mainly because it allows electrons to jump from one localized state to another (of different energy), and hence it delocalizes them. This delocalization appears only after an electron has travelled a distance comparable to the inelastic diffusion length L_T (7.122). As a result of this complication, localization effects become important only when L_T becomes comparable to or larger than any characteristic localization length.

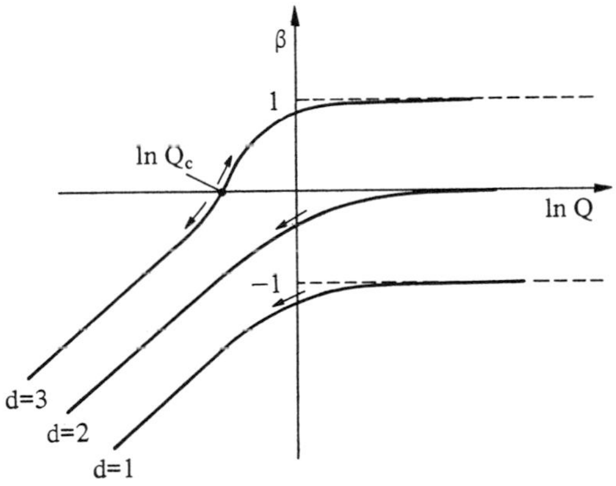

Fig.7.8. Plot of β vs. $\ell n\,Q$ for d = 1,2,3 dimensional disordered systems. Q is the dimensionless conductance of a d-dimensional cube of length L and $\beta = d\,\ell n\,Q/d\,\ell n\,L$. As L increases one moves along the curves in the direction indicated by the arrows

In the 3-D case there is a critical value Q_c of the conductance Q. For $Q < Q_c$, β is negative and hence, by the same arguments as in $d = 1,2$, the states are localized. On the other hand, for $Q > Q_c$, β is positive; as a result, by increasing the length, one increases Q and, hence, β itself. Thus one climbs up the β curve towards the flat metallic regime, associated with extended states of more or less uniform amplitude. It must be pointed out that the present analysis implies that the conductivity approaches zero continuously near the critical point. Indeed, for every length L, no matter how large, one can find a disorder such that Q is just above Q_c. Then the conductivity $\sigma \approx Q_c/L$ can become arbitrarily small by increasing the length L. There is currently considerable interest in finding how σ as a function of the energy or as a function of the disorder approaches zero. One assumes a power-law behavior and then the question of finding the exponents arises [7.127,129-141].

Let us now return to the weak scattering regime ($\ell n\, Q >> 1$) and let us assume that the classical limit d-2 is approached as 1/Q, i.e.,

$$\beta \approx d - 2 - \frac{A_d}{Q} \quad ; \quad Q \to \infty \quad . \tag{7.164}$$

One can then integrate (7.164) to find the first correction to the classical result for R^{-1} or for the DC conductivity σ. The final results are

$$\sigma = \sigma_0 - \frac{e^2 A_1}{\hbar} L \quad ; \quad d = 1 \quad , \tag{7.165}$$

$$= \sigma_0 - \frac{e^2 A_2}{\hbar} \ell n\, L \quad ; \quad d = 2 \quad , \tag{7.166}$$

$$= \sigma_0 + \frac{e^2 A_3}{\hbar} L^{-1} \quad ; \quad d = 3 \quad , \tag{7.167}$$

where σ_0 is the DC conductivity if vertex corrections other than the CPA were ignored. Equations (7.165-167) can also be obtained from (7.119). Thus (7.164) can be justified by the post-CPA vertex corrections. The latter, of course, would give explicit values for the constants A_d : $A_1 = 1/2\pi$, $A_2 = 1/\pi^2$, and $A_3 = \sqrt{2}/4\pi^2$. Here A_1 was obtained by expanding (7.133) with $|t|^2$ replaced by its geometric mean, i.e., by $\exp[-2L/\lambda]$. Such a replacement reproduces the classical result for σ_0.

As it was pointed out before, the theoretical results (7.165-167) have been checked extensively experimentally in quasi one- and two-dimensional systems and even in three-dimensional systems. For two-dimensional systems (thin films and interfaces) the experimental data [7.142-147] are in very good agreement with the theoretical results. For thin wires (quasi one-dimensional systems) the agreement is less impressive [7.148-150]. The situation is less clear for three-dimensional systems, where the theory is not well founded and the experimental data are limited to only impurity bands in semiconductors [7.136, 141,151]. Overall one can say that the scaling approach of THOULESS [7.80] and of ABRAHAMS et al. [7.127] has been at the center of the significant current developments in the field and it has received strong experimental evidence to its support. Further support was given to it by various independent approximate theoretical treatments. Thus, although it is not a rigorous theory, it seems to be well accepted as a tool for studying the role of disorder in transport properties. However, there are already indications [7.147,152,153] suggesting that some of the basic assumptions of the method may not be generally true: e.g., spin-orbit scattering seems to *change the sign* of A_2 in (7.164) and to reduce its magnitude by two [7.147,152,153].

7.4.3 Numerical Methods

The question of localization has been studied extensively by various numerical techniques. In this subsection no attempt will be made to cover this diverse subject. The interested reader is referred to some of the original literature. Most of the authors mentioned in Sect.7.4.1 (one-dimensional case) supplemented their theoretical analysis with numerical computations which are consistent with the picture we described in Sect.7.4.1. For higher-dimensional systems the numerical results can hardly be called conclusive because of the relatively small sizes of the samples; nevertheless they are not inconsistent with the prevailing ideas in the field.

YOSHINO and OKAZAKI [7.154], and YONEZAWA [7.155] have found eigenenergies and corresponding eigenfunctions of finite two-dimensional (100×100) TBM, by direct diagonalization of the matrix equation; they took advantage of the fact that most of the matrix elements are zero. LICCIARDELLO and THOULESS [7.128] examined numerically the sensitivity of the eigenenergies to changing boundary conditions. STEIN and KREY [7.156] implemented numerically the recursion method [7.59] according to which higher-dimensional disordered systems are mapped to one-dimensional systems with a much more complicated disorder; the 1-D systems are usually treated through the iteration (or continued-fractions) inherent in (B.12,13) (for $K = 1$). WEAIRE and SRIVASTAVA [7.157] examined numerically the time evolution of a random trial function and used their results to draw conclusions about localization. PRELOVSEK [7.158] employed a diffusion simulation technique. LEE [7.159] and SARKER and DOMANY [7.129] implemented numerically the scaling ideas presented in Sect.7.4.2; their results show that $d = 2$ seems to be a borderline dimensionality, but are nevertheless inconclusive regarding what is happening for $d = 2$. PICHARD and SARMA [7.160], MACKINNON and KRAMER [7.161], and SOUKOULIS et al. [7.162] studied the largest localization length λ of a strip of width M, or of a wire of square cross section M^2, as a function of M. For the wire case (which becomes 3-D as $M \rightarrow \infty$) the behavior depends on whether or not the disorder exceeds a critical value. For $W < W_c$, $\lambda(M)$ behaves as M^2/ξ (up to the largest M examined) and hence, it seems to approach infinity parabolically as $M \rightarrow \infty$; the characteristic length ξ (which depends on the disorder) can tentatively be given a physical interpretation as being the largest length beyond which the eigenfunctions look uniform in amplitude. Since a uniform amplitude implies a classical metallic behavior, the quantity ξ can also be interpreted as the length (for a given disorder) at which β is almost one. For $W > W_c$, $\lambda(M)$ increases with M slower than linearly, and it seems to saturate to a value $\lambda(\infty)$, which is the localization length. For the strip case (which becomes 2-D as $M \rightarrow \infty$) $\lambda(M)$ increases with M slower than linearly and it seems to saturate to a finite value at least to disorders down to $W/V = 4$ [7.161]. Around $W/V \approx 6$ (which early work identified as a critical point) the localization length seems to change very fast (e.g., $\lambda \approx 40$ for $W/V = 6$, while $\lambda \approx 500$ for $W/V = 4$). Such large localization lengths make conclusive numerical work extremely difficult.

7.4.4 Mapping the Problem to Other Problems

Recently there has been considerable activity in trying to find the connection between the localization question and other extensively studied problems, where the borderline dimensionality is $d = 2$. As in the previous subsection, we do not attempt to cover this diverse subject here; we only refer the interested reader to some of the original literature. ALLEN [7.163] connected the localization problem to the random walk problem, in particular to $S(N)/N$, where $S(N)$ is the mean number of distinct sites visited during a random walk of N steps. Then $S(N)/N$ goes to zero as $N \rightarrow \infty$ for $d \leq 2$, while it remains finite for $d > 2$. KAVEH and MOTT [7.92,130,164] obtained the vertex corrections to the conductivity (7.165-167) by mapping the problem to the classical diffusion process. WEGNER [7.132,133], HIKAMI [7.140], EFETOV et al. [7.135], and HOUGHTON et al. [7.134] have demonstrated that the localization problem is analoguous to the nonlinear σ-model; then they employ sophisticated field theoretical techniques to reach conclusions which are in agreement with the scaling approach. SCHUSTER [7.165] pursued an analogy with the XY model; such an analogy, although in agreement with the conclusions of PICHARD and SARMA [7.160], is inconsistent with the numerical results of MACKINNON and KRAMER [7.161] and of SOUKOULIS et al. [7.162], as well as with the conclusions of the scaling approach. The recursion method, which maps the problem to a one-dimensional one, produces results [7.166] which are consistent with the currently prevailing ideas in the field. We mention also the work by GÖTZE [7.167].

7.4.5 Green's Functions Approach I: RPE

Green's functions behave differently depending on whether the eigenstates are localized or extended. An example of such difference has already been encountered in 1-D, where for exponentially localized eigenstates $<\ell n|G(\ell,m)|> \sim -|\ell - m|/\lambda$, while for Bloch states (extended) $\ell n|G(\ell,m)|$ is independent of $|m-\ell|$, (5.29).

Before we proceed further let us clarify the concepts of extended and localized eigenstates. We call extended those eigenstates which do not decay to zero at infinity; furthermore we assume that these extended states look more or less uniform in amplitude at least beyond a characteristic length scale ξ. On the other hand, we define localized states as those which decay to zero "sufficiently fast". There is no rigorous examination of what decay law qualifies for this characterization. It seems that exponential decay is "fast enough"; it is even possible that all normalizable eigenfunctions are localized in the above sense. These definitions leave in principle the possibility of having eigenstates which are neither extended nor localized in the above sense. A decaying but not normalizable eigenstate is an example of such an intermediate situation. It is usually assumed that the eigenstates will be either exponentially localized or extended. In 1-D this assumption is true in almost all cases. For higher dimensionalities there is no proof.

Let us consider here a finite system consisting of N unit cells. The diagonal matrix element of $G(E;N)$ in a Wannier representation can be written as

$$G(\underline{m},\underline{m};z;N) = \sum_{\nu} \frac{f_{\nu}}{z - E_{\nu}} \quad , \tag{7.168}$$

where

$$f_\nu = \langle \underline{m}|\nu)(\nu|\underline{m}\rangle \tag{7.169}$$

and $|\nu)$ is an eigenfunction of the Hamiltonian H with eigenvalue E_ν; we have dropped the index $\underline{m}$ from f_ν for simplicity. Note that the symbol $\sum_\nu$ in (7.168) denotes a genuine summation since we are dealing with a finite system which possesses a discrete spectrum.

We consider now the limit $N \to \infty$, i.e., we allow the system to become infinite. A localized eigenstate $|\nu)$ tends to a well-defined limit as $N \to \infty$; hence, if $|\nu)$ is localized, $f_\nu \to f_\nu^\infty$ as $N \to \infty$ where f_ν^∞ is in general a positive quantity. The magnitude of f_ν^∞ (for localized eigenstates) depends strongly on how far away from the site $\underline{m}$ the localized state $|\nu)$ is. If we assume exponential decay of the localized eigenfunctions then $f_\nu \sim \exp(-2r_{\underline{m}\nu}/\lambda_\nu)$ as $r_{\underline{m}\nu} \to \infty$, where $r_{\underline{m}\nu}$ is the distance between the site $\underline{m}$ and the center of the localized state $|\nu)$ and λ_ν is its localization length; thus, the new f_ν's which appear as $N \to \infty$ are extremely small, since the distance $r_{\underline{m}\nu}$ is proportional to $N^{1/d}a$, where a is the linear dimension of the unit cell. On the other hand, for extended or propagating eigenstates $|\nu)$, the quantity f_ν behaves as $1/N$ in the limit $N \to \infty$; this behavior stems from the normalization factor $1/\sqrt{N}$ contained in every extended eigenstate $|\nu)$. This basic difference in the behavior of f_ν as $N \to \infty$ produce the following characteristic properties.

Consider the quantity $c_{\underline{m}}(t)$ which is the probability amplitude for finding a particle in the state $|\underline{m}\rangle$ at time t if initially (t = 0) the particle was in $|\underline{m}\rangle$. This quantity has been evaluated before and it is given by (6.34). Substituting (7.168) into (6.34) we obtain

$$c_{\underline{m}}(t) = \sum_\nu f_\nu \, e^{-iE_\nu t/\hbar} \quad . \tag{7.170}$$

The probability $|c_{\underline{m}}(t)|^2$ is

$$|c_{\underline{m}}(t)|^2 = \sum_{\nu\nu'} f_\nu f_{\nu'} \, e^{-i(E_\nu - E_{\nu'})t/\hbar} \quad . \tag{7.171}$$

Defining

$$|c_{\underline{m}}|^2 = \lim_{t\to\infty} \frac{1}{t} \int_0^t |c_{\underline{m}}(t')|^2 dt' \tag{7.172}$$

we obtain

$$|c_{\underline{m}}|^2 = \sum_\nu f_\nu^2 \quad . \tag{7.173}$$

Using (7.168), it is easy to show that

$$|c_{\underline{m}}|^2 = \frac{s}{\pi}\int_{-\infty}^{\infty} dEG(\underline{m},\underline{m};E + is)G(\underline{m},\underline{m};E - is) \quad . \tag{7.174}$$

Equation (7.173) shows that the localized eigenstates make a nonzero contribution to the probability $|c_{\underline{m}}|^2$ while the extended eigenstates make no contribution to $|c_{\underline{m}}|^2$ as $N \to \infty$. If all states were extended, then all $f_\nu \sim 1/N$, and consequently $\sum_\nu f_\nu^2 \sim \sum_\nu (1/N^2) \sim 1/N$ (there are N terms in the summation). On the other hand, for localized eigenstates each f_ν tends to a definite limit and the series (7.173) converges since f_ν^∞ becomes quite small for eigenstates $|\nu)$ appearing for large N. Equation (7.174), which is valid in the limit $s \to 0^+$ independently of whether or not the limit $N \to \infty$ has already been taken, shows that a branch cut in $G(E)$ makes no contribution to $|c_{\underline{m}}|^2$ and, hence, it corresponds to states which are not localized. It is probably true, although no rigorous proof has been given, that a branch cut in $G(E)$ indicates a decay slower than $r^{-d/4}$. Under rather general conditions, it is true that extended states produce a branch cut in G(E); this can be shown by taking the limit $N \to \infty$ in (7.168); one must keep in mind that the average spacing of the levels goes as 1/N and f_ν (for extended states) behaves as 1/N for $N \to \infty$. On the other hand, for localized eigenstates the quantities f_ν approach a finite limit as $N \to \infty$, which means that in the limit of $N \to \infty$ we will have a dense distribution of poles the residues of which remain finite; thus a line of singularity results which is not a branch cut and which is known as a natural boundary because the side limits $\lim G(E \pm is)$ do not exist as $s \to 0^+$.

Another way to distinguish between extended and propagating eigenstates is to consider the convergence of $G(E)$ as $N \to \infty$. Note first that each term $f_\nu/(z - E_\nu)$ in (7.168) is important as $z \to E$ (where E belongs to the continuous spectrum), only if $|E - E_\nu| \leq Af_\nu$, i.e., if $E \in P_\nu$ where P_ν is the interval $[E_\nu - Af_\nu, E_\nu + Af_\nu]$ with $A \gg 1$. Then the contribution of all the terms with $\nu > \nu_0$ is important only if E belongs to the union $\sum_0$ of P_ν's with $\nu > \nu_0$. Assume now that the terms in (7.168) have been arranged in order of decreasing f_ν. For localized eigenstates the extent of the interval P_ν decreases at least as $r_{m\nu}^{-d-\varepsilon}$ (because the states are normalizable) with $\varepsilon > 0$ as $\nu \to \infty$. Hence the extent of the union $\sum_0$ approaches $N_0^{-\varepsilon/d}$, i.e., zero as $\nu_0 \to \infty$. Consequently, the probability of E belonging to $\sum_0$ approaches zero as

$\nu_0 \to \infty$; hence the series (7.168) converges as $N \to \infty$ with probability 1. It has been assumed that this last statement is equivalent to saying that for localized states the probability distribution of $G(E)$ converges as $N \to \infty$. For extended eigenstates the union $\sum_0$ approaches a nonzero value as $\nu_0 \to \infty$; this implies that for extended eigenstates $G(E)$ diverges with probability 1 as $N \to \infty$.

On the basis of the above discussion one can state certain criteria to decide about the nature of the eigenstates at E. These statements, although very plausible, cannot be considered as rigorously proven.

(i) Consider the quantity

$$\lim \langle s\, G(\underline{m},\underline{m};E+is)G(\underline{m},\underline{m};E-is)\rangle, \text{ as } s \to 0^+$$

or more generally

$$\lim \langle s\, G(\underline{m},\underline{n};E+is)G(\underline{n},\underline{m};E-is)\rangle, \text{ as } s \to 0^+ \quad .$$

If the above quantities are nonzero the eigenstates at E are localized; if they approach zero linearly with s the eigenstates are extended (or possibly very slowly decaying as $r^{-\nu}$ with $\nu < d/4$); if they approach zero more slowly than linearly with s, they probably decay as $r^{-\nu}$ with $d/4 \leq \nu \leq d/2$ (with logarithmic factors in s appearing when $\nu = d/2$ or $d/4$).

(ii) Consider the probability distribution of $G(E;N)$. If it converges as $N \to \infty$ the eigenstates at E are localized. If it diverges the states are extended (or possibly slowly decaying but not normalizable).

(iii) Consider $G(\underline{m},\underline{m};E,N)$. If this quantity converges with probability one as $N \to \infty$ the states at E are localized. If it diverges the states are extended (or possibly slowly decaying but not normalizable).

The first criterion has been employed in 1-D by ECONOMOU and co-workers [7.126]. We shall return to it in the next subsection. The second criterion has been used in 1-D systems. The third criterion was first used in conjunction with the renormalized perturbation expansion (RPE) by ANDERSON [7.168] in his basic paper that marked the birth of the field of localization. ANDERSON's work has been further developed by ZIMAN [7.169], KIKUCHI [7.170], HERBERT and JONES [7.171], THOULESS [7.172], and by ECONOMOU and co-workers [7.91,173]. In all this work the RPE (presented in Appendix B) played a key role; the reason is that the RPE is a closed expression for a finite system. Hence, its convergence or divergence as $N \to \infty$ is equivalent to the convergence or divergence of $G(\underline{m},\underline{m};E,N)$ as $N \to \infty$; the latter, according to (iii) is connected directly with the nature of the eigenstates.

According to (B.9) the RPE for the quantity $\Delta(\underline{\ell})$ has the structure

$$\Delta(\underline{\ell}) = \sum_j t_j \quad , \tag{7.175}$$

$$= \sum_{N=1}^{\infty} \sum_j{}' \, t_j^{(N)} \quad , \tag{7.176}$$

where the summation in (7.175) is over all terms corresponding to self-avoiding paths starting from and ending at the site $\underline{\ell}$. In (7.176) we have rearranged the series so as to perform first the summation over all paths of $N+1$ steps and then sum from $N=1$ to ∞. We remind the reader that

$$t_j^{(N)} = V^{N+1} G(\underline{n}_1,\underline{n}_1[\ell])G(\underline{n}_2,\underline{n}_2[\underline{\ell},\underline{n}]) \dots \tag{7.177}$$

where the sites $\underline{\ell}$, $\underline{n}_1$, $\underline{n}_2$, ..., $\underline{\ell}$ belong to the self-avoiding path j of $N+1$ steps. The expansion (7.176) may diverge through two distinct mechanisms: one is the divergence of the series shown in (7.176); even if this series terminates (as in the 1-D case) the RPE may diverge as a result of the implicit iteration built in the RPE (Appendix B). Usually the assumption is made that the convergence of the series controls the convergence of the RPE; ABOU-CHACRA et al. [7.174] made the opposite assumption by terminating the series at $N=1$ and explicitly taking into account the iteration procedure. To study the convergence of the series one introduces the localization function $L(E)$ defined by

$$L(E) = \lim_{N\to\infty} [|\sum_j t_j^{(N)}|]^{1/N} \quad . \tag{7.178}$$

If $L(E)$ is less (more) than 1 with probability 1 the series converges (diverges) in probability. Here it must be pointed out that there is the possibility of convergence (not absolute convergence) even when $L(E)$ is larger than 1. This possibility has been disregarded. To evaluate (7.178) one needs to make several approximations. First we put

$$L(E) \approx L^*(E) \quad , \tag{7.179}$$

where

$$L^*(E) = \lim_{N\to\infty} (\sum_j |t_j^{(N)}|)^{1/N} \quad . \tag{7.180}$$

Equation (7.179) means that the sign fluctuations of the $t_j^{(N)}$'s are omitted. The second approximation is to replace the various G's in (7.177) by the first one

$$t_j^{(N)} \approx V^{N+1} G^N(n_1, n_1[\underline{\ell}]) \quad . \tag{7.181}$$

The third approximation replaces the $G(\underline{n}_1, \underline{n}_1[\ell])$ by its CPA average. Thus L(E) has been calculated for various 3-D and 2-D lattices and various probability distributions of $\{\varepsilon_{\underline{n}}'\}$. By putting $L(E) = 1$ one can obtain the trajectory of the critical point, i.e., the critical energy E_c separating extended from localized states as a function of the disorder, or, equivalently, the critical disorder as a function of the energy (Fig.7.7). The results were consistent with numerical data and with various limiting cases. The main drawback of the final result for L(E) is that it failed to predict any qualitative difference between $d = 2$ and $d = 3$. To check the origin of this failure SOUKOULIS and ECONOMOU [7.175] reexamined the three approximations mentioned above. They proved (B.19), which allowed them to relax approximation (7.181); they worked in the limit of zero disorder so that no average is needed; and finally they calculated both L(E) and $L^*(E)$. At the ends of the spectrum they found that $L^*(E) = L(E) \approx 1$ (with less than 1% error), which indicates that their calculations are quite accurate. For $d = 3$, they found that (7.179) is well satisfied and that L(E) is well above 1 within the band, which shows that a finite amount of disorder is required to make $L(E) < 1$ and localized the eigenstates. In contrast, for a 2-D lattice, it was found that while $L^*(E)$ is well above 1 within the band, L(E) is (within numerical uncertainties) equal to 1. This is a strong indication that any amount of disorder, no matter how small, is enough to localize the states in 2-D. It was suggested [7.175] that the equation $L^*(E) \approx 1$ indicates a crossover region where the localization length varies very fast.

7.4.6 Green's Functions Approach II: Vertex Corrections

In this last subsection on localization we show how the CPA and the post-CPA vertex corrections can be calculated; the result for the conductivity is the basic formula (7.119).

Any localization effects must appear as divergences of our expansions for the vertex correction. As can be seen from (7.113) (which is exact and it is also obeyed by the CPA) such divergences appear at least for the case where $B = 1$, and for z, z' approaching each other from *opposite* sides of the real axis. When $B = 1$, $\tilde{\gamma}_n(z,z')$ can be easily obtained from (7.107,109-111):

$$\tilde{\gamma}_n(z,z') = \frac{u(z,z')A(0;z,z')}{1 - u(z,z')A(0;z,z')} \quad . \tag{7.182}$$

Comparing (7.182) with (7.113) we see immediately that $u(z,z')A(0;z,z')$ approaches 1 as $z \to z'$ from opposite sides of the real axis. Under these conditions $\gamma(\underline{q};z,z')$ has a singularity when $\underline{q} \to 0$. Since $\gamma(\underline{q};z,z')$ is part of the integrand for the vertex corrections both within the CPA and beyond the CPA (if $A' = A$), it follows that the vertex corrections may become singular. This depends on the behavior of the integrand around the singularity, i.e., on the behavior of uA for small values of $z - z'$ and $\underline{q}$

$$u(z,z')A(\underline{q};z,z') \approx 1 + a_1(z - z') - a_2 q^2 \quad , \tag{7.183}$$

with $\text{Im}\{z\}\text{Im}\{z'\} < 0$. To be specific we take $\text{Im}\{z\} > 0$. The constant a_1 can be found by comparing (7.182) with (7.113). The result is

$$a_1 = \frac{i}{2|\Sigma_2|} \quad . \tag{7.184}$$

To find a_2 requires some algebra. We start with (7.110) and we expand $G_e(\underline{k} - \underline{q};z')$ in powers of $\underline{q}$. The coefficient of the first power is $1/N \sum_{\underline{k}} G_e(\underline{k}) \nabla_{\underline{k}} G_e(\underline{k})$, which is zero, because $G_e(\underline{k})$ is an even function of $\underline{k}$. The next term is

$$\frac{q^2}{2d} \frac{1}{N} \sum_{\underline{k}} G_e(\underline{k}) \nabla^2_{\underline{k}} G_e(\underline{k}) = -\frac{q^2}{2d} \frac{1}{N} \sum_{\underline{k}} \nabla_{\underline{k}} G_e(\underline{k}) \cdot \nabla_{\underline{k}} G_e(\underline{k}) \quad , \tag{7.185}$$

where cubic symmetry was assumed. Taking into account that $G_e(\underline{k};z) = [z - \Sigma - E(\underline{k})]^{-1}$, we see that $\nabla_{\underline{k}} G_e(\underline{k};z) = G_e^2 \cdot \nabla_{\underline{k}} E(\underline{k})$; hence the coefficient of q^2 is proportional to $\sigma^{(0)}/\Sigma_2^2$, where $\sigma^{(0)}$ is the conductivity without vertex corrections (7.93'). To find a_2 we also need $u(E^+,E^-) = A^{-1}(0;E^+,E^-) = [1/N \sum_{\underline{k}} G_e^+(\underline{k};E) G_e^-(\underline{k};E)]^{-1} = [(\Sigma^- - \Sigma^+)^{-1} 1/N \sum_k (G_e^- - G_e^+)]^{-1} = 2i|\Sigma_2| [2i\pi\rho'(E)]^{-1} = |\Sigma_2|/\pi\rho'$, where $\rho'(E)$ is the DOS per unit cell per spin. The coefficient a_2 can be expressed in a more compact way in terms of the diffusion coefficient $D^{(0)} = \sigma^{(0)}/2e^2\rho$:

$$a_2 = \frac{\hbar D^{(0)}}{2|\Sigma_2|} \quad . \tag{7.186}$$

As an application of the above let us first consider the quantity $s\langle G(\underline{0},\underline{0};E + is)G(\underline{0},\underline{0};E - is)\rangle$, from which conclusions can be drawn regarding the nature of the eigenstates. From (7.112) we have

$$s\langle G(\underline{0},\underline{0};E+is)G(\underline{0},\underline{0};E-is)\rangle = \frac{s}{N}\sum_{\underline{q}} \frac{A(q)}{1-uA(\underline{q})} \quad . \tag{7.187}$$

For $d=3$ the q^2 term in the denominator (in the limit $s \to 0^+$) is cancelled by a q^2 phase factor in the numerator and hence (7.187) approaches zero linearly in s, indicating thus the validity of the perturbation expansion. In contrast, for $d=2$ or $d=1$ the phase factor is q or 1, respectively, and the integral blows up. In these cases we have

$$s\langle G^+G^-\rangle = \frac{s}{N}\sum_{\underline{q}} \frac{\pi\rho'/|\Sigma_2|}{\frac{s}{|\Sigma_2|} + \frac{\hbar D^{(0)}}{2|\Sigma_2|}q^2} = \frac{s}{N}\sum_{\underline{q}} \frac{\pi\rho'}{s + \frac{\hbar D^{(0)}}{2}q^2} \tag{7.188}$$

Although the integral blows up as $s \to 0$ the quantity $s\langle G^+G^-\rangle$ still approaches zero: as $\sqrt{s/\hbar D(0)}$ in 1-D and as $(s/\hbar D^{(0)})\ell n(s/\hbar D^{(0)})$ in 2-D. If this result is accepted, one must conclude that the eigenstates are neither extended nor normalizable; instead, they decay very slowly to zero. However, we know that at least in 1-D the states are exponentially localized. Hence, we must conclude that the singularities of the vertex part obtained through the CPA or the post CPA, although indicative of localization effects, are not capable of fully describing the situation at all length scales. It is assumed that at short length scales, when one does not sample the extreme tails of the eigenfunctions, the CPA and post-CPA vertex corrections are satisfactory. One way of improving the present method is to replace in (7.188) $D^{(0)}$ by D, the corrected diffusion coefficient, which in general is ω and q dependent. If $D(\omega)$ is proportional to $i\omega$, then $s\langle G^+G^-\rangle$ goes to a nonzero limit as $s \to 0^+$, and thus regular localization is recovered. We discuss this idea a little further in connection with the results for the conductivity σ.

Expression (7.92) for the real part of the conductivity becomes at T = 0

$$\sigma_1(\omega) = \frac{e^2\hbar}{\pi\Omega}\langle \mathrm{Tr}\{\upsilon_x \mathrm{Im}\{G^+(E_F+\hbar\omega)\}\upsilon_x \mathrm{Im}\{G^+(E_F)\}\}\rangle \quad . \tag{7.189}$$

But $\mathrm{Im}\{G^+(E_F+\hbar\omega)\} = [G^+(E_F+\hbar\omega) - G^-(E_F+\hbar\omega)]/2i$ with a similar expression for $\mathrm{Im}\{G^+(E_F)\}$. Out of the four terms resulting from the multiplication of the two Im{G}'s, two have their arguments in the same side of the real axis and consequently they are not important as far as the vertex corrections are concerned. The other two make a significant contribution to $\delta\sigma_1$, i.e.,

$$\delta\sigma_1 = \varphi_1(E_F,\omega) + \varphi_1(E_F+\hbar\omega, -\omega) \tag{7.190}$$

where

$$\varphi_1(E_F,\omega) = \frac{e^2\hbar}{\pi\Omega}\frac{1}{4}\,\delta\,\langle \mathrm{Tr}\{v_x G^+(E+\hbar\omega) v_x G^-(E)\}\rangle \quad . \tag{7.191}$$

We consider here the post-CPA corrections shown in Fig.7.6, since the CPA corrections by themselves vanish. From Fig.7.6a, and transforming everything to the $\underline{k}$-representation we have

$$\varphi_1(E_F,\omega) = \frac{2e^2\hbar}{\pi\Omega}\frac{1}{4d}\frac{1}{N}\sum_{\underline{q}}\sum_{\underline{k}} \vec{v}_{\underline{k}}\cdot\vec{v}_{\underline{q}-\underline{k}} G^+(\underline{k};E+\hbar\omega)G^+(\underline{q}-\underline{k};E+\hbar\omega)$$

$$\cdot\, G^-(\underline{k};E)G^-(\underline{q}-\underline{k};E)\gamma'(\underline{q};\, E^+ + \hbar\omega^+, E^-) \quad . \tag{7.192}$$

Because the integral (for $d=2,1$) is dominated by the region of the singularity at $\underline{q}=0$ we can take $\underline{q}=0$ in the $\underline{v}_{q-k}$ and G's and thus separate the $\underline{k}$-integration from the q-integration. The former gives a result (as $\omega\to 0$) which is proportional to $-\sigma^{(0)}/\Sigma_2^2$ or $-D^{(0)}/\Sigma_2^2$. To perform the $\underline{q}$-integration we need γ', which we obtain from (7.116) and (7.183) ($A'=A$)

$$\gamma'(\underline{q},E^+ + \hbar\omega^+,E^-) = u/(1-uA) = \frac{|\Sigma_2|}{\pi\rho'}\,\frac{1}{-\dfrac{i\hbar\omega}{2|\Sigma_2|} + \dfrac{\hbar D^{(0)}}{2|\Sigma_2|}q^2}$$

$$= \frac{2|\Sigma_2|^2}{\pi\rho'\hbar D^{(0)}}\,\frac{1}{q^2 - i\dfrac{\omega}{D^{(0)}}} \quad . \tag{7.193}$$

Substituting in (7.192) we obtain for $\varphi_1(E_F,\omega)$

$$\varphi_1(E_F,\omega) = -\frac{e^2}{\pi\hbar}\frac{1}{(2\pi)^d}\int d\underline{q}\,\frac{1}{q^2 - i\omega/D^{(0)}} \quad , \tag{7.194}$$

which combined with $\varphi_1(E_F+\hbar\omega,-\omega)\approx\varphi_1(E_F,-\omega)$ gives the final result for $\delta\sigma_1$

$$\delta\sigma_1 = -\frac{e^2}{\pi\hbar}\frac{2}{(2\pi)^d}\int d\underline{q}\,\frac{q^2}{q^4+\omega^2/D^{(0)2}} \quad . \tag{7.195}$$

For $d=1$ no upper cutoff is needed; for $d=2,3$ we take as an upper cutoff L_m^{-1} as given by (7.120). By introducing the length $L_\omega = (D^{(0)}/\omega)^{\frac{1}{2}}$ (7.123) we obtain from (7.195)

$$\delta\sigma_1 = -\frac{e^2}{\sqrt{2}\pi\hbar} L_\omega \quad ; \quad 1-D \tag{7.196}$$

$$= -\frac{e^2}{\pi^2\hbar} \ln\frac{L_\omega}{L_m} \quad ; \quad 2-D \tag{7.197}$$

$$= \frac{e^2}{\pi^2\hbar 2\sqrt{2}} \frac{1}{L_\omega} - \frac{e^2}{\pi^3\hbar} \frac{1}{L_m} \quad . \tag{7.198}$$

Equations (7.196-198) coincide, as far as the dependence on L_ω, L_m is concerned, with (7.119) or with (7.164). The present approach allows, in addition, an explicit evaluation of the prefactors. Note though that the prefactor for $d = 1$ is by a factor $\sqrt{2}$ larger than what one obtains from (7.133). In obtaining (7.195) we have started from (7.92) for the real part of σ. It would be simpler if we had chosen as our starting point (7.90); in this case by a quite similar procedure we would have obtained the vertex corrections for both the real and the imaginary part of σ, i.e.,

$$\delta\sigma = -\frac{e^2}{\pi\hbar} \frac{2}{(2\pi)^d} \int d\underline{q} \frac{1}{q^2 - i\omega/D^{(0)}} \quad , \tag{7.199}$$

the real part of which coincides with (7.195), as it should. One can rewrite the lhs of (7.199) in terms of the diffusion coefficient; furthermore (7.199) can be improved if one replaces the $D^{(0)}$ inside the integral by the corrected $D = D^{(0)} + \delta D$. We have then

$$D(\omega) = D^{(0)} - \frac{1}{(2\pi)^d \pi\hbar\rho} \int d\underline{q} \frac{1}{q^2 - i(\omega/D(\omega))} \quad . \tag{7.200}$$

Equation (7.200) was obtained by VOLLHARDT and WÖLFLE [7.139]. Self-consistent equations for $D(\omega)$ were also obtained by GÖTZE [7.167] and by KAWABATA [7.131]. Equation (7.200) has been used [7.139] to derive the results of the scaling method and to provide explicit expressions for β versus $\ln Q$.

It must be pointed out that $\delta\sigma(\omega)$ as given by (7.199) [or $\delta D(\omega)$ as given by (7.200)] is, apart from a constant factor, the same as the unperturbed Green's function $G_0(\underline{r},\underline{r};z)$, with $z = i\omega/D$, see (1.37). Thus the divergence of $\delta\sigma$ as $\omega \to 0$ for $d \leq 2$ (which was interpreted as implying the localization of all eigenfunctions) stems from the divergence of the unperturbed Green's function at the band edge $z = 0$. We remind the reader that is is precisely this divergence which always allows a local potential fluctuation to trap a particle (for $d \le 2$). Hence a mathematical connection was established between the property that all states are localized in a d-dimensional disordered system ($d \le 2$) and the property that a potential well always traps a particle in d dimensions ($d \le 2$). This connection can be further quantified in terms of the localization length. Indeed, for the shallow potential well problem one can rewrite the basic equation (6.9) in terms of the effective mass m*, defined by $E(\underline{k}) = \hbar^2\underline{k}^2/2m^*$, and the localization length λ_p, connected to the binding energy E_b by the relation $E_b = \hbar^2/2m^*\lambda_p^2$:

$$\frac{\hbar^2}{2m^* a^d |\varepsilon|} = \frac{1}{(2\pi)^d} \int d\underline{q} \, \frac{1}{q^2 + \lambda_p^{-2}} \quad . \tag{7.201}$$

In the disordered case and for localized states the polarizability $\alpha(\omega)$ is finite as $\omega \to 0$. Hence the conductivity $\sigma(\omega) = -i\omega\alpha(\omega)$ and the diffusion coefficient $D(\omega)$ approach zero linearly in ω. As a result $-i\omega/D(\omega) \to \xi^{-2}$ as $\omega \to 0$. There is evidence that ξ equals to the localization length λ [7.107, 139]. Taking into account the above remarks we can recast (7.200) as follows:

$$\frac{\pi\hbar\sigma^{(0)}}{2e^2} = \frac{1}{(2\pi)^d} \int d\underline{q} \, \frac{1}{q^2 + \lambda^{-2}} \quad . \tag{7.202}$$

The similarity of (7.201,202) suggests that there must be a direct physical connection between the problem of localization in disordered systems and that of a bound level in a single potential well.

The CPA for the vertex corrections, if combined with electron-electron interactions, gives nonzero results, which, as was discussed before, are very similar with (7.199). The main difference from the post-CPA results is that the latter, in contrast with the former, are sensitive to an external magnetic field. This sensitivity is due to the fact that the magnetic field destroys the reality of the product $G^+(\underline{n},\underline{m};E)G^-(\underline{n},\underline{m};E)$. The detailed calculations for the combined effects of CPA vertex corrections and electron-electron interactions (known in the field as interaction effects) were done by ALTSHULER et al. [7.176]; we also mentioned earlier work by ALTSHULER and ARONOV [7.177] and by SCHMID [7.178].

The post-CPA terms for the vertex corrections (known as localization effects in the field) to the conductivity were first recognized as important by LANGER and NEAL [7.179]. They were utilized in connection to the localization problem by ABRAHAMS et al. [7.127] and by GOR'KOV et al. [7.180]. Important contributions were made by HIKAMI et al. [7.152], by MAEKAWA and FUKUYAMA

[7.153], and by ALTSHULER et al. [7.176]. We mention also the scaling approach of McMILLAN [7.138] and the work of FUKUYAMA [7.181] on the role of spins. Some of the above papers are reviewed in a recent book edited by NAGAOKA and FUKUYAMA [7.182].

We have already mentioned that the localization theory has found the most direct applications in the transport properties of thin wires, films, and interfaces. There are, however, classes of materials (such as quasi one-dimensional organic conductors [7.183], polymers [7.183], amorphous semiconductors [7.184], amorphous metals, etc. [7.185]) of high physical and/or technological importance, which are to some extent disordered. The ideas and theoretical techniques presented in this chapter are indispensable tools in studying these materials. However, due to their complexity, our understanding has not yet reached a quantitative stage of development comparable to that of crystalline materials.

7.5 Summary

In this chapter we examined first a system consisting of two "impurities" embedded in a periodic tight-binding Hamiltonian (TBH). The total t-matrix associated with the two impurities is not simply the sum of t-matrices of each impurity but includes infinitely many terms which correspond physically to multiple scattering events. A diagrammatic representation of this fact is shown in Fig.7.1, where the propagation from site $\underline{i}$ to site $\underline{j}$ [described by $G(\underline{j},\underline{i})$] is decomposed to unperturbed propagations [described by $G_0(\underline{n},\underline{n}')$] and scattering events (described by the t-matrices $t_{\underline{\ell}}$, $t_{\underline{m}}$) from the impurity sites $\underline{\ell}$ and $\underline{m}$.

In the case of two impurities the spectrum consists of a continuum (corresponding to extended scattering eigenstates) and no more than two discrete levels. The appearance and the positions of the discrete levels depend on the magnitude and the sign of the impurity potentials. A typical behavior is shown in Fig.7.3 for a 1-D system (Fig.7.3a) and a 3-D system (Fig.7.3b).

If the concentration of impurities is nonzero we have what is called random or disordered system. For such a system we are interested first in determining $\langle G\rangle$ where the symbol $\langle\rangle$ denotes averaging over the various configurations of impurities corresponding to the given concentration. In general, $\langle G\rangle$ cannot be determined exactly and consequently schemes of approximation are needed. Three such schemes are reviewed briefly in Sect.7.2. The most satisfactory scheme is the so-called coherent potential approximation (CPA) and its various extensions. The CPA, in its simplest form, determines $\langle G(E)\rangle$ as follows

$$\langle G(E)\rangle = G_0(E - \Sigma(E)) \quad , \tag{7.49}$$

where G_0 is the unperturbed periodic Green's function, and the so-called self-energy Σ obeys the equation

$$\Sigma(E) = \left\langle \frac{\varepsilon'_{\underline{m}}}{1 - (\varepsilon'_{\underline{m}} - \Sigma(E))G_0(\underline{m},\underline{m};E - \Sigma(E))} \right\rangle \quad . \tag{7.56}$$

The perturbation $H_1 = \sum_{\underline{m}} |\underline{m}\rangle\varepsilon'_{\underline{m}}\langle\underline{m}|$ with $\langle\varepsilon'_{\underline{m}}\rangle = 0$.

In Sect.7.3 the theory of electrical conductivity is developed. A general quantum mechanical expression for the conductivity $\sigma(\omega)$ is obtained (the Kubo-Greenwood formula) which can be recast in terms of Green's functions

$$\sigma_{\mu\nu}(\omega) = \frac{ie^2 n}{m\omega} + \frac{e^2}{\Omega m^2\omega} \int_{-\infty}^{\infty} \frac{dE}{d\pi} \mathrm{Tr}\{p_\mu g(E) p_\nu g(e + \hbar\omega)\} \quad , \tag{7.90}$$

$$\sigma_{1\mu\nu}(\omega) = \frac{e^2\hbar}{\pi\Omega m^2} \int_{-\infty}^{\infty} dE \, \frac{f(E) - f(E + \hbar\omega)}{\hbar\omega} \mathrm{Tr}\{p_\mu \mathrm{Im}\{G^+(E + \hbar\omega)\} p_\nu \mathrm{Im}\{G^+(E)\}\} \, . \tag{7.92}$$

In the above expressions $\sigma_1 = \mathrm{Re}\{\sigma\}$; the Tr operation includes a factor of 2 due to spin summation; p_ν is a cartesian component of the momentum operator. Equation (7.90) is valid at $T = 0$ and

$$\begin{aligned} g(E) &\equiv G^-(E) \quad ; \quad E < E_F \\ &\equiv G^+(E) \quad ; \quad E > E_F \quad . \end{aligned} \tag{7.88}$$

To obtain the ensemble average of $\sigma(\omega)$ we need the average of the product of two G's, which is in general different from the product of the averages. A first approximation for $\sigma(\omega)$ is obtained by ignoring this difference and then utilizing the CPA for $\langle G\rangle$. The result in the weak scattering limit is (for $\omega = 0$)

$$\sigma^{(0)} = \frac{2}{(2\pi)^d d} \frac{e^2}{\hbar} \ell S_F \quad , \tag{7.95''}$$

where $\ell = \upsilon\tau$ is the mean free path, υ is the average magnitude of the velocity over the Fermi surface S_F (which in the simplest case is $4\pi k_F^2$), d is the dimensionality, and τ is the relaxation time given by

$$\frac{1}{\tau} = \frac{2\pi}{\hbar}\rho_F n_{imp} \frac{1}{4\pi}\int d\Omega V^2(\underline{q}) \quad . \tag{7.75}$$

In (7.75) ρ_F is the DOS per volume per spin at E_F, n_{imp} is the concentration of the scatterers, each one of which is associated with a scattering potential the Fourier transform of which is $V(\underline{q})$. The integration is over all directions of the vector $\underline{q}$ ($E(\underline{q}) = E_F$).

Next we considered an extension of the CPA which allows the approximate evaluation of the difference between the average of the product of G's and the product of the average G's. The effect of this difference on σ (which is called vertex correction), according to the CPA, in the weak scattering limit, is to introduce a factor $(1-\cos\theta)$ in (7.75) where θ is the angle of $\underline{q}$ with respect to the direction of propagation. This factor by definition changes τ to τ_{tr} (transport relaxation time) and ℓ to ℓ_{tr}. Equation (7.95") with ℓ replaced by $\ell_{tr} = \upsilon\tau_{tr}$ is the semiclassical result for σ, obtained usually by employing Boltzmann's equation.

In addition to the CPA contribution (shown in Fig.7.5) to the vertex corrections, another contribution (shown in Fig.7.6) has been obtained. This post-CPA contribution to σ is of the form

$$\delta\sigma \sim -\int_{1/L_M}^{1/L_m} \frac{d\underline{q}}{q^2} \tag{7.119}$$

where the upper cutoff length L_M is dominated by the shortest of several lengths (7.122-126) and L_m is close to ℓ_{tr}. The contribution (7.119) is extremely important for two-dimensional systems, where $\delta\sigma$ diverges as $-\ell n(L_M)$ and for one-dimensional systems, where $\delta\sigma$ diverges as $-L_M$, in the limit $L_M \to \infty$. It was found recently that the CPA contribution to σ, if *combined* with *electron-electron* interaction processes, yields a form similar to (7.119) with a different prefactor, and somewhat different L_M (especially in the presence of external magnetic fields). These recent theoretical results have found impressive experimental confirmation in 2-D films or interfaces; in thin wires and in 3-D systems the agreement between theory and experiment is of qualitative or semiqualitative nature.

The physical interpretation of the divergence of $\delta\sigma$ as $L_M \to \infty$ in 1-D and 2-D is that all eigenstates are localized (i.e., they decay to zero for large distances) no matter how weak the disorder is. As $\delta\sigma$ increases (in absolute value) with increasing L_M it eventually becomes comparable to $\sigma^{(0)}$ and finally the total $\sigma = \sigma^{(0)} + \delta\sigma$ approaches zero as $L_M \to \infty$. Equation (7.119) is appropriate for describing $\delta\sigma$ only when $|\delta\sigma|/\sigma^{(0)}$ is considerably smaller than 1. Thus one- and two-dimensional systems never are truly metallic. This nonmetallic nature is revealed when L_M becomes comparable to or larger than the localization length of the eigenfunctions. The fact that for $d = 3$ (7.119) gives a $\delta\sigma \sim 1/L_M$ is interpreted as meaning that for weak disorder the eigenstates in 3-D systems are extended with uniform (more or less) amplitude at least for length scales exceeding a characteristic length ξ. When the disorder increases beyond a critical value the eigenstates become localized. Various approaches to this fundamental problem of disorder-induced localization are reviewed and an extensive list of references to the current literature is given.

Part III

Green's Functions in Many-Body Systems

8. Definitions

The Green's functions defined before are recast in a second quantized form. The resulting expressions can easily be generalized for the case where there are many *interacting* particles. The time evolution of the operators involves now the interaction terms in the Hamiltonian. As a result the generalized Green's functions obey differential equations containing extra terms which depend on more complicated Green's functions.

8.1 Single Particle Green's Functions in Terms of Field Operators

The Green's function formalism which was developed up to now is appropriate for the problem of a single quantum particle moving in an external potential. Quite often we are dealing with systems involving many quantum particles. It is possible to generalize the definition of the Green's functions to obtain from them important physical information about the properties of the many particle systems. Such generalization is achieved in two steps. We first re-express the single body Green's functions (which we studied in the previous chapters) in the second quantization formalism, which is the most convenient language for the description of many-body systems [4.2,8.1-5]. We can then generalize quite easily the definition of the Green's functions as we shall see below.

In Appendix C we present very briefly the highlights of the second quantization formalism for two characteristic and very important cases: 1) the case of the field $\psi(\underline{r},t)$ obeying Schrödinger's equation (first order in time); 2) the case of the field $u(\underline{r},t)$ obeying the wave equation (second order in time). We express first the various Green's functions for the wave equation in terms of the field operator $u(\underline{r},t)$. The field $u(\underline{r},t)$ can be expressed in terms of creation and annihilation operators $b^{\dagger}_{\underline{k}}$, $b_{\underline{k}}$ as follows (see Appendix C):

$$u(\underline{r},t) = \sum_{\underline{k}} \left(\frac{\hbar c^2}{2\omega_{k}0}\right)^{1/2} \left[b^{\dagger}_{\underline{k}} e^{i(\omega_k t - \underline{k}\underline{r})} + b_{\underline{k}} e^{-i(\omega_k t - \underline{k}\underline{r})}\right] \quad . \tag{8.1}$$

The above time dependence of the operator $u(\underline{r},t)$ results from the general relation

$$i\hbar \frac{\partial u(\underline{r},t)}{\partial t} = [u(\underline{r},t),H] \quad , \tag{8.2}$$

which can be solved formally to give

$$u(\underline{r},t) = e^{iHt/\hbar} u(\underline{r},0) e^{-iHt/\hbar} \quad ; \tag{8.3}$$

substituting in (8.3) the noninteracting Hamiltonian

$$H = \sum_{\underline{k}} \hbar\omega_k \left(b^{\dagger}_{\underline{k}} b_{\underline{k}} + \frac{1}{2}\right) \quad , \tag{8.4}$$

we obtain (8.1) by taking into account the commutation relations (C.15') obeyed by $b_{\underline{k}}$ and $b^{\dagger}_{\underline{k}}$. Since we ascribe all the time dependence to the operators and none to the state vectors, we are working within the so-called Heisenberg picture.

Now we express the various g's and $\tilde{g}$'s associated with the wave equation and defined in Chap.2 in terms of the operator $u(\underline{r},t)$ given by (8.1). We have

$$\begin{aligned}\tilde{g}(\underline{r},\underline{r}',t,t') &= -\frac{i}{\hbar} \langle 0|(u(\underline{r},t)u(\underline{r}',t') - u(\underline{r}',t')u(\underline{r},t))|0\rangle \\ &= -\frac{i}{\hbar} [u(\underline{r},t),u(\underline{r}',t')]\end{aligned} \tag{8.5}$$

$$\tilde{g}^{>}(\underline{r},\underline{r}',t,t') = -\frac{i}{\hbar} \langle 0|u(\underline{r},t)u(\underline{r}'t')|0\rangle \tag{8.6}$$

$$\tilde{g}^{<}(\underline{r},\underline{r}',t,t') = -\frac{i}{\hbar} \langle 0|u(\underline{r}'t')u(\underline{r},t)|0\rangle \tag{8.7}$$

$$g^{R}(\underline{r},\underline{r}',t,t') = -\frac{i}{\hbar} \theta(t-t')\langle 0|[u(\underline{r},t),u(\underline{r}'t')]|0\rangle \tag{8.8}$$

$$g^{A}(\underline{r},\underline{r}',t,t') = \frac{i}{\hbar} \theta(t'-t)\langle 0|[u(\underline{r},t),u(\underline{r}',t')]|0\rangle \tag{8.9}$$

$$g(\underline{r},\underline{r}',t,t') = -\frac{i}{\hbar} \langle 0|T(u(\underline{r},t)u(\underline{r}',t'))|0\rangle \quad . \tag{8.10}$$

The Green's function g^- can be expressed in terms of g^R, g^A and g as $g^- = g^R + g^A - g$. To prove the above relations we proceed as follows: for (8.5) we note that $g(\underline{r},\underline{r}',t,t')$ is uniquely defined by the wave equation it obeys,

$$\left(\nabla^2 - \frac{1}{c^2}\frac{\partial^2}{\partial t^2}\right)\tilde{g}(\underline{r},\underline{r}',t,t') = 0 \quad , \tag{8.11}$$

and the initial conditions $\tilde{g}(\underline{r},\underline{r}',t,t) = 0$ and $\dot{\tilde{g}}(\underline{r},\underline{r}',t,t) = -c^2\delta(\underline{r} - \underline{r}')$. These initial conditions follow immediately from (2.48). All we need to show is that the right-hand side of (8.5) obeys the same equation and the same initial conditions. The wave equation (8.11) is obeyed by the right-hand side of (8.5) because the operator $u(\underline{r},t)$ obeys it by definition. The initial conditions are obeyed as a consequence of the commutation relation (C.14'). Thus (8.5) is proved. Note that the unequal time commutator $[u(\underline{r},t),u(\underline{r}',t')]$, where $u(\underline{r},t)$ is given by (8.1), is a c-number, and as such there was no need to take its expected value in the vacuum state. We prefer to use the expected value of $[u(\underline{r},t),u(\underline{r}',t')]$ in defining $\tilde{g}$ first because we can then generalize easily for interacting fields and second because we conform more to the definition of the other g's and $\tilde{g}$'s. To prove (8.6) and (8.7) we use (8.1) in the rhs of (8.6,7), and we take into account that $b_{\underline{k}}|0\rangle = 0$ and $[b_{\underline{k}},b^\dagger_{\underline{k}'}] = \delta_{\underline{k}\underline{k}'}$; we find after some simple algebra that the right-hand sides of (8.6) and (8.7) are, respectively,

$$-\frac{ic}{2}\sum_{\underline{k}}\frac{(e^{i\underline{k}\underline{r}}/\sqrt{\Omega})(e^{-i\underline{k}\underline{r}'}/\sqrt{\Omega})}{k}e^{-ick(t-t')}$$

and

$$\frac{-ic}{2}\sum_{\underline{k}}\frac{(e^{i\underline{k}\underline{r}}/\sqrt{\Omega})(e^{-i\underline{k}\underline{r}}/\sqrt{\Omega})}{k}e^{ick(t-t')} \quad .$$

These expressions coincide with the expressions (2.45) and (2.46) which were obtained previously for $\tilde{g}^>$ and $\tilde{g}^<$. To prove this last statement take into account that for the present case $L = -\nabla^2$, $\lambda_n = k^2$ and $\phi_n(\underline{r}) = e^{i\underline{k}\underline{r}}/\sqrt{\Omega}$. The proof of (8.8) and (8.9) follows immediately from (8.5) and (2.42) and (2.43). Finally, the proof of (8.10) follows by recalling the definition of the chronological operator T

$$T[u(\underline{r},t)u(\underline{r}',t')] = u(\underline{r},t)u(\underline{r}',t') \quad , \quad t>t'$$
$$= u(\underline{r}',t')u(\underline{r},t) \quad , \quad t<t' \qquad (8.12)$$

and by taking into account (8.6,7) and (2.41). We have thus succeeded to express the various g's and $\tilde{g}$'s for the wave equation in terms of the vacuum expectation values of bilinear combinations of the field operator $u(\underline{r},t)$ which obeys the wave equation. The relations (8.5-10) are valid for the general case where the g's and $\tilde{g}$'s are associated with the general linear second-order equation $(-\partial^2/c^2\partial t^2 - L)\phi = 0$, and $u(\underline{r},t)$ obeys the same equation and the commutation relations (C.14').

Let us now try to express the g's and $\tilde{g}$'s of a first-order equation (like the Schrödinger equation) in terms of the field operators $\psi,\psi^\dagger$. In writing expressions analogous to (8.5-10) we must take into account that the operator ψ is not Hermitian and that it may satisfy an anticommutation (instead of commutation) relation; furthermore, its time dependence is

$$\psi(\underline{r},t) = \sum_n a_n\psi_n(\underline{r})\, e^{-iE_nt/\hbar} \quad , \qquad (8.13)$$

$$\psi^\dagger(\underline{r},t) = \sum_n a_n^\dagger\psi_n^*(\underline{r})\, e^{iE_nt/\hbar} \quad . \qquad (8.14)$$

Again this time dependence stems from the relation

$$\psi(\underline{r},t) = e^{iHt/\hbar}\,\psi(\underline{r},0)\, e^{-iHt/\hbar} \quad , \qquad (8.15)$$

$$\psi^\dagger(\underline{r},t) = e^{iHt/\hbar}\psi^\dagger(\underline{r},0)\, e^{-iHt/\hbar} \quad , \qquad (8.16)$$

where

$$H = \sum_n E_n a_n^\dagger a_n \qquad (8.17)$$

and the creation and annihilation operators $a_n^\dagger$ and a_n obey the relation (C.15).

In analogy to (8.5-10) we define for the fields $\psi(\underline{r},t)$ and $\psi^\dagger(\underline{r},t')$ the following $\tilde{g}$'s and g's:

$$\tilde{g}(\underline{r},\underline{r}',t,t') = -\frac{i}{\hbar} <0|(\psi(\underline{r},t)\psi^\dagger(\underline{r}',t')\mp\psi^\dagger(\underline{r}',t')\psi(\underline{r},t))|0>$$

$$= -\frac{i}{\hbar}[\psi(\underline{r},t),\psi^{\dagger}(\underline{r}',t']_{\mp} \tag{8.18}$$

$$\tilde{g}^{>}(\underline{r},\underline{r}',t,t') = -\frac{i}{\hbar}\langle 0|\psi(\underline{r},t)\psi^{\dagger}(\underline{r}',t')|0\rangle \tag{8.19}$$

$$\tilde{g}^{<}(\underline{r},\underline{r}',t,t') = \mp\frac{i}{\hbar}\langle 0|\psi^{\dagger}(\underline{r}'t')\psi(\underline{r},t)|0\rangle \tag{8.20}$$

$$g^{R}(\underline{r},\underline{r}',t,t') = -\frac{i}{\hbar}\theta(t-t')\langle 0|[\psi(\underline{r},t),\psi^{\dagger}(\underline{r}',t')]_{\mp}|0\rangle \quad , \tag{8.21}$$

$$g^{A}(\underline{r},\underline{r}',t,t') = \frac{i}{\hbar}\theta(t'-t)\langle 0|[\psi(\underline{r},t),\psi^{\dagger}(\underline{r}',t')]_{\mp}|0\rangle \quad , \tag{8.22}$$

$$g(\underline{r},\underline{r}',t,t') = -\frac{i}{\hbar}\langle 0|T(\psi(\underline{r},t)\psi^{\dagger}(\underline{r}',t'))|0\rangle \quad , \tag{8.23}$$

where

$$[A,B]_{\mp} = AB \mp BA \tag{8.24}$$

and

$$T[\psi(\underline{r},t)\psi^{\dagger}(\underline{r}',t')] = \psi(\underline{r},t)\psi^{\dagger}(\underline{r}',t') \qquad t > t'$$

$$= \pm\psi^{\dagger}(\underline{r}',t')\psi(\underline{r},t) \qquad t < t' \tag{8.25}$$

where the upper sign refers to bosons and the lower sign refers to fermions. It is easy to see that if the field operator ψ were Hermitian and were corresponding to bosons, the relations (8.18-23) would be identical to (8.5-10). We now show that the quantities defined by (8.18-23) are identical to the Green's functions g^{+}, g^{-} and $\tilde{g}$ defined in Chap.2 for a first-order equation. We note first that $\tilde{g}^{<} = 0$ because $\psi|0\rangle = 0$. Next $\tilde{g}^{>} = \tilde{g}$ for the same reason. We now show that $\tilde{g}$ as defined by (8.18) coincides with the $\tilde{g}$ defined in Chap.2 [see, e.g., (2.13)]. Both $\tilde{g}$'s satisfy the same equation (Schrödinger's equation) and obey the same initial condition $\tilde{g}(\underline{r},\underline{r}',t,t) = -i\delta(\underline{r}-\underline{r}')\hbar$ (for the Schrödinger case $c = 1/\hbar$, $L = H$, $\lambda_n = E_n$). We can easily see that $g(\underline{r},\underline{r}',t,t') = g^{R}(\underline{r},\underline{r}',t,t') = \theta(t-t')\tilde{g}(\underline{r},\underline{r}',t,t') = g^{+}(\underline{r},\underline{r}',t,t')$; the last relation follows from (2.10). Similarly, $g^{-}(\underline{r},\underline{r}',t,t') = g^{A}(\underline{r},\underline{r}',t,t')$. Hence, in the case of a noninteracting field obeying a first-order (in time) equation the $\tilde{g}$'s and g's defined in terms of the vacuum expectation values of bilinear combinations of field operators ψ and $\psi^{\dagger}$ are related to the g^{+}, g^{-} and $\tilde{g}$ associated with the same equation as follows

$$g^{+}(\underline{r},\underline{r}',t,t') = g(\underline{r},\underline{r}',t,t') = g^{R}(\underline{r},\underline{r}',t,t') \quad (8.26)$$

$$g^{-}(\underline{r},\underline{r}',t,t') = g^{A}(\underline{r},\underline{r}',t,t') \quad (8.27)$$

$$\tilde{g}(\underline{r},\underline{r}',t,t') = \tilde{g}(\underline{r},\underline{r}',t,t') = \tilde{g}^{>}(\underline{r},\underline{r}',t,t') \quad (8.28)$$

$$\tilde{g}^{<}(\underline{r},\underline{r}',t,t') = 0 \quad . \quad (8.29)$$

The reader may wonder why we have introduced three g's and three $\tilde{g}$'s since actually there are in the present case only two independent g's and one independent $\tilde{g}$. The reason is that when we generalize our definition to the many particle systems, all three g's and three $\tilde{g}$'s become independent.

8.2 Green's Functions for Interacting Particles

We now generalize the definition of Green's functions to systems of many interacting particles which are described by a Hamiltonian. We first define various g's and $\tilde{g}$'s for the Schrödinger case characterized by the field operators $\psi^{\dagger}(\underline{r})$ and $\psi(\underline{r})$. The time dependence of these operators in the Heisenberg picture is given by the general equations (8.15,16). In what follows we assume for simplicity that there is no external potential felt by the particles, i.e., the quantity V_e introduced in Appendix C is zero. We also take $\hbar = 1$ so that we can use frequencies or energies indiscriminately. Furthermore, we write x for the four vector $\underline{r}$, t, and we denote by f(x,x') the function $f(\underline{r},\underline{r}',t,t')$ where f is any of the g's or $\tilde{g}$'s. For the sake of simplicity we omit the spin indices throughout the discussion. Because there is no external potential, the Hamiltonian is invariant under translation in space or time. This implies that all Green's functions depend on the difference x - x'. The definitions of the various g's and $\tilde{g}$'s are the following

$$\tilde{g}(x,x') = \tilde{g}(x-x') \equiv -i\langle[\psi(x),\psi^{\dagger}(x')]_{\mp}\rangle \quad (8.30)$$

$$\tilde{g}^{>}(x,x') = \tilde{g}^{>}(x-x') \equiv -i\langle\psi(x)\psi^{\dagger}(x')\rangle \quad (8.31)$$

$$\tilde{g}^{<}(x,x') = \tilde{g}^{<}(x-x') \equiv \mp\, i\langle\psi^{\dagger}(x')\psi(x)\rangle \quad (8.32)$$

$$g^{R}(x,x') = g^{R}(x-x') \equiv -i\theta(t-t')\langle[\psi(x),\psi^{\dagger}(x')]_{\mp}\rangle \quad (8.33)$$

$$g^A(x,x') = g^A(x - x') \equiv i\theta(t' - t)\langle[\psi(x),\psi^\dagger(x')]_\mp\rangle \qquad (8.34)$$

$$g(x,x') = g(x - x') \equiv -i\langle T[\psi(x)\psi^\dagger(x')]\rangle \qquad (8.35)$$

where the upper sign refers to bosons and the lower to fermions. The symbol $\langle A\rangle$ denotes thermal average of the arbitrary quantity A over the grand canonical ensemble.

$$\langle A\rangle = \frac{\sum_i \langle i|A|i\rangle\, e^{-\beta(E_i-\mu N_i)}}{\sum_i e^{-\beta(E_i-\mu N_i)}} = \frac{\mathrm{Tr}\left\{A\, e^{-\beta(H-\mu N)}\right\}}{\mathrm{Tr}\left\{e^{-\beta(H-\mu N)}\right\}} \quad . \qquad (8.36)$$

H is the total Hamiltonian of the system, N is the operator of the total number of particles, μ is the chemical potential, $\beta = 1/k_BT$ is the inverse temperature and $\{|i\rangle\}$ are the common eigenfunctions (in the Heisenberg picture) of H and N with eigenvalues E_i and N_i, respectively. In the limit $T \to 0$ we have

$$\langle A\rangle \to \langle\Psi_0|A|\Psi_0\rangle \quad \text{as} \quad T \to 0^+ \quad , \qquad (8.37)$$

where $|\Psi_0\rangle$ is the ground state (in the Heisenberg picture) of the whole system. Thus the generalization from (8.18-23) to (8.30-35) consists in taking the actual time development of the operators $\psi(x)$ and $\psi^\dagger(x')$ [instead of (8.13)] and in calculating the averages over the actual state of the system (instead of the vacuum). Obviously if no particles are present in the system the g's and g's defined by (8.30-35) reduce to those defined by (8.18-25) which as we have seen are identical with the g's and $\tilde{g}$'s defined in Chap.2. In other words, we can say by inspection of (8.18-25) that the $\tilde{g}$'s and g's defined in Chap.2 describe the propagation of a single particle from x' to x *in the absence of other particles*. On the other hand, the $\tilde{g}$'s and g's defined by (8.30-35) describe the propagation of a *single* particle (or hole) from x' to x (or x to x') *in the presence of other particles*. For example, $\tilde{g}^>(x,x')$ from (8.31) can be interpreted as the probability amplitude to find at x an extra particle which was added in our system at x' without any other modification in our system; thus $\tilde{g}^>$ describes the propagation of an extra particle added to our system (such propagation has physical meaning for t>t'). Similarly, $\tilde{g}^<(x,x')$ describes the propagation of a hole from x to x'. It follows that g(x,x') describes the propagation of an extra particle when t>t' and the propagation of a hole when t'>t. If the system is the vacuum, as in (8.18-23), we cannot have holes, and as a result $\tilde{g}^< = 0$. On the other hand, for the general case (8.32) a hole can be created (i.e., a particle can be eliminated) and hence $\tilde{g}^<$ is not, in general, zero.

The various $\tilde{g}$'s and g's defined by (8.30-35) are related with each other in exactly the same way as the $\tilde{g}$'s and g's introduced in Chap.2 for the second-order (in time) equation, i.e., by (2.38-43).

We introduce also the Fourier transform of $f(x - x')$ with respect to the variable $\underline{\rho} = \underline{r} - \underline{r}'$ and with respect to the four vector $x - x'$, where f is any of the $\tilde{g}$'s or g's, i.e.,

$$f(\underline{k},\tau) = \int d^3\rho e^{-i\underline{k}\underline{\rho}} f(x - x') \tag{8.38}$$

and

$$f(\underline{k},\omega) = \int d^3\rho d\tau e^{-i\underline{k}\underline{\rho} + i\omega\tau} f(x - x') \tag{8.39}$$

where $\tau = t - t'$. The Fourier transform $f(\underline{k},\omega)$ is the most widely used. The $\underline{k},\tau$ Fourier transforms can be expressed in terms of the $a^\dagger_{\underline{k}}$ and $a_{\underline{k}}$ operators creating and annihilating a particle with momentum $\underline{k}$ as follows

$$\tilde{g}^>(\underline{k},\tau) = -i\langle a_{\underline{k}}(\tau + t')a^\dagger_{\underline{k}}(t')\rangle \tag{8.40}$$

with similar expressions for the other $\tilde{g}$'s and g's. Equation (8.40) can be proved using (C.20), which relates $\psi(\underline{r})$ and $\psi^\dagger(\underline{r}')$ with $a_{\underline{k}'}$, and $a^\dagger_{\underline{k}''}$. Again $a_{\underline{k}}(t) = \exp(i\mathcal{H}t)a_{\underline{k}} \exp(-i\mathcal{H}t)$ and $a^\dagger_{\underline{k}}(t) = \exp(i\mathcal{H}t)a^\dagger_{\underline{k}} \exp(-i\mathcal{H}t)$. One can use the relations (8.30-35) to define the Green's functions associated with any field. In particular, for the case of longitudinal acoustic (LA) phonons it is customary to use the scalar field $\phi(\underline{r},t)$ which is real and depends on the LA phonon creation and annihilation operators $b^\dagger_{\underline{k}}$ and $b_{\underline{k}}$ as in (C.37). We have again that

$$\phi(\underline{r},t) = \exp(i\mathcal{H}t)\phi(\underline{r},0)\exp(-i\mathcal{H}t) \tag{8.41}$$

Denoting by D the Green's functions associated with the LA phonon field, we have, e.g.,

$$D(x,x') = D(x - x') = -i\langle T[\phi(x)\phi(x')]\rangle \tag{8.42}$$

with similar expressions for D^R, D^A, $\tilde{D}$, $\tilde{D}^>$ and $\tilde{D}^<$. Again the $\tilde{D}$'s and D's are related to each other by (2.38-43), i.e., in exactly the same way as the single particle Green's functions.

In the case where the total Hamiltonian H involves interactions, there is a basic difference between the Green's function defined in this section and those defined in Chap.2. In the presence of interaction the field ψ(or ϕ) does not obey the same equation as in the noninteracting case. As a result the differential equations obeyed by the $\tilde{g}$'s or g's are more complicated than those examined in Chap.2. To be more specific, consider the Schrödinger case with a total Hamiltonian $H = T + V_i$ where T is the kinetic energy given by (C.23) and V_i is the interaction part given by (C.26). In this case we have shown in Appendix C that

$$\left(i \frac{\partial}{\partial t} + \frac{\nabla^2}{2m}\right)\psi(\underline{r},t) = \int d^3\underline{r}'v(\underline{r} - \underline{r}')\psi^\dagger(\underline{r}',t)\psi(\underline{r}',t)\psi(\underline{r},t) \quad . \tag{8.43}$$

Thus, the application of the operator $i\partial/\partial t + \nabla^2/2m$ on $\psi(\underline{r},t)$ does not produce zero as in the noninteracting case but an extra term involving the interaction potential v and three ψ (or $\psi^\dagger$). As a result of this extra term application of $i\partial/\partial t + \nabla_r^2/2m$ on g(x,x') will give

$$\left(i \frac{\partial}{\partial t} + \frac{\nabla_r^2}{2m}\right)g(x,x') = \delta(x - x') \pm i \int d^4x_1v(\underline{r} - \underline{r}_1)g_2(x,x_1;x',x_1^+)\Big|_{t_1 = t} \tag{8.44}$$

where x_1^+ means $\underline{r}_1$, $t_1 + s$ as $s \to 0^+$ and g_2, the two particle Green's function, is defined as

$$g_2(x_1,x_2;x_1',x_2') = (-i)^2\langle T[\psi(x_1)\psi(x_2)\psi^\dagger(x_2')\psi^\dagger(x_1')]\rangle \quad , \tag{8.45}$$

where the chronological operator T arranges the operators in chronological order so that the earliest time appears on the right and the latest on the left. In addition, for fermions only, we introduce a factor ± 1 depending on whether the time-ordered product is an even or odd permutation of the original ordering. Thus for $t_1>t_2>t_1'>t_2'$ we have that $g_2(x_1,x_2;x_1',x_2') = \langle\psi(x_1)\psi(x_2)\psi^\dagger(x_1')\psi^\dagger(x_2')\rangle$; in this case g_2 describes the propagation of *two* extra particles added to our system. Equation (8.44) shows that in the presence of pairwise interactions the differential equation obeyed by

$g(x,x')$ involves an extra term depending on the two-particle Green's function g_2, which is unknown. In a similar way one finds that the differential equation for g_2 involves the three-particle Green's function g_3 and so on, where the n-particle Green's function is defined by

$$g_n(x_1 \dots x_n;x_1' \dots x_n') = (-i)^n \langle T[\psi(x_1) \dots \psi(x_n)\psi^\dagger(x_n') \dots \psi^\dagger(x_1')]\rangle \quad . \tag{8.46}$$

Thus, the existence of interaction complicates the calculation of g in an essential way: while in the noninteracting case g was determined by a single differential equation (and the appropriate initial conditions), in the presence of interactions we have an infinite hierarchy of equations each connecting a Green's function of order n to one of order n + 1. We postpone the question of calculation of g until Chap.10. In the next section we calculate g for the special case where there are no interactions among the particles.

Since the various g's and $\tilde{g}$'s are so closely interrelated, there is no need to consider all of them. Most of the monographs on the subject develop the whole formalism by using the causal Green's function g defined by (8.35) or (8.42). Some other authors (see, e.g., [8.6,7]) prefer to work with g^R.

In this section we have defined the various g's and $\tilde{g}$'s for a system consisting of many particles. From these Green's functions one can obtain quantities of physical importance. We will discuss this question in the next chapter.

8.3 Green's Functions for Noninteracting Particles

Consider first the Schrödinger case $\psi(x)$. The Hamiltonian for noninteracting particles is, [see (C.23)],

$$H = \sum_{\underline{k}} \varepsilon_{\underline{k}} a_{\underline{k}}^\dagger a_{\underline{k}} \tag{8.47}$$

where $\varepsilon_{\underline{k}} = k^2/2m$. From the general equation $ida_{\underline{k}}(t)/dt = [a_{\underline{k}}(t),H]$ we have in the present case $ida_{\underline{k}}(t)/dt = \varepsilon_{\underline{k}} a_{\underline{k}}(t)$, which leads to

$$a_{\underline{k}}(t) = \exp(-i\varepsilon_k t)a_{\underline{k}} \quad . \tag{8.48}$$

Substituting this into (8.40) we obtain

$$\tilde{g}^{>}(\underline{k},t) = -i\langle e^{-i\varepsilon_k t}a_{\underline{k}}a_{\underline{k}}^{\dagger}\rangle = -ie^{-i\varepsilon_k t}(1\pm\langle a_{\underline{k}}^{\dagger}a_{\underline{k}}\rangle) \quad . \tag{8.49}$$

Similarly,

$$\tilde{g}^{<}(\underline{k},t) = \mp\, ie^{-i\varepsilon_k t}\langle a_{\underline{k}}^{\dagger}a_{\underline{k}}\rangle \quad . \tag{8.50}$$

The average number operator $\langle a_{\underline{k}}^{\dagger}a_{\underline{k}}\rangle$ is given by the Bose or Fermi function for noninteracting particles;

$$\langle a_{\underline{k}}^{\dagger}a_{\underline{k}}\rangle = f_{\mp}(\varepsilon_k) = \frac{1}{e^{\beta(\varepsilon_k-\mu)}\mp 1} \tag{8.51}$$

where μ is the chemical potential. For the other g's and $\tilde{g}$'s we have [using (2.40-43)]

$$\begin{aligned} g(\underline{k},t) &= -ie^{-i\varepsilon_k t}(1\pm f_{\mp}(\varepsilon_k)) \quad , \qquad t>0 \\ &= \mp\, ie^{-i\varepsilon_k t}f_{\mp}(\varepsilon_k) \quad , \qquad t<0 \end{aligned} \tag{8.52}$$

$$\tilde{g}(\underline{k},t) = -\, ie^{-i\varepsilon_k t} \tag{8.53}$$

$$g^{R}(\underline{k},t) = -i\theta(t)e^{-i\varepsilon_k t} \tag{8.54}$$

$$g^{A}(\underline{k},t) = i\theta(-t)e^{-i\varepsilon_k t} \quad . \tag{8.55}$$

We see that for noninteracting systems the quantities $\tilde{g}$, g^R and g^A do not involve the temperature or μ and that they have exactly the same form as in the case of a single particle moving in the vacuum. On the other hand, the quantities $\tilde{g}^{>}$, $\tilde{g}^{<}$ and g involve information pertaining not only to the motion of the added particle (or hole) but to the state of the system as well.

For fermions in the limit $T \to 0$ we have $\mu = \varepsilon_F = k_F^2/2m$; thus $g(\underline{k},t)$ becomes

$$g(\underline{k},t) = -ie^{-i\varepsilon_k t}\theta(k - k_F) \quad , \qquad t>0$$

$$= ie^{-i\varepsilon_k t}\theta(k_F - k) \quad , \qquad t<0 \quad . \tag{8.56}$$

For bosons the limit $T \to 0$ is more complicated because of the phenomenon of Bose condensation (see [8.3]).

Let us now calculate the Fourier transforms with respect to t. It is easy to show that

$$\tilde{g}(\underline{k},\omega) = -2\pi i\delta(\omega-\varepsilon_k) \tag{8.57}$$

$$g^R(\underline{k},\omega) = \lim_{s\to 0^+} \frac{1}{\omega+is-\varepsilon_k} \tag{8.58}$$

$$g^A(\underline{k},\omega) = \lim_{s\to 0^+} \frac{1}{\omega-is-\varepsilon_k} \quad . \tag{8.59}$$

For $T = 0$ and for fermions we obtain from (8.56)

$$g(\underline{k},\omega) = \lim_{s\to 0^+} \left[\frac{\theta(k-k_F)}{\omega+is-\varepsilon_k} + \frac{\theta(k_F-k)}{\omega-is-\varepsilon_k}\right]$$

$$= \lim_{s\to 0^+} \frac{1}{\omega-\varepsilon_k+is\bar{\varepsilon}(k-k_F)} \quad , \tag{8.60}$$

where $\bar{\varepsilon}(x) = 1$ for $x>0$ and -1 for $x<0$. By defining the function

$$G(\underline{k},z) = \frac{1}{z-\varepsilon_k} \quad , \tag{8.61}$$

we can rewrite (8.58-61) as

$$g^R(\underline{k},\omega) = \lim_{s\to 0^+} G(\underline{k},\omega+is) \tag{8.62}$$

$$g^A(\underline{k},\omega) = \lim_{s\to 0^+} G(\underline{k},\omega-is) \tag{8.63}$$

$$g(\underline{k},\omega) = \lim_{s\to 0^+} G[\underline{k},\omega+is\bar{\varepsilon}(\omega-\varepsilon_F)] \quad . \tag{8.64}$$

Equations (8.58-60) can be proved by applying the residue theorem to show that the inverse Fourier transform of g^R, g^A and g give (8.54-56).

We consider next the Green's functions for the LA phonon field $\phi(x)$. As is discussed in Appendix C, $\phi(x)$ can be expressed in terms of the phonon creation and annihilation operators $b_{\underline{k}}^\dagger$ and $b_{\underline{k}}$ as follows

$$\phi(x) = \sum_{\underline{k}} \left(\frac{\omega_{\underline{k}}}{2\Omega}\right)^{1/2} \left[b_{\underline{k}}^\dagger e^{ikx} + b_{\underline{k}} e^{-ikx}\right] \tag{8.65}$$

where $kx = \omega_k t - \underline{k}\underline{r}$ and the summation over $\underline{k}$ is restricted by $|\underline{k}| \leq k_D$. In obtaining (8.65) we have used the fact that the phonons do not interact, i.e., that their Hamiltonian is $H = \sum_{\underline{k}} \omega_k (b_{\underline{k}}^\dagger b_{\underline{k}} + 1/2)$. Substituting (8.65) in the definition of $\tilde{D}^>(x,x')$, we have

$$\tilde{D}^>(x,x') = -i\langle\phi(x)\phi(x')\rangle = \frac{-i}{2\Omega} \sum_{\underline{k}\underline{k}'} \sqrt{\omega_{\underline{k}}\omega_{\underline{k}'}} \left[e^{-ikx+ik'x'} \langle b_{\underline{k}} b_{\underline{k}'}^\dagger\rangle + e^{ikx-ik'x'} \langle b_{\underline{k}}^\dagger b_{\underline{k}'}\rangle\right] \quad . \tag{8.66}$$

Taking into account that $\langle b_{\underline{k}}^\dagger b_{\underline{k}'}\rangle = \langle b_{\underline{k}'} b_{\underline{k}}^\dagger\rangle - \delta_{\underline{k}\underline{k}'} = \delta_{\underline{k}\underline{k}'} n_{\underline{k}} = \delta_{\underline{k}\underline{k}'} (e^{\beta\omega_k} - 1)^{-1}$, we obtain

$$\tilde{D}^>(x,x') = \frac{-i}{2\Omega} \sum_{\underline{k}} \omega(k) \left[e^{ik(x-x')} n_{\underline{k}} + e^{-ik(x-x')}(1+n_{\underline{k}})\right] \; ; \quad \omega(k) \equiv \omega_k \; ; \tag{8.67}$$

in the T = 0 limit we have

$$\tilde{D}^>(x,x') = \frac{-i}{2\Omega} \sum_{\underline{k}} \omega(k) e^{-ik(x-x')} \quad . \tag{8.68}$$

The quantity $\tilde{D}^<(x,x')$ can be obtained immediately from the relation $\tilde{D}^<(x,x') = \tilde{D}^>(x',x)$. We have then for $\tilde{D}(x-x')$

$$\tilde{D}(x-x') = \frac{-i}{2\Omega} \sum_{\underline{k}} \omega(k) [e^{-ik(x-x')} - e^{ik(x-x')}] = -\frac{1}{\Omega} \sum_{\underline{k}} \omega(k) \sin[k(x-x')] \quad . \tag{8.69}$$

The Green's functions D, D^R and D^A can then be obtained by applying (2.41-43). Having obtained the various D(x - x')'s and $\tilde{D}$(x - x')'s, we can calculate their Fourier transforms easily. We find

$$\tilde{D}(\underline{k},\omega) = -\pi i\omega_k[\delta(\omega - \omega_k)-\delta(\omega + \omega_k)]\theta(k_D - k) \quad ; \qquad k \equiv |\underline{k}| \tag{8.70}$$

$$D^R(\underline{k},\omega) = \frac{\omega_k^2}{(\omega+is)^2-\omega_k^2}\,\theta(k_D - k) \tag{8.71}$$

$$D^A(\underline{k},\omega) = \frac{\omega_k^2}{(\omega-is)^2-\omega_k^2}\,\theta(k_D - k) \quad . \tag{8.72}$$

In the T → 0 limit we obtain for $D(\underline{k},\omega)$

$$D(\underline{k},\omega) = \frac{\omega_k^2}{\omega^2-\omega_k^2+is}\,\theta(k_D - k) \quad . \tag{8.73}$$

The factor $\theta(k_D - k) = \theta(\omega_D - \omega_k)$, where $\omega_D = ck_D$ comes from the restriction $k \leq k_D$ in the basic expression (8.65); physically it is justified by the requirement that there are as many degrees of freedom in our continuous elastic model as there are degrees of freedom in an actual (discrete) solid.

8.4 Summary

Let $\psi(\underline{r})$ and $\psi^\dagger(\underline{r})$ be the field operators resulting by quantizing (second-quantization) the wave function (and its complex conjugate) corresponding to the Schrödinger equation. We take $\hbar = 1$, $\underline{r}$, t = x, and we omit any external potential; we also omit for simplicity any spin indices. $\psi(\underline{r},t) = \exp(iHt)\psi(\underline{r})\exp(-iHt)$ with an identical expression for $\psi^\dagger(\underline{r},t)$, where H is the total Hamiltonian describing our system. We define the one particle Green's functions for our many-body system as follows:

$$\tilde{g}(x,x') = -i\langle[\psi(x), \psi^\dagger(x')]_\mp\rangle \tag{8.30}$$

$$\tilde{g}^>(x,x') = -i\langle\psi(x)\psi^\dagger(x')\rangle \tag{8.31}$$

$$\tilde{g}^<(x,x') = \mp i\langle\psi^\dagger(x')\psi(x)\rangle \quad (8.32)$$

$$g^R(x,x') = \theta(t-t')\tilde{g}(x,x') \quad (8.33)$$

$$g^A(x,x') = -\theta(t'-t)\tilde{g}(x,x') \quad (8.34)$$

$$g(x,x') = -i\langle T[\psi(x)\psi^\dagger(x')]\rangle \quad (8.35)$$

where $[A,B]_\mp = AB \mp BA$; the upper sign refers to bosons and the lower to fermions; the chronological operator T is defined as

$$\begin{aligned} T[\psi(x)\psi^\dagger(x')] &= \psi(x)\psi^\dagger(x') \qquad t>t' \\ &= \pm\psi^\dagger(x')\psi(x) \qquad t<t' \quad ; \end{aligned} \quad (8.25)$$

the symbol <> denotes thermal average over the grand canonical ensemble, i.e.,

$$\langle A\rangle = \mathrm{Tr}\left\{Ae^{-\beta(H-\mu N)}\right\}/\mathrm{Tr}\left\{e^{-\beta(H-\mu N)}\right\} \quad . \quad (8.36)$$

For zero temperature $\langle A\rangle = \langle\Psi_0|A|\Psi_0\rangle$ where $|\Psi_0\rangle$ is the ground state (in the Heisenberg picture) of our system.

The Green's functions defined above describe the propagation of an extra particle (or hole) added to our system which is in thermal equilibrium. If our system contains no particles at all, the Green's functions defined by (8.30-35) reduce to those introduced in Chap.2, which describe the propagation of a particle in the vacuum. Using the commutation (anticommutation) relations of the field operators, one can prove that the $\tilde{g}$'s and g's defined by (8.30-35) obey (2.38-43) like the Green's functions introduced in Chap.2. Because the various g's and $\tilde{g}$'s are interrelated, one can limit oneself to one of them; the most commonly used is the causal g defined by (8.35); a few authors use the retarded g^R.

The time evolution of the field operator $\psi(x)$ depends on the total Hamiltonian which involves the other particles of the system if interactions are present. Thus, for interacting systems the differential equations obeyed by the $\tilde{g}$'s or g's contain extra terms which involve more complicated Green's functions.

In practical calculations it is more convenient to work with the Fourier transforms of the various $\tilde{g}$'s and g's with respect to the four-variable $x - x'$.

The definitions (8.30-35) can be applied to any quantized field. An example is given for the scalar field of the longitudinal acoustic phonons in a continuous elastic medium.

Finally, we calculated the Green's function for the simple case of non-interacting Schrödinger or phonon fields.

9. Properties and Use of the Green's Functions

The Green's functions defined in Chap.8 have similar but not identical analytical properties as the Green's functions defined in Chap.2 corresponding to second-order (in time) differential equations. They can all be expressed in terms of a generalized DOS and the Fermi or Bose thermal equilibrium distributions. From the Green's functions (or the generalized DOS) one can easily obtain all thermodynamic quantities and linear response functions, like the conductivity. The poles of an appropriate analytic continuation of G in the complex E-plane can be interpreted as the energy (the real part of the pole) and the inverse life time (the imaginary part of the pole) of quasi-particles. The latter are entities which allow us to map an interacting system to a non-interacting one.

9.1 Analytical Properties of the g's and $\tilde{g}$'s

Let us consider first the quantity $\tilde{g}^{>}(\underline{k},t)$. We have

$$\begin{aligned}\tilde{g}^{>}(\underline{k},\tau) &= -i\langle a_{\underline{k}}(t)a_{\underline{k}}^{\dagger}(t')\rangle = -i\sum_{m}\rho_m\langle m|a_{\underline{k}}(t)a_{\underline{k}}^{\dagger}(t')|m\rangle \\ &= -i\sum_{m\ell}\rho_m\langle m|e^{iHt}a_{\underline{k}}^{\dagger}\,e^{-iHt}|\ell\rangle\langle\ell|e^{iHt'}a_{\underline{k}}^{\dagger}\,e^{-iHt'}|m\rangle \\ &= -i\sum_{m\ell}\rho_m\,e^{-i\tau(E_\ell-E_m)}|\langle\ell|a_{\underline{k}}^{\dagger}|m\rangle|^2 \quad ,\end{aligned} \tag{9.1}$$

where $|m\rangle$ and $|\ell\rangle$ are eigenfunctions of the total H and N, i.e., $H|m\rangle = E_m|m\rangle$. $H|\ell\rangle = E_\ell|\ell\rangle$, $N|m\rangle = N_m|m\rangle$ and $N|\ell\rangle = N_\ell|\ell\rangle$; $\rho_m = \exp[-\beta(E_m-\mu N_m)]/\mathrm{Tr}\{\exp[-\beta(H-\mu N)]\}$; $\tau = t - t'$; the summation extends over all eigenstates. By taking the Fourier transform with respect to τ, we find

$$\tilde{g}^{>}(\underline{k},\omega) = -2\pi i \sum_{m\ell} \rho_m |\langle\ell|a_{\underline{k}}^{\dagger}|m\rangle|^2 \delta(\omega + E_m - E_\ell) \quad . \tag{9.2}$$

In a similar way we can show that

$$\tilde{g}^{<}(\underline{k},\omega) = \mp 2\pi i \sum_{mn} \rho_m |\langle n|a_{\underline{k}}|m\rangle|^2 \delta(\omega + E_n - E_m) \quad . \tag{9.3}$$

Equations (9.2,3) become in the T = 0 limit

$$\tilde{g}^{>}(\underline{k},\omega) = -2\pi i \sum_{\ell} |\langle\ell|a_{\underline{k}}^{\dagger}|\Psi_o\rangle|^2 \delta(\omega + E_g - E_\ell) \tag{9.2'}$$

$$\tilde{g}^{<}(\underline{k},\omega) = \mp 2\pi i \sum_{n} |\langle n|a_{\underline{k}}|\Psi_o\rangle|^2 \delta(\omega + E_n - E_g) \tag{9.3'}$$

where $|\Psi_o\rangle$ is the ground state and E_g the ground state energy.

One can easily show from (9.2,3) that

$$\tilde{g}^{<}(\underline{k},\omega) = \pm e^{-\beta(\omega-\mu)}\tilde{g}^{>}(\underline{k},\omega) \quad . \tag{9.4}$$

In the T = 0 limit, (9.4) becomes

$$\tilde{g}^{<}(\underline{k},\omega) = 0 \quad \text{for} \quad \omega>\mu$$

$$\tilde{g}^{>}(\underline{k},\omega) = 0 \quad \text{for} \quad \omega<\mu \quad . \tag{9.4'}$$

Equation (9.4') can be proved directly from (9.2',3') by taking into account that $\mu = E_g(N + 1) - E_g(N) = E_g(N) - E_g(N - 1)$.

Let us now define the quantity $A(\underline{k},\omega)$ as follows

$$A(\underline{k},\omega) \equiv i\tilde{g}(\underline{k},\omega) = i[\tilde{g}^{>}(\underline{k},\omega) - \tilde{g}^{<}(\underline{k},\omega)] \quad . \tag{9.5}$$

Taking into account (9.4,5) we can express both $\tilde{g}^{>}(\underline{k},\omega)$ and $\tilde{g}^{<}(\underline{k},\omega)$ in terms of $A(\underline{k},\omega)$. We find

$$\tilde{g}^{>}(\underline{k},\omega) = -iA(\underline{k},\omega)[1 \pm f_{\mp}(\omega)] \tag{9.6}$$

$$\tilde{g}^{<}(\underline{k},\omega) = \mp iA(\underline{k},\omega)f_{\mp}(\omega) \tag{9.7}$$

where

$$f_{\mp}(\omega) = \frac{1}{e^{\beta(\omega-\mu)}\mp 1} \qquad (9.8)$$

is the Bose (Fermi) distribution. Equations (9.6,7) become in the T = 0 limit and for fermions

$$\tilde{g}^{>}(\underline{k},\omega) = -iA(\underline{k},\omega)\theta(\omega-\mu) \qquad (9.6')$$

$$\tilde{g}^{<}(\underline{k},\omega) = iA(\underline{k},\omega)\theta(\mu-\omega) \quad ; \qquad (9.7')$$

for bosons the T = 0 limit is more complicated as a result of the phenomenon of Bose condensation.

From (9.2,3) and (9.5) it is easy to see that the quantity $A(\underline{k},\omega)$ is real. For fermions it is always nonnegative. For bosons it is nonnegative for $\omega>\mu$ and nonpositive for $\omega<\mu$. One can show further that $A(\underline{k},\omega)$ obeys the following sum rule

$$\int \frac{d\omega}{2\pi} A(\underline{k},\omega) = 1 \quad . \qquad (9.9)$$

Equation (9.9) can be proved with the aid of (9.2,3) and (9.5) or as follows

$$\int \frac{d\omega}{2\pi} A(\underline{k},\omega) = i \int \frac{d\omega}{2\pi} \int dt\, e^{i\omega t}[\tilde{g}^{>}(\underline{k},t) - \tilde{g}^{<}(\underline{k},t)]$$

$$= i[\tilde{g}^{>}(\underline{k},0) - \tilde{g}^{<}(\underline{k},0)] = \langle a_{\underline{k}}a^{\dagger}_{\underline{k}} \mp a^{\dagger}_{\underline{k}}a_{\underline{k}}\rangle = 1 \quad .$$

Consider now the function $G(\underline{k},\omega)$ defined by

$$G(\underline{k},\omega) = \int_{-\infty}^{\infty} \frac{d\omega'}{2\pi} \frac{A(\underline{k},\omega)}{\omega-\omega'} \quad . \qquad (9.10)$$

The function $G(\underline{k},\omega)$ is analytic in the complex ω-plane and has singularities (branch cuts, in general) along those portions of the real axis where $A(\underline{k},\omega) \neq 0$.

From (9.10) and (9.5) it follows that

$$G(\underline{k},\omega+is) - G(\underline{k},\omega-is) = -iA(\underline{k},\omega) = \tilde{g}(\underline{k},\omega) \qquad (9.11)$$

where ω is real and $s \to 0^{+}$.

In view of (9.11) it follows immediately that $\tilde{g}(\underline{k},\tau)$ is given by integrating $\exp(-i\omega\tau)G/2\pi$ along the contour shown in Fig.9.1f.

Now we can show that $g^R(\underline{k},\tau)$ can be written as

$$g^R(\underline{k},\tau) = \int_C \frac{d\omega}{2\pi} e^{-i\omega\tau} G(\underline{k},\omega) \tag{9.12}$$

where the integration path C is shown in Fig.9.1b. The proof is as follows: For negative τ we can close the integration path by a semicircle in the upper half plane, and consequently $g^R(\underline{k},\tau) = 0$ for $\tau<0$ as it should be. For $\tau>0$ we can close the path by a semicircle in the lower half plane and the resulting contour can be deformed as in Fig.9.1f.

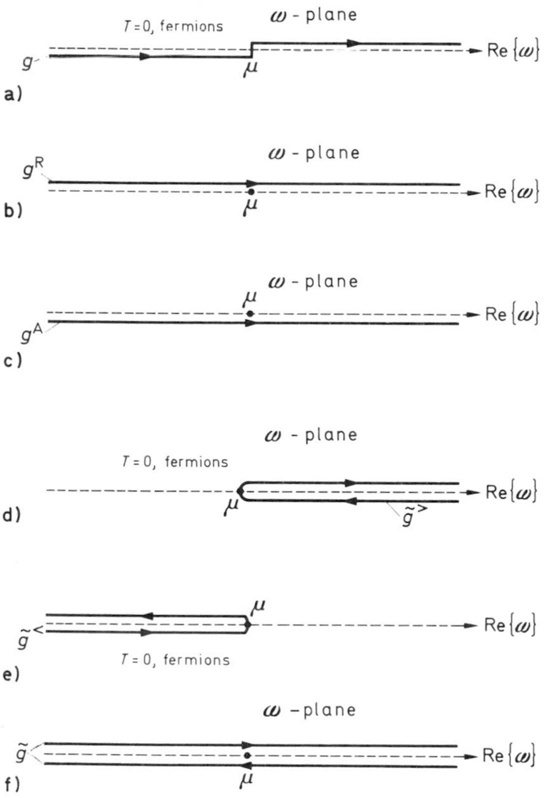

Fig.9.1. Integration paths in the complex ω-plane for obtaining the various $g(\tau)$'s and $\tilde{g}(\tau)$'s. The quantities g, $\tilde{g}^>$ and $\tilde{g}^<$ are given by the corresponding integrals only for T = 0. The singularities of the integrand are located on the real ω-axis

Hence for $\tau>0$

$$\int_C \frac{d\omega}{2\pi} e^{-i\omega\tau} G(\underline{k},\omega) = \int_{-\infty}^{\infty} \frac{d\omega}{2\pi} e^{-i\omega\tau}[G(\underline{k},\omega+is) - G(\underline{k},\omega-is)]$$

$$= \int_{-\infty}^{\infty} \frac{d\omega}{2\pi} e^{-i\omega\tau} \tilde{g}(\underline{k},\omega) = \tilde{g}(\underline{k},\tau) = \tilde{g}^R(\underline{k},\tau) \quad .$$

In a similar way we can show that

$$g^A(\underline{k},\tau) = \int_C \frac{d\omega}{2\pi} e^{-i\omega\tau} G(\underline{k},\omega) \tag{9.13}$$

where the path C is shown in Fig.9.1c.
For the Fourier transforms we have

$$g^R(\underline{k},\omega) = \lim_{s\to 0^+} G(\underline{k},\omega+is) \tag{9.14}$$

$$g^A(\underline{k},\omega) = \lim_{s\to 0^+} G(\underline{k},\omega-is) \tag{9.15}$$

where ω is real.

Equations (9.14,15) can be written by taking into account (9.10) as follows

$$\mathrm{Re}\{g^R(\underline{k},\omega)\} = \mathrm{Re}\{g^A(\underline{k},\omega)\} = P\int_{-\infty}^{\infty} \frac{d\omega'}{2\pi} \frac{A(\underline{k},\omega')}{\omega-\omega'} \quad , \tag{9.16}$$

$$\mathrm{Im}\{g^R(\underline{k},\omega)\} = -\mathrm{Im}\{g^A(\underline{k},\omega)\} = -\frac{1}{2} A(\underline{k},\omega) \quad . \tag{9.17}$$

From the general relation $g = g^R + \tilde{g}^<$ [see (2.39)] with the aid of (9.7) and (9.17) we obtain for real ω

$$\mathrm{Re}\{g(\underline{k},\omega)\} = \mathrm{Re}\{g^R(\underline{k},\omega)\} = \mathrm{Re}\{g^A(\underline{k},\omega)\} \tag{9.18}$$

$$\mathrm{Im}\{g(\underline{k},\omega)\} = -\frac{1}{2} A(\underline{k},\omega) \mp A(\underline{k},\omega) f_{\mp}(\omega) = -A(\underline{k},\omega)\left[\frac{1}{2} \pm f_{\mp}(\omega)\right]$$

$$= \mathrm{Im}\{g^R(\underline{k},\omega)\}[1 \pm 2f_{\mp}(\omega)]$$

$$= \mathrm{Im}\{g^R(\underline{k},\omega)\} \begin{matrix}\coth\\ \tanh\end{matrix} \left[\frac{\beta(\omega-\mu)}{2}\right] \quad . \tag{9.19}$$

Equation (9.19) in the limit when $T = 0$ for fermions becomes

$$\begin{aligned} \mathrm{Im}\{g(\underline{k},\omega)\} &= \bar{\varepsilon}(\omega - \mu)\, \mathrm{Im}\{g^R(\underline{k},\omega)\} \\ &= -\frac{1}{2}\bar{\varepsilon}(\omega - \mu)A(\underline{k},\omega) = \lim_{s \to 0^+} \mathrm{Im}\{G(\underline{k},\omega + is\bar{\varepsilon}(\omega - \mu))\} \quad . \end{aligned} \tag{9.19'}$$

From (9.19') it follows that for $T = 0$ and for fermions $g(\underline{k},\omega)$ can be obtained by integrating $G(\underline{k},\omega)\cdot \exp(-i\omega\tau)/2\pi$ along the path shown in Fig.9.1a. By taking into account (9.6',7') and (9.11), we see that $\tilde{g}^>(\underline{k},\tau)$ and $\tilde{g}^<(\underline{k},\tau)$ (for $T = 0$ and for fermions) are given by integrating $G(\underline{k},\omega)\cdot \exp(-i\omega\tau)/2\pi$ along the contours shown in Fig.9.1d and e, respectively. With the aid of (9.18,16, and 19), we can express the $\mathrm{Re}\{(g(\underline{k},\omega)\}$ in terms of $\mathrm{Im}\{g(\underline{k},\omega)\}$ as follows

$$\mathrm{Re}\{g(\underline{k},\omega)\} = -2P \int_{-\infty}^{\infty} \frac{d\omega'}{2\pi} \frac{\mathrm{Im}\{g(\underline{k},\omega')\}}{\omega - \omega'} \begin{matrix}\tanh \\ \coth\end{matrix}\left[\frac{\beta(\omega-\mu)}{2}\right] \quad . \tag{9.20}$$

We conclude this section with some remarks:

1) We see that all g's and $\tilde{g}$'s can be expressed in terms of a single quantity, namely $A(\underline{k},\omega) = i\tilde{g}(\underline{k},\omega)$, or in terms of the limiting values as $s \to 0^+$ of an analytic function $G(\underline{k},\omega+is)$. One must note that $\tilde{g}(\underline{k},\omega)$, $g^R(\underline{k},\omega)$ and $g^A(\underline{k},\omega)$ depend only on $A(\underline{k},\omega)$ [or $\lim G(\underline{k},\omega+is)$]; on the other hand, $g(\underline{k},\omega)$, $\tilde{g}^>(\underline{k},\omega)$ and $\tilde{g}^<(\underline{k},\omega)$ involve both $A(\underline{k},\omega)$ and universal thermal factors; in other words, $\tilde{g}^>$, $\tilde{g}^<$ and g are not simply the limits of $G(\underline{k},\omega\pm is)$ as $s \to 0^+$ (except when $T = 0$ and for fermions). This means that $\tilde{g}^>(\underline{k},\omega)$, $\tilde{g}^<(\underline{k},\omega)$ and $\tilde{g}(\underline{k},\omega)$ cannot be continued off the real ω-axis to produce functions which are analytic everywhere off the real ω-axis. This could be a serious drawback when one uses the residue theorem to evaluate integrals or when one attempts to uniquely determine $g(\underline{k},\omega)$ by continuing off the real ω-axis the equation obeyed by $g(\underline{k},\omega)$. We shall return to this last point in the next chapter.

2) One must note the close analogy of the analytic properties of the various g's and $\tilde{g}$'s for a many-body system and the g's and $\tilde{g}$'s defined in Chap.2. Thus the quantity $A(\underline{k},\omega)$ is the analog of $2\pi\rho(\underline{k},\omega)$ where $\rho(\underline{k},E)$ is the density of states per unit volume in the $\underline{k}$-space. The function $G(\underline{k},z) = \int d\omega'(z - \omega')^{-1}A(\underline{k},\omega')/2\pi$ is the analog of the diagonal matrix element (in the $\underline{k}$-representation) of the Green's function $G(\underline{k},z) = \langle\underline{k}|(z - \mathcal{H})^{-1}|\underline{k}\rangle$. Similarly, the quantities $\tilde{g}$, $\tilde{g}^>$, $\tilde{g}^<$, g^R, g^A and g have their counterparts for the single-particle case. One must keep in mind that

there are differences between the g's defined in Chap.2 and those defined in Chap.8. We mention some: the $\omega = 0$ point in the Chap.2 case corresponds to the $\omega = \mu$ point in the Chap.8 case (compare Figs.2.2 and 9.1). This difference can be easily eliminated by slightly modifying the definition of Green's function, i.e., by replacing H by $H - \mu N$ in (8.15,16). This replacement implies that $\omega \to \omega' = \omega - \mu$. Many authors use this modified definition (see, e.g., [8.3,9.1]). The quantities $\tilde{g}^>$, $\tilde{g}^<$ and g have the same analytic structure as the corresponding functions defined in Chap.2 only for fermions at T = 0. The quantity $\rho(E)$ can become a δ-function in a proper representation, while $A(\omega)$ cannot, in general, become a δ-function no matter what representation we choose. Finally, we mention once more that the equations obeyed by the many-body Green's functions are not as simple as those obeyed by the g's (or $\tilde{g}$'s) of Chap.2 because an extra term is present involving higher-order Green's functions.

3) In Table 9.1 we present the connection of our notation with that used in various books [8.3, 9.1-5] on the subjects. All these monographs deal mainly with the causal Green's function g.

9.2 Physical Significance and Use of the g's and $\tilde{g}$'s

In this section we examine the question of what physical information can be extracted from the various Green's functions.

The thermodynamic average of any quantity which is the sum of one-body terms can be expressed immediately in terms of the Green's functions. As we have mentioned in Appendix C such quantities can be written in the second quantized formalism as

$$F = \int d^3r \psi^\dagger(\underline{r}) F(\underline{r}) \psi(\underline{r}) \quad , \tag{9.21}$$

where $F(\underline{r})$ is the first quantized one-particle operator. Examples of operators F are the kinetic energy T with $T(\underline{r}) = -\nabla^2/2m$, the total number of particles N with $N(\underline{r}) = 1$ and the density operator $n(\underline{r}_0)$ with $n(\underline{r}) = \delta(\underline{r}-\underline{r}_0)$. It follows from the definition of $\tilde{g}^<$ that

$$\begin{aligned} \langle F \rangle &= \pm i \int d^3r F(\underline{r}) \tilde{g}^<(\underline{r},t,\underline{r},t) \\ &= \pm i \int d^3r F(\underline{r}) \lim_{t' \to t^+} g(\underline{r},t,\underline{r},t') \quad . \end{aligned} \tag{9.22}$$

Table 9.1. Connection of the notation used in the present work with that used in other books on the subject. The symbol (fT = 0) denotes the corresponding quantity for fermions at T = 0

Our notation	[8.3]	[9.1]	[9.2]	[9.3]	[9.4]	[9.5]
$g(\underline{k},\omega)$	$\bar{G}(\underline{k},\omega-\mu)$	$G(\underline{k},\omega-\mu)$	$G(\underline{k},\omega)$	-	$G(\underline{k},\omega)$	$G^T(\underline{k},\omega)$
$g(\underline{k},\omega)(fT = 0)$	$G(\underline{k},\omega)$	$G(\underline{k},\omega-\mu)$	-	$-G(\underline{k},\omega)$	$G(\underline{k},\omega)$	$G(\underline{k},\omega)$
$g^R(\underline{k},\omega)$	$\bar{G}^R(\underline{k},\omega-\mu)$	$G^R(\underline{k},\omega-\mu)$	-	-	$G^R(\underline{k},\omega)$	-
$g^R(\underline{k},\omega)(fT = 0)$	$G^R(\underline{k},\omega)$	$G_R(\underline{k},\omega-\mu)$	-	-	$G^R(\underline{k},\omega)$	-
$g^A(\underline{k},\omega)$	$\bar{G}^A(\underline{k},\omega-\mu)$	$G^A(\underline{k},\omega-\mu)$	-	-	-	-
$g^A(\underline{k},\omega)(fT = 0)$	$G^A(\underline{k},\omega)$	$G_A(\underline{k},\omega-\mu)$	-	-	-	-
$\tilde{g}(\underline{k},\omega)$	$-i\rho(\underline{k},\omega-\mu)$	$2\pi i\rho(\underline{k},\omega-\mu)$	$-iA(\underline{k},\omega)$	-	-	-
$\tilde{g}(\underline{k},\omega)(fT = 0)$	-	-	-	-	-	-
$A(\underline{k},\omega)$	$\rho(\underline{k},\omega-\mu)$	$-2\pi\rho(\underline{k},\omega-\mu)$	$A(\underline{k},\omega)$	-	-	-
$A(\underline{k},\omega)(fT = 0)$	-	-	-	-	-	-
$\tilde{g}^>(\underline{k},\omega)$	-	-	$G^>(\underline{k},\omega)$	-	$-iJ_1(\underline{k},\omega)$	-
$\tilde{g}^>(\underline{k},\omega)(fT = 0)$	$-2\pi iA(\underline{k},\omega-\mu)$	$-2\pi iA(\underline{k},\omega-\mu)$	-	$-2\pi iA_+(\underline{k},\omega-\mu)$	-	$-2\pi iA^+(\underline{k},\omega-\mu)$
$\tilde{g}^<(\underline{k},\omega)$	-	-	$G^<(\underline{k},\omega)$	-	$\mp iJ_2(\underline{k},\omega)$	-
$\tilde{g}^<(\underline{k},\omega)(fT = 0)$	$2\pi iB(\underline{k},\mu-\omega)$	$2\pi iB(\underline{k},\mu-\omega)$	-	$2\pi iA_-(\underline{k},\mu-\omega)$	-	$+2\pi iA^-(\underline{k},\mu-\omega)$
$G(\underline{k},z)$	$\Gamma(\underline{k},z-\mu)$	-	$G(\underline{k},z)$	-	$G(\underline{k},z)$	-
$G(\underline{k},z)(fT = 0)$	-	-	-	-	$G(\underline{k},z)$	-

More specifically, we have the kinetic energy $<T>$ and the density $<n(\underline{r})>$

$$<T> = \pm i \int d^3r \lim_{\substack{\underline{r}'\to\underline{r} \\ t'\to t^+}} \left[-\frac{\nabla^2}{2m} g(x,x')\right]$$

$$= \pm i \sum_{\underline{k}} \int_{-\infty}^{\infty} \frac{d\omega}{2\pi} \tilde{g}^<(\underline{k},\omega) \frac{k^2}{2m} = \sum_{\underline{k}} \int_{-\infty}^{\infty} \frac{d\omega}{2\pi} A(\underline{k},\omega) f_{\mp}(\omega) \frac{k^2}{2m} \tag{9.23}$$

$$<n(\underline{r})> = \pm i \lim_{t'\to t^+} g(\underline{r},t,\underline{r},t')$$

$$= \int \frac{d^3k}{(2\pi)^3} \int_{-\infty}^{\infty} \frac{d\omega}{2\pi} A(\underline{k},\omega) f_{\mp}(\omega) \quad . \tag{9.24}$$

The density in $\underline{k}$-space is

$$<n(\underline{k})> = <a^\dagger_{\underline{k}} a_{\underline{k}}> = \int_{-\infty}^{\infty} \frac{d\omega}{2\pi} A(\underline{k},\omega) f_{\mp}(\omega) \quad . \tag{9.25}$$

Operators which involve summation over pairs of particles such as the interaction energy V_i (see Appendix C) contain four ψ's (two ψ's and two $\psi^\dagger$'s). These operators can be expressed in terms of g_2 and not g. However, the interaction Hamiltonian V_i, which is connected through the Schrödinger equation with T, can be written in terms of g. To show this we use the basic equation (see Appendix C)

$$\left(i \frac{\partial}{\partial t} + \frac{\nabla^2}{2m}\right) \psi(\underline{r},t) = \int d^3r_1 v(\underline{r} - \underline{r}_1)\psi^\dagger(\underline{r}_1,t)\psi(\underline{r}_1,t)\psi(\underline{r},t) \tag{9.26}$$

and its adjoint

$$\left(-i \frac{\partial}{\partial t'} + \frac{\nabla'^2}{2m}\right)\psi^\dagger(\underline{r}',t') = \psi^\dagger(\underline{r}',t') \int d^3r_2 v(\underline{r}' - \underline{r}_2)\psi^\dagger(\underline{r}_2,t')\psi(\underline{r}_2,t') \quad . \tag{9.27}$$

Multiplying (9.26) from the left by $\psi^\dagger(\underline{r}',t')/4$ and (9.27) from the right by $\psi(\underline{r},t)/4$, subtracting the resulting equations, putting $t' = t$ and $\underline{r}' = \underline{r}$ and integrating over $\underline{r}$ we obtain

$$\frac{1}{4}\int d^3r \left[\left(i\frac{\partial}{\partial t} - i\frac{\partial}{\partial t'}\right)\psi^\dagger(\underline{r},t')\psi(\underline{r},t)\right]_{t=t'} = \frac{1}{2}\langle T\rangle + \langle V_i\rangle \quad . \tag{9.28}$$

Using (9.23) and the definition of g, we obtain from (9.28)

$$\langle V_i\rangle = \pm\frac{i}{2}\int d^3r \lim_{\substack{t'\to t^+\\ \underline{r}'\to \underline{r}}} \left[i\frac{\partial}{\partial t} + \frac{\nabla_r^2}{2m}\right] g(x,x')$$

$$= \sum_{\underline{k}} \int \frac{d\omega}{2\pi}\frac{1}{2}\left(\omega - \frac{k^2}{2m}\right)A(\underline{k},\omega)f_{\mp}(\omega) \quad . \tag{9.29}$$

Adding (9.23) and (9.29) we obtain

$$\langle H\rangle = \pm\frac{i}{2}\int d^3r \lim_{\substack{t'\to t^+\\ \underline{r}'\to \underline{r}}} \left(i\frac{\partial}{\partial t} - \frac{\nabla_r^2}{2m}\right) g(x,x')$$

$$= \sum_{\underline{k}} \int \frac{d\omega}{2\pi}\frac{1}{2}\left(\omega + \frac{k^2}{2m}\right) A(\underline{k},\omega)f_{\mp}(\omega) \quad . \tag{9.30}$$

To obtain all the other thermodynamic quantities it is enough to calculate the grand partition function $Z_G \equiv \mathrm{Tr}\{\exp[-\beta(H-\mu N)]\}$ as a function of the volume Ω, the chemical potential μ, and the temperature $T \equiv 1/k_B\beta$. Z_G is directly related to the pressure by the general thermodynamic equation

$$Z_G = e^{\beta\Omega P} \quad . \tag{9.31}$$

Furthermore, the pressure can be expressed in terms of the density as [9.2]

$$P(\beta,\mu) = \int_{-\infty}^{\mu} d\mu' n(\beta,\mu') \quad . \tag{9.32}$$

Substituting in (9.32) from (9.24) we have

$$P(\beta,\mu) = \int_{-\infty}^{\mu} d\mu' \int \frac{d^3k}{(2\pi)^3} \int \frac{d\omega}{2\pi} A(\underline{k},\omega)f_{\mp}(\omega) \quad , \tag{9.33}$$

where both $A(\underline{k},\omega)$ and $f_{\mp}(\omega)$ depend, in general, on the inverse temperature and the chemical potential μ'. Unfortunately, the integral over μ' can rarely be performed explicitly [9.2].

Another way to calculate Z_G (which avoids this difficulty) is based on the general thermodynamic relation [9.6]

$$\left\langle \frac{\partial H}{\partial \alpha} \right\rangle = - \Omega \left. \frac{\partial P}{\partial \alpha} \right|_{T,\mu} \tag{9.34}$$

where α is any parameter in the total Hamiltonian H.

We write for H

$$H = T + \alpha V_i \quad , \tag{9.35}$$

so that $\alpha = 0$ corresponds to the noninteracting case, and $\alpha = 1$ corresponds to the actual case. Taking into account (9.31,34,35) we obtain

$$-\beta \langle V_i \rangle = \frac{\partial}{\partial \alpha} [\ln(Z_G)] \quad . \tag{9.36}$$

Integrating (9.36) over α and using (9.29), we have finally

$$\ln(Z_G) = \beta P \Omega = \beta P_0 \Omega - \beta \int_0^1 \frac{d\alpha}{\alpha} \sum_{\underline{k}} \int \frac{d\omega}{2\pi} \frac{1}{2} \left(\omega - \frac{k^2}{2m} \right) A_\alpha(\underline{k},\omega) f_{\mp}(\omega) \quad , \tag{9.37}$$

where the subscript α in $A(\underline{k},\omega)$ denotes that $A_\alpha(\underline{k},\omega)$ corresponds to the Hamiltonian (9.35), and P_0 is the pressure for the noninteracting system.

Another class of important physical quantities which are related to the Green's functions are the so-called linear response functions.

A system is perturbed from equilibrium at $t = t_0$ by an external Hamiltonian $H_e(t)$ which has the form

$$H_e(t) = \int d^3r B(\underline{r},t) f(\underline{r},t) \quad , \tag{9.38}$$

where $B(\underline{r},t)$ is an operator and $f(\underline{r},t)$ is a c-number such that $f(\underline{r},t) = 0$ for $t < t_0$. We wish to find the first-order (in H_e) change $\delta\langle B(\underline{r},t)\rangle$ of the average value of the operator $B(\underline{r},t)$. The result is [8.3]

$$\delta\langle B(\underline{r},t)\rangle = \int_{-\infty}^{\infty} dt' \int d^3r' D^R(\underline{r},t,\underline{r}',t') f(\underline{r}',t') \quad . \tag{9.39}$$

Equation (9.39) is the analog of (2.52) with $\psi - \phi$ replaced by $\delta\langle B(\underline{r},t)\rangle$. The quantity $D^R(x,x')$ is

$$D^R(x,x') = -i\theta(t - t')\langle [B(x), B(x')] \rangle \quad , \tag{9.40}$$

i.e., it has the form of a retarded Green's function with the field operator ψ replaced by the operator B.

A specific example of the above is the case of an external electrostatic potential $\phi_e(\underline{r},t)$ applied to a system of electrons (each of charge e) placed in a uniform positive background (to ensure overall electrical neutrality). In this case, $B(\underline{r},t) = \delta n(\underline{r},t)$, where $\delta n(\underline{r},t)$ is the deviation of the electronic density from its equilibrium value n_0, $f(\underline{r},t) = e\phi_e(\underline{r},t)$ and $\delta\langle B(\underline{r},t)\rangle$ is $\langle\delta n(\underline{r},t)\rangle$. Thus the Fourier transform of (9.39) takes the form

$$\langle\delta n(\underline{k},\omega)\rangle = eD^R(\underline{k},\omega)\phi_e(\underline{k},\omega) \quad , \tag{9.41}$$

where $D^R(\underline{k},\omega)$ is the Fourier transform of $D^R(x,x')$ with respect to the variable $x - x'$ and

$$\begin{aligned} D^R(x,x') &= -i\theta(t - t')\langle[\delta n(x) \; , \; \delta n(x')]\rangle \\ &= -i\theta(t - t)\langle[(\psi^\dagger(x)\psi(x) - n_0), (\psi^\dagger(x')\psi(x') - n_0)]\rangle \\ &= -i\theta(t - t')\langle[\psi^\dagger(x)\psi(x), \psi^\dagger(x')\psi(x')]\rangle \quad . \end{aligned} \tag{9.42}$$

Equation (9.42) shows that $D^R(x,x')$ is related with the two-particle Green's function g_2. Taking into account that the potential ϕ_e is related to the external density n_e with Poisson's equation

$$\phi_e(\underline{k},\omega) = 4\pi e n_e(\underline{k},\omega)/k^2 \tag{9.43}$$

and that the longitudinal dielectric function $\varepsilon(\underline{k},\omega)$ is defined by

$$\varepsilon(\underline{k},\omega) \equiv \frac{en_e(\underline{k},\omega)}{en_e(\underline{k},\omega) + e\langle\delta n(\underline{k},\omega)\rangle} \quad , \tag{9.44}$$

we can express $\varepsilon(\underline{k},\omega)$ in terms of D^R as follows

$$\frac{1}{\varepsilon(\underline{k},\omega)} - 1 = \frac{4\pi e^2}{k^2} D^R(\underline{k},\omega) \quad . \tag{9.45}$$

Since the longitudinal conductivity $\sigma(\underline{k},\omega)$ is related with $\varepsilon(\underline{k},\omega)$ by [9.4]

$$\varepsilon(\underline{k},\omega) = 1 - \frac{4\pi}{i\omega}\sigma(\underline{k},\omega) \quad , \tag{9.46}$$

we conclude that the longitudinal conductivity can be expressed in terms D^R, i.e., in terms of g_2. The quantity $\delta n(x)$ in (9.42) is related to the current operator by the continuity equation; as a result D^R can be expressed in terms of the current - current commutator. This is a special case of the fluctuation-dissipation theorem [9.7] which connects the imaginary part of a response function (which determines the dissipation of energy) to the square of the fluctuation of the corresponding physical quantity. For more details on the evaluation of the transverse and longitudinal conductivity the reader is referred to [9.1,4] and [7.90].

As we can see from (9.23,24,29,30,33,37), the various thermodynamic quantities involve an integral of the type

$$I = \lim_{\sigma\to 0^-} I(\sigma) = \lim_{\sigma\to 0^-} \int \frac{d\omega}{2\pi} e^{-\omega\sigma} F(\omega) A(\underline{k},\omega) f_{\mp}(\omega) \quad , \tag{9.47}$$

where $F(\omega)$ is a polynomial; the factor $\exp(-\omega\sigma)$ was introduced in order to allow the transformations which follow. Taking into account (9.11) and subtracting the contribution of the pole of $f_-(\omega)$ at $\omega=\mu$ we can write for $I(\sigma)$

$$I(\sigma) + \frac{1}{\beta} e^{-z_0\sigma} F(z_0) G(\underline{k},z_0) = i \int_C \frac{d\omega}{2\pi} e^{-\omega\sigma} F(\omega) G(\underline{k},\omega) f_{\mp}(\omega)$$

$$= i \int_{C_R} \frac{d\omega}{2\pi} e^{-\omega\sigma} F(\omega) G(\underline{k},\omega) f_{\mp}(\omega) - i \int_{C_A} \frac{d\omega}{2\pi} e^{-\omega\sigma} F(\omega) G(\underline{k},\omega) f_{\mp}(\omega) \; , \tag{9.48}$$

where the contour C and the paths C_R and C_A are shown in Figs.9.1f,b and c, respectively. For fermions the last term of the lhs of (9.48) must be taken as zero. The integral over the path C_R, I_R, can be transformed as shown in Fig.9.2. The new path $C_1 + C_2 + C_3 + C_4$ avoids the poles of the integrand in the upper half ω-plane, these poles come from $f_{\mp}(\omega)$ and are given by

$$z_\nu = \mu + \frac{i\pi\nu}{\beta} \quad ; \quad \nu = 2n \text{ for bosons}$$
$$= 2n-1 \text{ for fermions} \tag{9.49}$$

where n is a positive integer. Assuming that

$$-\beta<\sigma<0 \quad , \tag{9.50}$$

we can easily show that the contributions from C_1 and C_4 vanish. Thus

$$I_R(\sigma) = \mp \frac{1}{\beta} \sum_{\nu>0} e^{-z_\nu\sigma} F(z_\nu) G(\underline{k},z_\nu) \quad ; \tag{9.51}$$

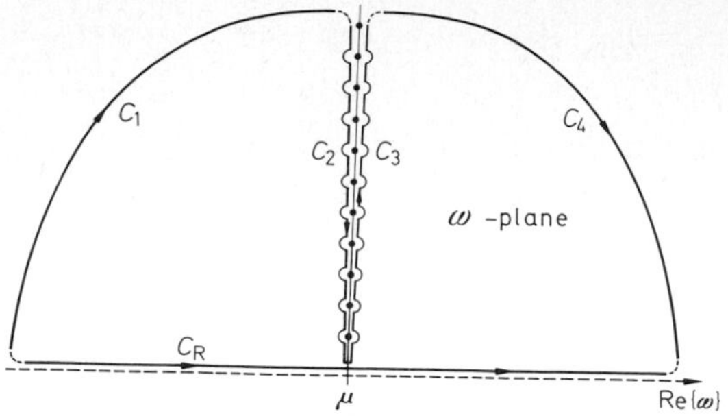

Fig.9.2. The integration of $\exp(-\omega\sigma)F(\omega)G(\underline{k},\omega)f_{\mp}(\omega)/2\pi$ (see text) along the path C_R can be equivalently performed along the composite path $C_1 + C_2 + C_3 + C_4$

in a similar way one shows that

$$I_A(\sigma) = \pm \frac{1}{\beta} \sum_{\nu<0} e^{-z_\nu \sigma} F(z_\nu) G(\underline{k}, z_\nu) \quad . \tag{9.52}$$

Combining (9.48) and (9.51,52) we have

$$I(\sigma) = \mp \frac{1}{\beta} \sum_{\nu} e^{-z_\nu \sigma} F(z_\nu) G(\underline{k}, z_\nu)$$

$$I = \mp \frac{1}{\beta} \sum_{\nu} F(z_\nu) G(\underline{k}, z_\nu) \quad . \tag{9.53}$$

Equation (9.53) shows that the thermodynamic quantities can be obtained from the values of $G(\underline{k},\omega)$ at the special points z_ν given by (9.49). The importance of this result stems from the fact that the quantities $G(\underline{k},z_\nu)$ are easier to calculate than $A(\underline{k},\omega)$. We shall return to this point in the next chapter.

The results of this section show that the quantity $A(\underline{k},\omega)/2\pi$ not only reduces in the absence of interactions to the density of states $\rho(\underline{k},\omega)$ per unit volume in $\underline{k}$-space, but it also allows the calculation of various thermodynamic quantities in a way similar to the way that $\rho(\underline{k},\omega)$ gives the thermodynamics of noninteracting particles. Because of this property we can interpret the quantity $A(\underline{k},\omega)/2\pi$ as a generalized density of states per unit frequency and unit volume in $\underline{k}$-space. Then with the aid of (9.7), $\tilde{g}^{<}(\underline{k},\omega)$ can be interpreted (apart from a factor $\pm i$) as the average number of particles per unit volume in $\underline{k}$-ω-space; similarly, with the aid of (9.6), $\tilde{g}^>(\underline{k},\omega)$ can be interpreted as the average number of states (per unit volume in $\underline{k}$-ω-space) *available* for the addition of an extra particle to the system.

In the absence of interactions and for a translationally invariant system, we have

$$A(\underline{k},\omega)/2\pi = \rho(\underline{k},\omega) = \delta(\omega - \varepsilon_{\underline{k}}^{0}) \tag{9.54}$$

where for Schrödinger particles $\varepsilon_{\underline{k}}^{0} = k^2/2m$. Equation (9.54) expresses simply the fact that in the absence of interactions a particle of momentum $\underline{k}$ necessarily has energy $\varepsilon_{\underline{k}}^{0}$. A formal proof of (9.54) was given in Chap.8. Substituting (9.54) in the basic formulas of Sects.9.1 and 9.2 we rediscover the results of Sect.8.3 and the well-known thermodynamic expressions for noninteracting particles.

9.3 Quasi-Particles

Let us assume that $A(\underline{k},\omega)$ has a sharp peak at $\omega = \varepsilon_{\underline{k}}$: for simplicity we assume further that the peak has a lorentzian shape, i.e.,

$$A(\underline{k},\omega) = A_b(\underline{k},\omega) + \frac{2|\gamma_{\underline{k}}|w_{\underline{k}}}{(\omega-\varepsilon_{\underline{k}})^2+\gamma_{\underline{k}}^2} \quad . \tag{9.55}$$

Here $A_b(\underline{k},\omega)$ is the smooth background, $\varepsilon_{\underline{k}}$ is the position of the maximum, $|\gamma_{\underline{k}}|$ is the width, and $w_{\underline{k}}$ $(0<w_{\underline{k}}\leq 1)$ is the weight of the peak. It is natural to attempt to interpret the peak as representing a quasi-particle of momentum $\underline{k}$, energy $\varepsilon_{\underline{k}}$ and lifetime $\tau_{\underline{k}} = 1/|\gamma_{\underline{k}}|$; $w_{\underline{k}}$ can be interpreted as the percentage of a real particle participating in the creation of the corresponding quasi-particle. The quasi-particle can be thought of as a dressed particle consisting in part $(w_{\underline{k}})$ by a bare (real) particle and in part $(1-w_{\underline{k}})$ by a cloud of other particles surrounding the bare one. As the interactions approach zero each quasi-particle tends to the corresponding bare particle, i.e.,

$$\varepsilon_{\underline{k}} \to \varepsilon_{\underline{k}}^{0} \quad , \quad \tau_{\underline{k}} \to \infty \, , \quad |\gamma_{\underline{k}}| \to 0^+ \quad , \quad w_{\underline{k}} \to 1 \quad , \quad A_b \to 0 \quad .$$

The concept of quasi-particle is of fundamental importance in analyzing the behavior of many-body systems; it allows us to replace a system of strongly interacting particles by an equivalent system of weakly interacting quasi-particles. Then in a first approximation the interactions among the quasi-particles can be omitted and the problem reduces to one which can be solved. A straightforward method to obtain the weakly interacting quasi-particles out of the strongly interacting particles is by a canonical transformation; this method, although conceptionally simple, is extremely difficult in practice since for every particular system one must find the appropriate canonical transformation. On the other hand, the Green's function method provides a general way to recognize the existence of weakly interacting quasi-particles as peaks in $A(\underline{k},\omega)$. Furthermore, it gives basic information about the quasi-particles, i.e., it gives its energy as a function of its momentum; it gives its finite lifetime resulting from the weak interactions with the other quasi-particles; and finally, it gives what percentage of the quasi-particle is made up of bare particle and what percentage is made up of cloud. For a more precise statement of the last sentence the reader is referred to NOZIERES [9.3].

A peak in $A(\underline{k},\omega)$ of the type shown in (9.55) will appear as a pole in the analytic continuation of $G(\underline{k},\omega)$ across the discontinuity along the real ω-axis. The analytic continuation of $G(\underline{k},\omega)$, a.c.$\{G(\underline{k},\omega)\}$, to the lower ω-half plane is, according to (9.11),

$$\text{a.c.}\{G(\underline{k},\omega)\} = G(\underline{k},\omega) - i\,\text{a.c.}\{A(\underline{k},\omega)\} \;;\quad \text{Im}\{\omega\}<0 \;; \tag{9.56}$$

similarly,

$$\text{a.c.}\{G(\underline{k},\omega)\} = G(\underline{k},\omega) + i\,\text{a.c.}\{A(\underline{k},\omega)\} \;;\quad \text{Im}\{\omega\}>0 \;. \tag{9.57}$$

Substituting (9.55) in (9.56,57) we obtain

$$\text{a.c.}\{G(\underline{k},\omega)\} = G(\underline{k},\omega)\} \pm i\,\text{a.c.}\{A_b(\underline{k},\omega)\} \pm \frac{2\gamma_{\underline{k}} w_{\underline{k}} i}{(\omega-\varepsilon_{\underline{k}})^2+\gamma_{\underline{k}}}$$

$$= G(\underline{k},\omega) \pm i\,\text{a.c.}\{A_b(\underline{k},\omega)\} \pm \frac{2\gamma_{\underline{k}} w_{\underline{k}} i}{(\omega-z_{\underline{k}})(\omega-z^*_{\underline{k}})} \;;\quad \text{Im}\{\omega\}\gtrless 0 \tag{9.58}$$

where

$$z_{\underline{k}} = \varepsilon_k + i\gamma_{\underline{k}} \quad . \tag{9.59}$$

Thus, a sharp peak of lorentzian shape in $A(\underline{k},\omega)$ appears as a pole (near the real axis) of the analytical continuation a.c.$\{G(\underline{k},\omega)\}$ of $G(\underline{k},\omega)$ across the real axis. The real part of the pole gives the energy of the quasi-particle, the imaginary part gives the inverse lifetime and the residue equals the weight $w_{\underline{k}}$. Taking into account (9.14,15) and (9.58) we have

$$\begin{aligned} \text{a.c.}\{g^R(\underline{k},\omega)\} &= G(\underline{k},\omega) && ; \quad \text{Im}\{\omega\}>0 \\ &= G(\underline{k},\omega) - i\,\text{a.c.}\{A_b\} - \frac{2\gamma_{\underline{k}} w_{\underline{k}} i}{(\omega-z_{\underline{k}})(\omega-z^*_{\underline{k}})} && ; \quad \text{Im}\{\omega\}<0 \end{aligned} \tag{9.60}$$

$$\begin{aligned} \text{a.c.}\ g^A(\underline{k},\omega) &= G(\underline{k},\omega) + i\,\text{a.c.}\{A_b\} + \frac{2\gamma_{\underline{k}} w_{\underline{k}} i}{(\omega-z_{\underline{k}})(\omega-z^*_{\underline{k}})} && ; \quad \text{Im}\{\omega\}>0 \\ &= G(\underline{k},\omega) && ; \quad \text{Im}\{\omega\}<0 \quad . \end{aligned} \tag{9.61}$$

Equations (9.60,61) mean that a pole in the analytic continuation of $g^R(\underline{k},\omega)[g^A(\underline{k},\omega)]$ in the lower (upper) ω-half plane describes a quasi-particle. The closer the pole to the real ω-axis the better the quasi-particle is defined.

For fermions at $T = 0$ we obtain with the aid of (9.19') that the pole of the analytic continuation of $g(\underline{k},\omega)$, a.c.$\{g(\underline{k},\omega)\}$, appears if at all, in the upper ω-half plane when $\text{Re}\{\omega\}<\mu$ and in the lower ω-half plane when $\text{Re}\{\omega\}>\mu$, i.e.,

$$\begin{aligned} \text{a.c.}\{g(\underline{k},\omega)\} &= G(\underline{k},\omega) \quad ; \quad \text{Re}\{\omega\}>\mu \ , \quad \text{Im}\{\omega\}>0 \\ &= \text{a.c.}\{G(\underline{k},\omega)\} \quad ; \quad \text{Re}\{\omega\}<\mu \ , \quad \text{Im}\{\omega\}>0 \\ &= G(\underline{k},\omega) \quad ; \quad \text{Re}\{\omega\}<\mu \ , \quad \text{Im}\{\omega\}<0 \\ &= \text{a.c.}\{G(\underline{k},\omega)\} \quad ; \quad \text{Re}\{\omega\}>\mu \ , \quad \text{Im}\{\omega\}<0 \end{aligned} \tag{9.62}$$

where a.c.$\{G(\underline{k},\omega)\}$ is given by (9.58). The function a.c.$\{g(\underline{k},\omega)\}$ as defined by (9.62) has a branch cut along the line $\text{Re}\{\omega\} = \mu$.

Furthermore, a peak in $A(\underline{k},\omega)$ of the type shown in (9.55) appears as a peak in $\text{Im}\{\tilde{g}^<(\underline{k},\omega)\}$when $\varepsilon_{\underline{k}}<\mu$ or as a peak in $-\text{Im}\{\tilde{g}^>(\underline{k},\omega)\}$ when $\varepsilon_{\underline{k}}>\mu$. Since $\tilde{g}^<(\tilde{g}^>)$ describes the propagation of a hole (particle), it follows that when

$\varepsilon_{\underline{k}}<\mu$, the pole in a.c.$\{g(\underline{k},\omega)\}$ which lies in the upper half plane describes a quasi-hole; when $\varepsilon_{\underline{k}}>\mu$, the pole in a.c.$\{g(\underline{k},\omega)\}$ which lies in the lower ω-half plane describes a quasi-particle. This situation is summarized in Fig.9.3. When k is very small the elementary excitation is expected to be quasi-hole; this means that for very small k the pole $z_{\underline{k}}$ lies in region II of Fig.9.3. On the other hand, for very large k the pole lies in the region IV. Assuming that the trajectory defined by $z_{\underline{k}}$ as k increases from zero is a continuous one, it follows that this trajectory must cross the real ω-axis at the point $\omega = \mu$. Thus, a characteristic surface in k-space, which is called the Fermi surface, can be defined by the relation

$$\varepsilon_{\underline{k}_F} = \mu \quad . \tag{9.63}$$

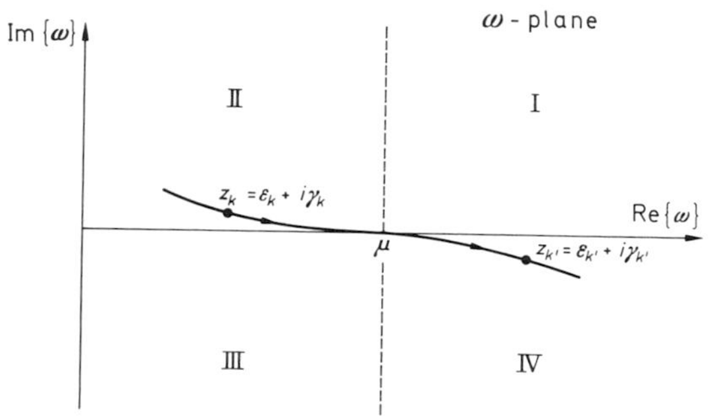

Fig.9.3. Analytic structure in the complex ω-plane of the analytic continuation of $g(\underline{k},\omega)$, a.c.$\{g(\underline{k},\omega)\}$, as given by (9.62) for fermions at T=0. A pole or any other singularity will appear either in region II (corresponding to a quasi-hole) or in region IV (corresponding to a quasi-particle). As k increases the pole moves along a trajectory an example of which is shown. The function a.c.$\{g(\underline{k},\omega)\}$ has a branch cut along the line $\mathrm{Re}\{\omega\} = \mu$

For a rotationally invariant system the Fermi surface is a sphere of radius k_F; k_F is called the Fermi momentum. The Fermi surface as defined above has (under certain conditions) one property of the Fermi surface of a system of noninteracting particles, namely the quantity $\langle n(\underline{k})\rangle$ is discontinuous at $k = k_F$. To show this, use (9.25) and (9.55) and assume that the slope of the pole trajectory is continuous at μ and of magnitude $s = \lim(|\gamma_{\underline{k}}|/|\varepsilon_{\underline{k}}-\mu|)$ as $k \to k_F$. Then we obtain

$$\langle n(k_F^-)\rangle - \langle n(k_F^+)\rangle = \frac{2w_{k_F}}{\pi} \tan^{-1} \frac{1}{s} \quad . \tag{9.64}$$

Thus, if the slope is zero (as in Fig.9.3), the discontinuity equals to w_{k_F}. If the slope is infinite or if the trajectory is discontinuous at k_F, the difference $\langle n(k_F^-)\rangle - \langle n(k_F^+)\rangle$ is zero and the Fermi surface loses its usual physical meaning. In Fig.9.4 we plot $\langle n(k)\rangle$ versus k for the case where s = 0. Systems for which the trajectory of the pole is continuous with a zero slope at μ are called normal.

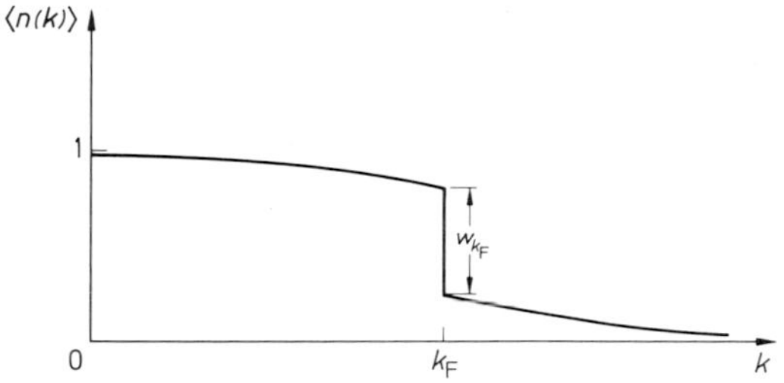

Fig.9.4. The average particle number per unit volume in $\underline{k}$-space for a normal interacting Fermi system at T = 0; k_F is the Fermi momentum

We show now that for normal systems the quasi-particles (or quasi-holes) near the Fermi surface can be used to analyze not only the thermodynamic behavior of the system but the dynamic behavior as well. For this purpose we try to calculate $g(\underline{k},t)$, which for t>0 describes the propagation of an extra particle added to the system at t = 0. As was mentioned before, for fermions at T = 0

$$g(\underline{k},t) = \int_C \frac{d\omega}{2\pi} e^{-i\omega t} G(\underline{k},\omega) = \int_C \frac{d\omega}{2\pi} e^{-i\omega t} \, \text{a.c.}\{g(\underline{k},\omega)\} \tag{9.65}$$

where the path C is shown in Fig.9.1a and (9.62) was used. For t>0 the path can be closed by a semicircle in the lower ω-half plane. The resulting contour can be deformed as shown in Fig.9.5. We have assumed that there is a pole at z_k, and the rest of the singularities in region IV of the a.c.$\{g(\underline{k},\omega)\}$, if any, are below the line $\text{Im}\{\omega\} = -\Gamma$. The contribution from this line contain a factor $\exp(-\Gamma t)$, and, consequently, it is negligible for $t \gg 1/\Gamma$. The contribution from the pole is

$$-iw_{\underline{k}}\, e^{(-i\varepsilon_{\underline{k}}t - |\gamma_{\underline{k}}|t)} \tag{9.66}$$

and the contribution from the contour around the branch cut is according to (9.62) and (9.58)

$$\int_{\mu}^{\mu-i\infty} \frac{d\omega}{2\pi} e^{-i\omega t}[\text{a.c.}\{G(\underline{k},\omega)\} - G(\underline{k},\omega)]$$

$$\simeq \int_{\mu}^{\mu-i\infty} \frac{d\omega}{2\pi} e^{-i\omega t} \frac{-2\gamma_{\underline{k}}w_{\underline{k}}i}{(\omega-z_{\underline{k}})(\omega-z_{\underline{k}}^*)} \quad . \tag{9.67}$$

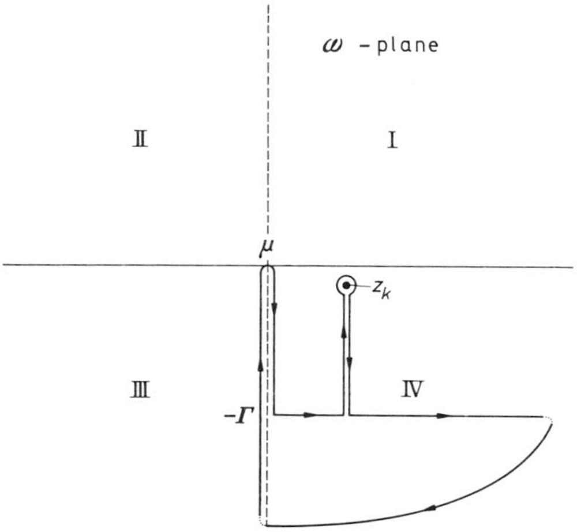

Fig.9.5. Contour in ω-plane for evaluating $g(\underline{k},t)$ for $t>0$

In obtaining the last expression we have omitted the quantity $-i\text{a.c.}\{A_b(\underline{k},\omega)\}$ in (9.58) because it is small in comparison with the last term in (9.58) when the pole $z_{\underline{k}}$ is close to the point $\omega = \mu$. The contribution (9.67) is comparable to the contribution (9.66) when $t \lesssim 1/(\varepsilon_{\underline{k}} - \mu)$. For $t \gg 1/(\varepsilon_{\underline{k}} - \mu)$ the contribution (9.67) becomes about equal to

$$\frac{\gamma_{\underline{k}}w_{\underline{k}}e^{-i\mu t}}{\pi t(\varepsilon_{\underline{k}}-\mu)^2} = \frac{w_{\underline{k}}e^{-i\mu t}}{\pi t(\varepsilon_{\underline{k}}-\mu)} \frac{\gamma_{\underline{k}}}{\varepsilon_{\underline{k}}-\mu} \quad . \tag{9.68}$$

Since $t(\varepsilon_{\underline{k}} - \mu) \gg 1$ and $\gamma_{\underline{k}}/(\varepsilon_{\underline{k}} - \mu) \to 0$ as $k \to k_F$, it follows that the contribution (9.68) is negligible. We can thus conclude that for

$$\max\left\{\frac{1}{\Gamma}\ ,\ \frac{1}{|\varepsilon_{\underline{k}}-\mu|}\right\} \ll t \ll \frac{1}{|\gamma_{\underline{k}}|} \tag{9.69}$$

the propagator $g(\underline{k},t)$ is given by

$$g(\underline{k},t) \simeq -iw_k\, e^{(-i\varepsilon_{\underline{k}}t - |\gamma_{\underline{k}}|t)} \quad , \tag{9.70}$$

which is the propagator of an entity of weight w_k, of energy ε_k and of lifetime $1/|\gamma_k|$, i.e., of the quasi-particle we have already introduced. We can interpret the results (9.69,70) physically as follows: following the addition of a new bare particle in our system, a time or order $1/|\varepsilon_k - \mu|$ is required for its dressing, i.e., for the appearance of the quasi-particle. After the quasi-particle is formed, it behaves as an independent entity for a period of the order $1/|\gamma_k|$. For a normal system $1/|\varepsilon_k - \mu| \ll 1/|\gamma_k| \to \infty$ as $k \to k_F$. Hence, for a normal system there is considerable time span when the quasi-particles (and quasi-holes) lying close to the Fermi surface are well-defined independent elementary excitations of the system. Further considerations supporting this last statement can be found in [9.3].

Physically we can understand the existence of long-lived quasi-particles because the available phase space for decaying processes is proportional to $(k - k_F)^2$ as $k \to k_F$ [9.1] [while $(\varepsilon_k - \mu) \sim (k - k_F)$ as $k \to k_F$]. This $(k-k_F)^2$ result for the available phase space is based upon the conservation of energy and momentum and upon the assumption that the quasi-particles are fermions with a well-defined Fermi surface. Because of this last assumption the argument above only shows that the normal state is a consistent state of an interacting Fermi system and not necessarily the actual one. However, if this consistent state is the actual one for very small α(as in the case where the perturbation expansion in powers of αV_i converges) by a continuity argument, one expects that it may remain so for the real value $\alpha = 1$.

Examples of normal systems are: a dense system of electrons repelling each other via Coulomb forces and moving in a positive background to ensure overall electrical neutrality; a system of fermions interacting through short range repulsive forces. On the other hand, a system of fermions attracting each other is not a normal one. We have seen in Chap.6 that no matter how weak the attraction is, the system rearranges itself to a new state, the superconducting state, consisting of bound pairs.

Although the poles of a.c.$\{G(\underline{k},\omega)\}$ lying near the real ω-axis give the quasi-particles (or quasi-holes), which are independent elementary excitations of the system, not all the elementary excitations appear as poles of a.c.$\{G(\underline{k},\omega)\}$. Excitations made out of pairs of particles must appear as poles of a.c.$\{g_2\}$, which describes the propagation of two particles. Also, even in normal systems there are excitations, such as plasmons or zero sound, which are elementary density waves. Such collective excitations must appear as poles of a.c.$\{D(\underline{k},\omega)\}$, where $D(x,x') \equiv -i\langle T[n(x),n(x')]\rangle$ ($n(x) = \psi^\dagger(x)\psi(x)$ is the density operator). Thus, in a typical normal system the role of the interactions is twofold: first, the bare particles are dressed and they appear as independent quasi-particles characterized by a new dispersion relation $\varepsilon = f(\underline{k})$; second, new collective density wave-type excitations are created. In a system that is not normal the interactions play a more drastic role creating a new ground state, which, in general, is fundamentally different from the ground state of the noninteracting system.

9.4 Summary

Properties

The Green's functions defined in Chap.8 for a many-body system have similar but not identical analytical properties to the Green's functions defined in Chap.2 corresponding to a second-order (in time) differential equation. For a translationally invariant system the Fourier transforms of the g's and $\tilde{g}$'s can be expressed in terms of a single real quantity $A(\underline{k},\omega)$ which can be interpreted as 2π times a generalized density of states in $\underline{k}$-ω-space. We have for real ω

$$\tilde{g}(\underline{k},\omega) = -iA(\underline{k},\omega) \quad , \tag{9.5}$$

$$g^R(\underline{k},\omega) = \lim_{s\to 0^+} G(\underline{k},\omega+is) \quad , \tag{9.14}$$

$$g^A(\underline{k},\omega) = \lim_{s\to 0^+} G(\underline{k},\omega-is) \quad , \tag{9.15}$$

$$\tilde{g}^>(\underline{k},\omega) = -iA(\underline{k},\omega)[1 \pm f_{\mp}(\omega)] \quad , \tag{9.6}$$

$$\tilde{g}^<(\underline{k},\omega) = \mp iA(\underline{k},\omega)f_{\mp}(\omega) \quad , \tag{9.7}$$

$$g(\underline{k},\omega) = g^R(\underline{k},\omega) + \tilde{g}^<(\underline{k},\omega) \quad , \qquad (9.18,19)$$

where, for a complex ω,$G(\underline{k},\omega)$ is defined as

$$G(\underline{k},\omega) = \int_{-\infty}^{\infty} \frac{d\omega'}{2\pi} \frac{A(\underline{k},\omega')}{\omega - \omega'} \qquad (9.10)$$

and

$$f_{\mp}(\omega) = [e^{\beta(\omega-\mu)} \mp 1]^{-1} \quad . \qquad (9.8)$$

$A(\underline{k},\omega)$ can be expressed in terms of G as follows

$$A(\underline{k},\omega) = i \lim_{s\to 0^+}[G(\underline{k},\omega+is) - G(\underline{k},\omega-is)] \;;\quad \omega \text{ real} \quad . \qquad (9.11)$$

Equations (9.5,14,15,10, and 11) are the analogs of (1.22), (2.34), (2.35), (1.30), and (1.21), respectively. Equations (9.18,19) reduce to (2.33) only for fermions at T = 0 and with $\omega = \mu$ instead of $\omega = 0$; similarly, for (9.6,7). The analogy can be seen also by comparing Fig.9.1 with Fig.2.2.

Use

1) Having the generalized density of states $A(\underline{k},\omega)/2\pi$, one can obtain several quantities of physical interest, such as the density of particles in $\underline{k}$-space, $\langle n(\underline{k})\rangle$, the total kinetic energy $\langle T\rangle$, the total potential energy $\langle V_i\rangle$, the total Hamiltonian $\langle H\rangle$, and the grand partition function Z_G. We have explicitly

$$\langle n(\underline{k})\rangle = \int_{-\infty}^{\infty} \frac{d\omega}{2\omega} A(\underline{k},\omega) f_{\mp}(\omega) \qquad (9.25)$$

$$\langle T\rangle = \sum_{\underline{k}} \int_{-\infty}^{\infty} \frac{d\omega}{2\pi} \frac{k^2}{2m} A(\underline{k},\omega) f_{\mp}(\omega) \qquad (9.23)$$

$$\langle V_i\rangle = \sum_{\underline{k}} \int_{-\infty}^{\infty} \frac{d\omega}{2\pi} \frac{\omega - k^2/2m}{2} A(\underline{k},\omega) f_{\mp}(\omega) \qquad (9.29)$$

$$\langle H\rangle = \sum_{\underline{k}} \int_{-\infty}^{\infty} \frac{d\omega}{2\omega} \frac{\omega + (k^2/2m)}{2} A(\underline{k},\omega) f_{\mp}(\omega) \qquad (9.30)$$

$$\ln Z_G = \ln Z_{Go} - \beta \int_0^1 \frac{d\alpha}{\alpha} \sum_{\underline{k}} \int_{-\infty}^{\infty} \frac{d\omega}{2\pi} \frac{\omega - k^2/2m}{2} A_\alpha(\underline{k},\omega) f_{\mp}(\omega) \quad , \tag{9.37}$$

where $A_\alpha(\underline{k},\omega)$ corresponds to an interaction part in the Hamiltonian equal to αV_i, and Z_{Go} is the grand partition function for $\alpha = 0$.

2) The poles of the analytic continuation of $G(\underline{k},\omega)$, a.c. $\{G(\underline{k},\omega)\}$, in the complex ω-plane can be interpreted as representing quasi-particles or dressed particles, i.e., weakly interacting entities determining the low-lying excitation spectrum of the many body system. The real part of the pole gives the energy of the quasi-particle, the inverse of the imaginary part gives its lifetime, and the residue the percentage of the dressed particle consisting of a bare (actual) particle.

3) Green's functions give also the linear response of the system to an external perturbation. For example, the dielectric function $\varepsilon(\underline{k},\omega)$ is given by

$$\frac{1}{\varepsilon(\underline{k},\omega)} - 1 = \frac{4\pi e^2}{k^2} D^R(\underline{k},\omega) \tag{9.45}$$

where D^R is a retarded Green's function.

10. Calculational Methods for g

There are two basic approaches to the approximate calculation of Green's functions. One is based upon the differential equation obeyed by g. In the other a perturbation expansion is employed where g is expressed as a series, the terms of which involve the unperturbed g_0 and the interaction potential $v(\underline{r}-\underline{r}')$.

10.1 Equation of Motion Method

As was mentioned in Chap.8, the Green's functions for an interacting many-body system obey a hierarchy of equations the first of which has the form

$$\left(i\frac{\partial}{\partial t}+\frac{\nabla_r^2}{2m}\right)g(x,x')=\delta(x-x')\pm i\int d^3r_1 v(\underline{r}-\underline{r}_1)g_2(x,x_1;x',x_1^+)_{t_1=t} \tag{10.1}$$

and connects the 1-particle causal Green's function with the 2-particle causal Green's function; the next equation in the hierarchy connects g_2 with g_3; and so on. A similar hierarchy of equations is obeyed by the retarded and advanced Green's functions [8.6]. The infinite number of coupled differential equations shows clearly the essential complication introduced by the interparticle interactions. It is obvious that in order to obtain a solution, one has to terminate the hierarchy at some point by employing an approximate relation connecting g_n with g_{n-1}, g_{n-2}, etc. The simplest such approximation attempts to express g_2 in terms of g and then substitute in (10.1). Thus a nonlinear integro-differential equation for g will result. We consider first the simplest approximate relation expressing g_2 in terms of g, which is equivalent to the so-called Hartree approximation,

$$g_2(x_1,x_2;x_1',x_2') \simeq g(x_1,x_1')g(x_2,x_2') \quad . \tag{10.2}$$

Remembering the physical interpretation of g_2 as describing the propagation of two additional particles from x_1, x_2 to x_1',x_2', we can see that (10.2) means that the two particles propagate independently, one from x_1 to x_1' and the other from x_2 to x_2'. Note that (10.2) does not satisfy the basic symmetry property demanding that g_2^2 is invariant under the exchange $x_1 \rightleftharpoons x_2$ or under the exchange $x_1' \rightleftharpoons x_2'$. Thus (10.2) is a quite drastic approximation indeed.

Substituting (10.2) in (10.1) we obtain

$$\left(i\frac{\partial}{\partial t} + \frac{\nabla_r^2}{2m}\right)g(x,x') = \delta(x - x') \pm ig(x,x') \int d^3r_1 v(\underline{r} - \underline{r}_1)g(x_1,x_1^+)$$

$$= \delta(x - x') + g(x,x') \int d^3r_1 v(\underline{r} - \underline{r}_1)\langle n(\underline{r}_1)\rangle \tag{10.3}$$

where (9.24) was used. Introducing the effective one-body potential

$$V(\underline{r}) \equiv \int d^3r_1 v(\underline{r} - \underline{r}_1)\langle n(\underline{r}_1)\rangle \quad , \tag{10.4}$$

we can rewrite (10.3) as

$$\left[i\frac{\partial}{\partial t} + \frac{\nabla_r^2}{2m} - V(\underline{r})\right]g(x,x') = \delta(x - x') \quad , \tag{10.5}$$

which shows that the added particle (or hole) moves (independently) in the average potential $V(\underline{r})$ created by the particles of the system. For a translationally invariant system, the density $\langle n(\underline{r})\rangle$ is a constant n_0, and as a result $V(\underline{r})$ is a constant $n_0 v_0$ where $v_0 = \int d^3r\, v(\underline{r})$. Then (10.5) becomes a simple differential equation which by a Fourier transformation can be written as

$$(\omega - k^2/2m - n_0 v_0)g(\underline{k},\omega) = 1 \quad . \tag{10.6}$$

The general solution of (10.6) is the sum of a particular solution plus the general solution of the corresponding homogeneous equation, which is proportional to $\delta(\omega - k^2/2m - n_0 v_0)$. In Chap.1, to get rid of this indeterminacy, we have allowed ω to become complex, in which case the homogeneous equation has no solution, and then we have taken the limit of the solution as ω approaches the real axis. This method works here for *fermions* at $T = 0$ where $g(\underline{k},\omega)$ is the limit of an analytic function as can be seen from (9.19'). Thus for fermions at T = 0 we have by combining (10.6) with (9.19') that

$$g(\underline{k},\omega) = \lim_{s \to 0^+} \frac{1}{\omega-k^2/2m-n_o v_o+is\bar{\varepsilon}(\omega-\mu)} \quad . \tag{10.7}$$

For $T \neq 0$, $g(\underline{k},\omega)$ is not the limit of an analytic function, and consequently one cannot continue (10.6) in the complex ω-plane without further analysis. This point shows why the equation of motion method works well for g^R (or g^A), since the latter is the limit of an analytic function for all temperatures [see (9.14,15)]. There is also a trick which allows us to use the equation of motion method for the "imaginary time" causal Green's functions. Consider the quantity

$$\begin{aligned} g(\underline{r}_1, -i\sigma_1,\underline{r}_2, -i\sigma_2) &= \tilde{g}^>(\underline{r}_1, -i\sigma_1,\underline{r}_2, -i\sigma_2) \quad , \quad \sigma_1>\sigma_2 \\ &= \tilde{g}^<(\underline{r}_1, -i\sigma_1,\underline{r}_2, -i\sigma_2) \quad , \quad \sigma_2>\sigma_1 \end{aligned} \tag{10.8}$$

resulting by replacing in the causal Green's function $g(\underline{r}_1,t_1,\underline{r}_2,t_2)$ the time t_1 by $-i\sigma_1$, the time t_2 by $-i\sigma_2$, and the time ordering by σ-ordering; the quantities σ_1, σ_2 are confined in the interval $[0,\beta]$ where β is the inverse temperature. For the systems we consider, $g(\underline{r}_1,-i\sigma_1,\underline{r}_2,-i\sigma_2)$ is a function of $\underline{r} = \underline{r}_1 - \underline{r}_2$ and $\sigma = \sigma_1 - \sigma_2$. Using the relation

$$\tilde{g}^<(\underline{r}_1,t_1,\underline{r}_2,t_2) = \int_{-\infty}^{\infty} \frac{d\omega}{2\pi} \int \frac{d^3k}{(2\pi)^3} e^{-i\omega(t_1-t_2)+i\underline{k}(\underline{r}_1-\underline{r}_2)}\tilde{g}^<(\underline{k},\omega)$$

together with (10.8) and (9.7), we obtain

$$\tilde{g}^<(\underline{r}_1,-i\sigma_1,\underline{r}_2,-i\sigma_2) = \mp i \int_{-\infty}^{\infty} \frac{d\omega}{2\pi} \int \frac{d^3k}{(2\pi)^3} e^{-\omega\sigma} e^{i\underline{k}(\underline{r}_1-\underline{r}_2)} A(\underline{k},\omega) f_{\mp}(\omega) \tag{10.9}$$

when $\sigma<0$. From the general equation (9.53) we have that

$$g(\underline{r}_1,-i\sigma_1,\underline{r}_2,-i\sigma_2) = \mp i \int \frac{d^3k}{(2\pi)^3} e^{i\underline{k}(\underline{r}_1-r_2)} \frac{\mp 1}{\beta} \sum_{\nu} e^{-z_\nu\sigma} G(\underline{k},z_\nu) \quad ,$$

which shows that the Fourier transform of $g(\underline{r}_1,-i\sigma_1,\underline{r}_2,-i\sigma_2)$ with respect to $\underline{r}_1 - \underline{r}_2$, $g(\underline{k},-i\sigma)$, obeys the relation

$$g(\underline{k}, -i\sigma) = \frac{i}{\beta}\sum_{\nu} e^{-z_\nu\sigma}G(\underline{k},z_\nu) \tag{10.10}$$

where z_ν is given by (9.49). In a similar way one obtains that (10.10) is valid also for $\sigma>0$. Equation (10.10) shows that the imaginary time Green's function $g(\underline{k},-i\sigma)$ can be expanded in a Fourier series the coefficients of which are the values of the analytic function $G(\underline{k},\omega)$ at the points z_ν. Inverting (10.10) we have

$$G(\underline{k},z_\nu) = -\frac{i}{2}\int_{-\beta}^{\beta} g(\underline{k},-i\sigma)\, e^{z_\nu\sigma}d\sigma \quad . \tag{10.11}$$

From (10.10) it follows immediately that

$$g(\underline{k},-i\sigma) = \pm\, e^{\beta\mu}\, g(\underline{k},-i(\sigma+\beta)) \quad ; \quad \sigma<0 \quad . \tag{10.12}$$

From the definition (10.8), it follows that $g(\underline{r}_1,-i\sigma_1,\underline{r}_2,-i\sigma_2)$ obeys the basic equation (10.1) with each t replaced by $-i\sigma$ and $\delta(t_1 - t_2)$ replaced by $\delta(\sigma_1 - \sigma_2)$. Using (10.2) for the imaginary time Green's function, substituting in (10.1) and employing (10.10), we obtain

$$(z_\nu - k^2/2m - n_o v_o)G(\underline{k},z_\nu) = 1 \quad . \tag{10.13}$$

Because of the analyticity of $G(\underline{k},\omega)$ in the complex ω-plane, it follows immediately from (10.13) that

$$G(\underline{k},z_\nu) = \frac{1}{z_\nu-k^2/2m-n_o v_o} \quad , \tag{10.14}$$

which by analytic continuation gives

$$G(\underline{k},\omega) = \frac{1}{\omega-k^2/2m-n_o v_o} \quad . \tag{10.15}$$

From $G(\underline{k},\omega)$ one can obtain all the real time Green's functions by employing the general relations of Sect.9.1. The above analysis shows that the imaginary time causal Green's functions are very convenient tools because, on the one hand, they obey the hierarchy of equations of motion and, on the other hand, their Fourier coefficients are the values of the analytic function $G(\underline{k},\omega)$ at the points $\omega = z_\nu$.

To summarize the calculational procedure for $T \neq 0$:

1) The imaginary time causal Green's functions (resulting by substituting $t_1 = i\sigma_1$, etc., in g, g_2, ...) obey a hierarchy of equations the first of which is (10.1) with $t = -i\sigma$, ... and $\delta(t - t') = \delta(\sigma - \sigma')$.

2) The hierarchy is approximately terminated usually by expressing g_2 in terms of g. There is a systematic procedure for obtaining increasingly more accurate expressions of g_2 in terms of g [9.2].

3) Combining (10.1) and the approximate relation $g_2 \simeq f(g)$ one obtains an equation for g, which in view of (10.10) becomes an equation for $G(\underline{k},z_\nu)$. The solution of this equation determines the analytic function $G(\underline{k},\omega)$ at the points $\omega = z_\nu$. As we have seen in Sect.9.2, the values of $G(\underline{k},z_\nu)$ is all we need in order to calculate the thermodynamic quantities.

4) If the real time Green's functions are needed, one has to obtain $G(\underline{k},\omega)$ by analytically continuing $G(\underline{k},z_\nu)$ and by taking into account that $G(\underline{k},\omega) \to 1/\omega$ as $\omega \to \infty$ [8.3]. This asymptotic behavior is a consequence of (9.9,10). Determining $G(\underline{k},\omega)$ from its values at the points $\omega = z_\nu$ may be a difficult problem; however, in most practical cases it involves no more than substituting z_ν by ω. Having $G(\underline{k},\omega)$, one can obtain all Green's functions by employing the equations of Sect.9.1.

The above procedure was illustrated in the simple case of the Hartree approximation (10.2). As can be seen from (10.15), within the framework of this approximation a quasi-particle of momentum $\underline{k}$ has energy $\varepsilon_k = k^2/2m + n_o v_o$, infinite lifetime and weight equal to 1, i.e., it is a free particle with the added energy $n_o v_o$. The average energy is obtained by combining (9.30,11 and 10.15); so

$$\langle H \rangle = \sum_{\underline{k}} \varepsilon_k f_{\mp}(\varepsilon_k) - \frac{1}{2} \Omega n_o^2 v_o \qquad (10.16)$$

where the last term corrects for the double counting of the interaction $n_o v_o$. The equation of state in the low-density limit is [9.2]

$$P - \frac{1}{2} n_o^2 v_o = n_o k_B T \quad , \qquad (10.17)$$

which is of the van der Waals type without the volume-exclusion effect [9.2].

An improvement over the Hartree approximation can be obtained by taking into account the symmetry (or antisymmetry) or g_2 under the exchange $x_1 \rightleftharpoons x_2$ or under the exchange $x_1' \rightleftharpoons x_2'$ while still considering the added particles as moving independently of each other. Under these conditions we have the so-called Hartree-Fock approximation

$$g_2(x_1,x_2;x_1',x_2') \simeq g(x_1,x_1')g(x_2,x_2') \pm g(x_1,x_2')g(x_2,x_1') \quad . \tag{10.18}$$

By following the procedure outlined above, one obtains within the framework of the Hartree-Fock approximation

$$G(\underline{k},\omega) = \frac{1}{\omega - \varepsilon_k} \tag{10.19}$$

where

$$\varepsilon_{\underline{k}} = k^2/2m + n_o v_o \pm \int \frac{d^3k'}{(2\pi)^3} v(\underline{k} - \underline{k}')\langle n(\underline{k}')\rangle \tag{10.20}$$

and $v(\underline{k}) = \int d^3r \exp(-i\underline{k}\underline{r})\, v(\underline{r})$ is the Fourier transform of the potential $v(\underline{r})$. Note that the quasi-particle energy $\varepsilon_{\underline{k}}$ depends implicitly on the temperature T through the density $\langle n(\underline{k}')\rangle$. In [9.2] more complicated $g_2 \simeq f(g)$ are examined.

10.2 Diagrammatic Method for Fermions at T = 0

This method is applicable to the important case where the total Hamiltonian can be decomposed as

$$H = H_0 + H_1 \quad , \tag{10.21}$$

where H_0, the unperturbed part, is such that the Green's functions corresponding to H_0 can be easily calculated. The method here is analogous to that presented in Chap.4 where the eigenvalues and eigenfunctions of the one body H were determined from $G = (E - H)^{-1}$ and G was expressed as a perturbation series in terms of H_1 and $G_0 = (E - H_0)^{-1}$. In the present many-body case, working with g is much more advantageous than in the one-body case. The reason is that the causal Green's function g not only has the simplest perturbation expansion than any other quantity, but it also provides a host of important physical information without unimportant details of the many-body system. For comparison we mentioned that while the perturbation expansion for the ground-state energy is more complicated than the expansion for g, the information we obtain is clearly less.

To obtain the perturbation expansion for g, we need to work within the interaction picture according to which the time development of the operators is determined by the unperturbed part H_0 while the time development of the states is determined by the perturbation H_1. More explicitly we have for any operator A

$$A_I(t) = \exp(iH_{0I}t)A_S \exp(-iH_{0I}t) \quad , \tag{10.22}$$

which is equivalent to

$$idA_I(t)/dt = [A_I(t), H_{0I}] \quad . \tag{10.23}$$

We also have

$$|\Psi_I(t)\rangle = \exp(iH_{0I}t)|\Psi_S(t)\rangle \quad , \tag{10.24}$$

which leads immediately to

$$id|\Psi_I(t)\rangle/dt = H_{1I}(t)|\Psi_I(t)\rangle \quad . \tag{10.25}$$

The subscript "I" denotes the interaction picture; the subscript "S" denotes the Schrödinger picture according to which

$$dA_S/dt = 0 \tag{10.26}$$

and

$$id|\Psi_S(t)\rangle/dt = H_S|\Psi_S(t)\rangle \quad , \tag{10.27}$$

or, equivalently,

$$|\Psi_S(t)\rangle = \exp(-iHt)|\Psi_S(0)\rangle \quad . \tag{10.28}$$

We consider also the Heisenberg picture where

$$A_H(t) = \exp(iHt)A_S \exp(-iHt) \tag{10.29}$$

or, equivalently,

$$i dA_H(t)/dt - [A_H(t), H_H] \tag{10.29'}$$

and

$$|\Psi_H(t)> = |\Psi_S(0)> \quad . \tag{10.30}$$

From the above relations it is easy to see that the observable matrix elements are the same in all pictures:

$$<\Phi_S(t)|A_S(t)|\Psi_S(t)> = <\Phi_H|A_H(t)|\Psi_H> = <\Phi_I(t)|A_I(t)|\Psi_I(t)> \quad . \tag{10.31}$$

By integrating (10.25) or by employing (4.39,42) and (10.24) we obtain

$$|\Psi_I(t)> = S(t,t_0)|\Psi_I(t_0)> \quad , \tag{10.32}$$

where $S(t,t_0)$ is given by (4.55) with all the integration limits being t_0 and t; as can be seen from (4.51) and (10.22), we have $H_1^I(t) \equiv H_{1I}(t)$. In what follows we assume that H_1 contains a factor $\exp(-s|t|)$ and that after the calculations are done, we will take the limit $s \to 0^+$. This means that the interaction is turned on adiabatically at $t = -\infty$ and is turned off adiabatically as $t = \infty$. We can now prove that

$$<\Psi_H|A_H(t)|\Psi_H> = \frac{<\Phi|S(\infty,t)A_I(t)S(t,-\infty)|\Phi>}{<\Phi|S|\Phi>} \quad , \tag{10.33}$$

where $|\Psi_H>$ is the normalized ground state of the total Hamiltonian H, $|\Phi>$ is the ground state of the unperturbed Hamiltonian H_0 and $S \equiv S(\infty,-\infty)$. The proof is as follows:

$$<\Psi_H|A_H(t)|\Psi_H> = <\Psi_I(t)|A_I(t)|\Psi_I(t)> = <\Psi_I(t_0)|S(t_0,t)A_I(t)S(t,t_0)|\Psi_I(t_0)>$$

$$= <\Phi|S(-\infty,\infty)S(\infty,t)A_I(t)S(t,-\infty)|\Phi> \quad . \tag{10.34}$$

The second step follows from (10.32) and the third step from (10.24) and the fact that $|\Psi_I(t_0)> = \exp(-iH_0 t_0)|\Phi>$ as $t_0 \to -\infty$ since $H_1 \to 0$ as $t \to -\infty$: we have also used the basic property $S(t_1,t_2) = S(t_1,t_3)S(t_3,t_2)$

with $t_1 = -\infty$, $t_3 = \infty$, $t_2 = t$. Assuming that the ground state $|\Phi\rangle$ is nondegenerate and taking the limit $s \to 0^+$, we can see from (4.46) that $S|\Phi\rangle$ is proportional to $|\Phi\rangle$, i.e.,

$$S|\Phi\rangle = e^{i\varphi}|\Phi\rangle \quad . \tag{10.35}$$

Equation (10.35) means that an interaction adiabatically switched on and off does not produce any transition from a nondegenerate state. From (10.35) we have

$$\langle\Phi|S(-\infty,\infty) = e^{-i\varphi}\langle\Phi| \tag{10.36}$$

and

$$e^{-i\varphi} = \frac{1}{\langle\Phi|S|\Phi\rangle} \quad . \tag{10.37}$$

Combining (10.37) and (10.36) with (10.34), we obtain (10.33). With the help of (4.55) we can rewrite (10.33) in a more explicit way as follows:

$$\langle\Psi_H|A_H(t)|\Psi_H\rangle = \frac{1}{\langle\Phi|S|\Phi\rangle}\langle\Phi|\sum_{n=0}^{\infty}\frac{(-i)^n}{n!}\int_{-\infty}^{\infty}dt_1 \ldots \int_{-\infty}^{\infty}dt_n T[H_{1I}(t_1)$$

$$\ldots H_{1I}(t_n)A_I(t)]|\Phi\rangle \quad . \tag{10.38}$$

In a similar way we can prove that

$$\langle\Psi_H|T[A_{1H}(t)A_{2H}(t') \ldots]|\Psi_H\rangle = \frac{1}{\langle\Phi|S|\Phi\rangle}\langle\Phi|\sum_{n=0}\frac{(-i)^n}{n!}\int_{-\infty}^{\infty}dt_1 \ldots \int_{-\infty}^{\infty}dt_n$$

$$T[H_{1I}(t_1)\ldots H_{1I}(t_n)A_{1I}(t)A_{2I}(t')\ldots]|\Phi\rangle \quad . \tag{10.39}$$

Equation (10.39) is the basis for the perturbative-diagrammatic expansion of the causal Green's function. We consider the case where the interaction part H_1 is given by

$$H_1 = \frac{1}{2}\int d^3r d^3r' \psi^\dagger(\underline{r})\psi^\dagger(\underline{r}')v(\underline{r}-\underline{r}')\psi(\underline{r}')\psi(\underline{r}) \quad , \tag{10.40}$$

which can be rewritten as

$$\int_{-\infty}^{\infty} H_{1I}(t)dt = \frac{1}{2} \int dxdx' \psi_I^\dagger(x)\psi_I^\dagger(x')v(x - x')\psi_I(x')\psi_I(x) \tag{10.41}$$

with

$$v(x - x') = v(\underline{r} - \underline{r}')\delta(t - t') \tag{10.42}$$

and

$$\psi_I(x) = \exp(iH_0 t)\psi(\underline{r})\exp(-iH_0 t) \quad . \tag{10.43}$$

Using the definition of g(x,x'), (10.39), and (10.41) we obtain the following expansion

$$g(x,x') = \frac{N}{D} \tag{10.44}$$

$$N = -i \sum_{n=0}^{\infty} (\tfrac{1}{2})^n \frac{(-i)^n}{n!} \int dx_1dx_1' \ldots \int dx_ndx_n' <\Phi|T[\psi_I^\dagger(x_1)\psi_I^\dagger(x_1')\psi_I(x_1')\psi_I(x_1)$$

$$\ldots \psi_I^\dagger(x_n)\psi_I^\dagger(x_n')\psi_I(x_n')\psi_I(x_n)\psi_I(x)\psi_I^\dagger(x')]|\Phi>v(x_1-x_1') \ldots v(x_n-x_n') \tag{10.45}$$

$$D = \sum_{n=0}^{\infty} (\tfrac{1}{2})^n \frac{(-i)^n}{n!} \int dx_1dx_1' \ldots \int dx_ndx_n' <\Phi|T[\psi_1^\dagger(x_1)\psi_1^\dagger(x_1')$$

$$\ldots \psi_I(x_n')\psi_I(x_n)]\Phi>v(x_1-x_1') \ldots v(x_n-x_n') \quad . \tag{10.46}$$

Taking into account (8.46) we can write (10.45,46) as follows:

$$N = \sum_{n=0}^{\infty} (\tfrac{1}{2})^n \frac{i^n}{n!} \int dx_1dx_1' \ldots \int dx_ndx_n' \; g_{2n+1,o} \; (x,x_1,x_1',\ldots;x',x_1,x_1',\ldots)$$

$$v(x_1 - x_1') \ldots v(x_n - x_n') \tag{10.47}$$

$$D = \sum_{n=0}^{\infty} (\tfrac{1}{2})^n \frac{i^n}{n!} \int dx_1dx_1' \ldots \int dx_ndx_n' \; g_{2n,o} \; (x_1,x_1',\ldots;x_1,x_1',\ldots)$$

$$v(x_1 - x_1') \ldots v(x_n - x_n') \quad ; \tag{10.48}$$

the subscript "o" denotes that the Green's functions g_{2n+1} and g_{2n} correspond to the unperturbed Hamiltonian H_o (since all ψ_I's, $\psi_I^\dagger$'s and $|\Phi>$ refer to H_o). Remembering the physical interpretation of g_m as describing the propagation of m additional particles and taking into account that the unperturbed part H_o does not include any interparticle interactions, we can conclude that the m added particles propagate independently of each other, and hence $g_{m,o}$ can be written as a product of one-particle propagators. This product must be symmetrized (or antisymmetrized) in order to take into account the invariance of physical quantities under particle exchange. Thus, we have

$$g_{2n+1,o}(x,x_1,x_1', \ldots; x',x_1,x_1', \ldots) = \sum (-1)^P g_o(x,\tilde{x}') \ldots g_o(x_n',\tilde{x}_n') \tag{10.49}$$

where $\{\tilde{x}',\tilde{x}_1,\tilde{x}_1', \ldots \tilde{x}_n,\tilde{x}_n'\}$ is an arbitrary permutation of the set $\{x',x_1,x_1', \ldots x_n,x_n'\}$, the summation is over all permutations, and P is even (odd) when the permutation is even (odd). Equal time g_o must be interpreted as $\tilde{g}_o^<$; this can be seen from the starting equations (10.45,46) where the creation operator precedes the equal time annihilation operator. Equation (10.49) can be proved formally [8.3]; it is known as Wick's theorem. Combining (10.49) with (10.47,48) and (10.44) we immediately see that we have succeeded in expressing g in powers of g_o and $v(x - x')$; this expansion is substantially more complicated than the analogous expansion in the single-particle case.

It is obvious that the zero-order contributions to N and D are $N^{(0)} = g(x,x')$ and $D^{(0)} = 1$, so that, to zero order, $g(x,x') = g_o(x,x')$ as expected. The first-order contribution to $N, N^{(1)}$, contains 3! = 6 terms; $N^{(2)}$ contains 5! = 120 terms, and so on. At this point we introduce a set of diagrams each of which is in one-to-one correspondence with each term in $N^{(n)}$ via certain well-defined rules to be presented below. We introduce also diagrams for each term contributing to $D^{(n)}$. The introduction of these diagrams, called Feynman diagrams, was done for the following reasons:

1) to facilitate the task of keeping track of the enormous number of terms contributing to N or D. We will also see that the diagrams have some additional very important advantages;

2) certain cancelations are revealed by the introduction of diagrams;

3) one can ascribe a physical meaning to each diagram as representing a particular process associated with the propagation of the added particle in our system.

To be specific, consider the terms contributing to $N^{(1)}$; there are six terms in the summation (10.49) for n = 1. These six terms are in one-to-one correspondence with the six ways one can connect by directed lines the points x, x', x_1, x_1' in Fig.10.1a. These six ways give the six diagrams shown in Fig.10.1b. The rules for finding the contribution to $N^{(1)}$ from the corresponding diagrams can be obtained by simple inspection of (10.49) and (10.47). Thus each directed line starting at x_λ and ending at x_μ corresponds to $g_0(x_\mu, x_\lambda)$; the wavy line connecting x_1 with x_1' corresponds to $v(x_1 - x_1')$. We multiply these factors times $(i/2)(-1)^P$, and we integrate over the inter-

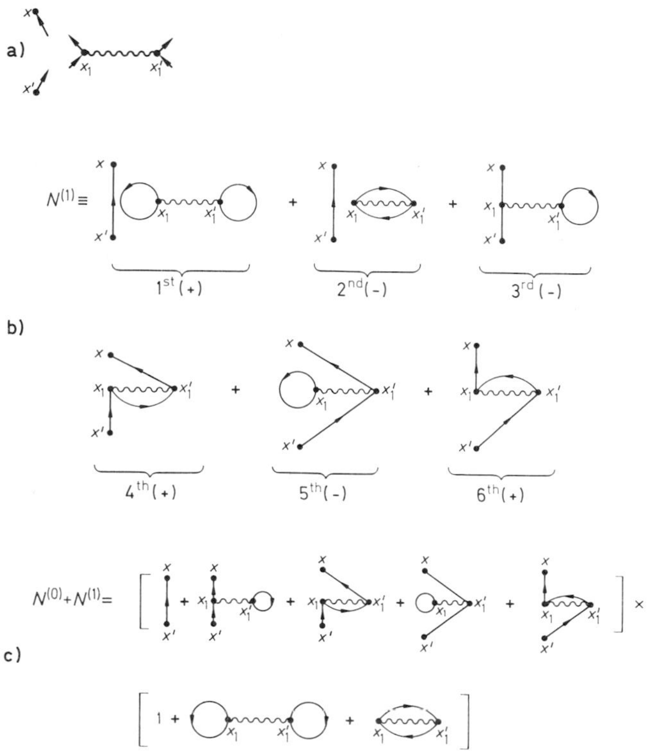

Fig.10.1. The six terms contributing to $N^{(1)}$ (see text) are in one-to-one correspondence with the six diagrams (b) resulting from all possible ways of connecting the two external points x, x' and the two internal points x_1, x_1' (a) according to the following rules: a directed line must start from x'; a directed line must end at x; one directed line must start from and one must end at each internal point; and no directed line must start from or end at any other point except x, x', x_1, x_1'. The plus or minus sign in parentheses is the sign of $(-1)^P$ [see (10.49)] for the corresponding term. The terms $N(0) + N(1)$ can be written as in (c) where up to first-order terms must be kept

nal variables x_1 and x_1'. Note that $(-1)^P$ equals $(-1)^m$ where m is the number of closed loops. The same rules allow us to calculate $D^{(1)}$ from the two diagrams shown in Fig.10.2.

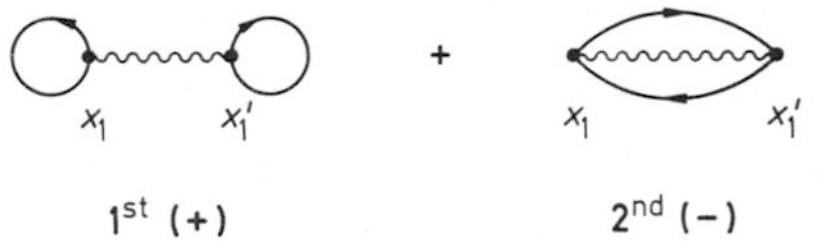

Fig.10.2. The two diagrams contributing to $D^{(1)}$(see text)

As one can see from Fig.10.1b, the diagrams can be classified as connected (such as the 3rd to 6th) and disconnected (such as the 1st and the 2nd). The contribution of a disconnected diagram consisting of two or more connected subdiagrams can be written as a product of terms each one corresponding to each subdiagram. The reason is that the subdiagrams have no common integration variables, and consequently the integral of the corresponding products is the product of the corresponding integrals. Thus, the contribution to N up to first order can be written as in Fig.10.1c. Taking into account that $D^{(0)} = 1$ and $D^{(1)}$ is given by the diagrams in Fig.10.2, we see that the quantity in the second parenthesis in Fig.10.1c equals to the contribution to D up to first order. It follows that the contributions to g up to first order are given by the *connected* diagrams up to first order contributing to N and shown in the first parenthesis in Fig.10.1c. It turns out [8.3] that this feature is correct to all orders, so that g(x,x') is given by the sum of *connected* diagrams contributing to N.

Because we are left with connected diagrams only, we can make some further simplifications. We observe that n^{th}-order diagrams resulting from each other by permutation of the pairs, (x_1,x_1'), (x_2,x_2'), ... , (x_n,x_n') are equal since this permutation is equivalent to renaming the integration variables. There are n! such diagrams. Thus we can keep only one of them and drop the factor 1/n! in (10.47). Similarly, interchanging x_λ with x_λ' leaves the contribution of the diagram unchanged. Thus we can consider only one of the set of all diagrams resulting from interchanges of the type $x_\lambda \rightleftharpoons x_\lambda'$ and drop the factor $(1/2)^n$ in (10.47). We are now in a position to give the final rules for calculating the nth-order contribution to g(x,x'):

1) Draw all topologically distinct connected diagrams with n interaction (wavy) lines, two external points, and 2n+1 directed g_0 lines.

2) Label each vertex with a fourth-dimensional point $x_i \equiv \underline{r}_i, t_i$.

3) For each directed line starting from x_ν and ending at x_μ write a factor $g_0(x_\mu, x_\nu)$.

4) For each interaction (wavy) line between x_i and x_i' write a factor $iv(x_i - x_i')$.

5) Integrate over all internal variables x_i, x_i'.

6) Multiply the expression by $(-1)^m$ where m is the number of closed loops.

7) Interpret $g_0(\underline{r}_i, t_i, \underline{r}_i', t_i)$ as being equal to $\tilde{g}_0^<(\underline{r}_i, t_i, \underline{r}_i', t_i)$.

In Fig.10.3 we plot all the diagrams for g up to second order. The contribution of the two first-order diagrams is, according to the rules above,

$$- i \int g_0(x,x_1) g_0(x_1,x') g_0(x_1',x_1') v(x_1 - x_1') dx_1 dx_1'$$

$$+ i \int g_0(x,x_1') g_0(x_1',x_1) g_0(x_1,x') v(x_1 - x_1') dx_1 dx_1' \quad .$$

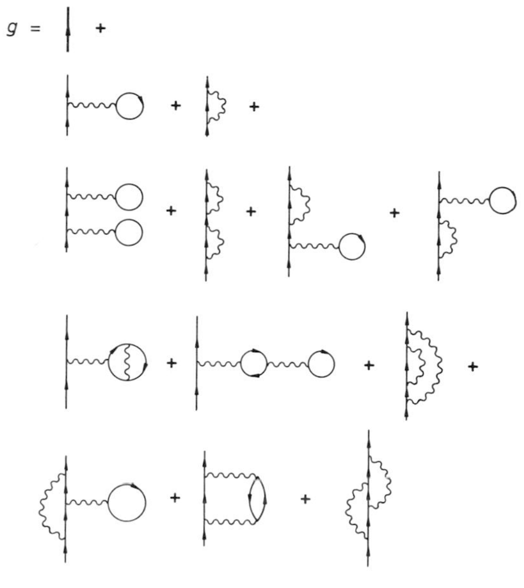

Fig.10.3. Feynman diagrams for g of zero order (first line), first order (second line), and second order (third, fourth and fifth lines)

For translationally invariant systems the calculations are facilitated by working in momentum-frequency space. This is achieved by expressing all

$g(x_\nu - x_\mu)$ and $v(x_i - x_i')$ in terms of their Fourier transforms with respect to the variables $x_\nu - x_\mu$ and $x_i - x_i'$, respectively. Then the integration over the internal variables x_i can be performed explicitly giving δ-functions expressing energy-momentum conservation at each vertex. Thus, in momentum space, we have rules resulting from the previous ones by the following replacement: 2 → 2', 3 → 3', 4 → 4', 5 → 5', 7 → 7' where

2') Label each line with a four momentum $q \equiv \underline{k},\omega$; conserve energy-momentum at each vertex.

3') For each directed line labeled with a four momentum $\underline{k},\omega$ write a factor

$$g_0(\underline{k},\omega) = \lim_{s \to 0^+} \frac{1}{\omega - \varepsilon_k^0 + is\bar{\varepsilon}(\omega-\mu)} \quad .$$

4') For each interaction (wavy) line labeled by $\underline{k},\omega$ write a factor

$$iv(\underline{k}) = i \int d^3r v(\underline{r})\, e^{-i\underline{k}\underline{r}} \quad .$$

5') Integrate over all internal independent four momenta (with a factor $1/2\pi$ for each single integration).

7') Interpret each $g_0(\underline{k},\omega)$ corresponding to a line starting from and ending at the same point (or linked by the same interaction line) as being $\tilde{g}_0^<(\underline{k},\omega) = 2\pi i\delta(\omega - \varepsilon_k^0)\theta(k_F - k)$.

According to the above rules the contribution to $g(q)$ from the first-order diagrams shown in Fig.10.4 is

$$- i \int \frac{d^4q'}{(2\pi)^4} \tilde{g}_0^<(q')v(0)g_0(q)g_0(q) + i \int \frac{d^4q'}{(2\pi)^4} g_0(q)g_0(q)\tilde{g}_0^<(q')v(\underline{k} - \underline{k}')$$

$$= g_0^2(q)\left[v(0) \int \frac{d^3k'}{(2\pi)^3} \theta(k_F - k') - \int \frac{d^3k'}{(2\pi)^3} v(\underline{k} - \underline{k}')\theta(k_F - k')\right] \quad .$$

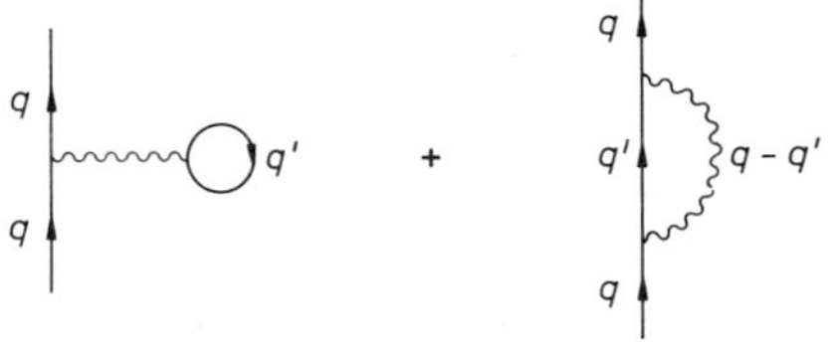

Fig.10.4. First-order Feynman diagrams for g in the four momentum space

It should be noted that the above rules for obtaining g are appropriate for an interaction term H_1 of the form given by (10.40). Similar but not identical rules would apply for the case where $H_0 = H_{oe} + H_{op}$ and $H_1 = \gamma \int d^3r\psi^\dagger(\underline{r})\psi(\underline{r})\phi(\underline{r})$, where H_{oe} describes noninteracting electrons, H_{op} describes noninteracting phonons and H_1 is the electron-phonon interaction. In this case we have a causal Green's functions g for the electron field, and another causal Green's function D for the phonon field, for each of which we can write perturbative expansions involving both the unperturbed Green's functions g_o, D_o, and the interaction γ. For details the reader is referred to [8.3] or [9.1].

10.3 Diagrammatic Method for T ≠ 0

In the last section we succeeded in expanding the T = 0 fermion g in terms of g_o and v. This expansion was based: 1) upon (10.39), expressing the average (at T = 0) of a chronological product of Heisenberg operators as N/D, where N is the unperturbed average (at T = 0) of the chronological product of $S(\infty, -\infty)$ and the same operators in the interaction picture, and D is the unperturbed average (at T = 0) of $S(\infty, -\infty)$; 2) upon Wick's theorem (10.49), expressing $g_{m,o}$ in terms of products of g_o's. For finite temperatures an equation analogous to (10.39) does not exist. Hence, the real time g for T ≠ 0 cannot be expanded in terms of g_o and v. However, it turns out that such an expansion for T ≠ 0 is possible for the imaginary time Green's functions introduced in Sect.10.1. The basis for such an expansion is the following equation

$$\langle T[A_{1H}(-i\sigma)A_{2H}(-i\sigma') \ldots] \rangle = \frac{\langle T[S(-i\beta,0)A_{1I}(-i\sigma)A_{2I}(-i\sigma') \ldots] \rangle_o}{\langle S(-i\beta,0) \rangle_o} \quad (10.50)$$

where $\langle A \rangle \equiv \mathrm{Tr}\{A \exp[-\beta(H-\mu N)]\}/\mathrm{Tr}\{\exp[-\beta(H-\mu N)]\}$; in $\langle A \rangle_o$ H has been replaced by H_o. Equation (10.50) is the analog of (10.39). To prove (10.50) we use the relation

$$|\Psi_I(0)\rangle = |\Psi_S(0)\rangle = |\Psi_H\rangle \quad (10.51)$$

which follows from (10.24) and (10.30). From (10.51), (10.32) and (10.31) it follows that

$$A_H(t) = S(0,t)A_I(t)S(t,0) \quad . \tag{10.52}$$

We have also that

$$e^{-iHt} = e^{-iH_0t}S(t,0) \quad . \tag{10.53}$$

Equation (10.53) follows from (10.28), (10.24), (10.32) and (10.51). Replacing t by $-i\beta$ in (10.53) and multiplying by $\exp(\beta\mu N)$, we obtain

$$e^{-\beta(H-\mu N)} = e^{-\beta(H_0-\mu N)}S(-i\beta,0) \quad . \tag{10.54}$$

To arrive at (10.54) we need to assume that N commutes with H and H_0. If this is not the case, then $\mu = 0$; consequently, (10.54) is always valid. Combining (10.54) with (10.52) and the property $S(t_1,t_2) = S(t_1,t_3)S(t_3,t_2)$, we obtain (10.50).

Taking into account (10.50), (10.49) (which holds for imaginary times as well [8.3]) and the definition of $g(\underline{r}_i,-i\sigma_1,\underline{r}_2,-i\sigma_2)$, we have a diagrammatic expansion of the latter in terms of $g_0(\underline{r}_i,-i\sigma_i,\underline{r}_i',-i\sigma_i')$ and $v(\underline{r}_i - \underline{r}_i')$ based on rules resulting from the previous ones by the substitutions: 2 by II, 4 by IV, 5 by V, and 7 by VII, where

II) Label each vertex with a four-dimensional point $x_i \equiv \underline{r}_i,-i\sigma_i$.

IV) For each interaction (wavy) line between x_i and x_i' write a factor $v(\underline{r}_i - \underline{r}_i')\delta(\sigma_i - \sigma_i')$.

V) Integrate over all internal variables $\underline{r}_i,\sigma_i$: $\int d^3r_i \int_0^\beta d\sigma_i$.

VII) Interpret $g_0(\underline{r}_i,-i\sigma_i,\underline{r}_i',-i\sigma_i)$ as being equal to $\tilde{g}_0^<(\underline{r}_i,-i\sigma_i,\underline{r}_i',-i\sigma_i)$.

The calculational effort is greatly simplified if we work in the $\underline{k}$, z_ν-space, i.e., if we try to calculate $G(\underline{k},z_\nu)$ given by (10.11). The final rules for obtaining the nth-order contribution to $G(\underline{k},z_\nu)$ are the following:

I') Draw all topologically district connected diagrams with n interaction lines, two external points, and 2n+1 directed lines.

II') Label each line with a four momentum $\underline{k}',z_\nu'$; conserve momentum and $\mathrm{Im}\{z_\nu'\}$ at each vertex.

III') For each directed line labeled with a four momentum $\underline{k},z_\nu$ write a factor

$$G_0(\underline{k},z_\nu) = \frac{1}{z_\nu-\varepsilon_k^0} \quad .$$

IV') For each interaction (wavy) line labeled by $\underline{k}, z_\nu$ write a factor $-v(\underline{k})$.

V') Integrate over all internal independent momenta $\underline{k}[(2\pi)^{-3} \int d^3k]$ and sum over all internal independent discrete frequencies $(\beta^{-1}\sum_\nu)$.

VI') Multiply by $(-1)^m$ where m is the number of closed fermion loops.

VII') Whenever a directed line either closes on itself or is joined by the same interaction line, insert a convergence factor $\exp(-z_\nu\sigma)$ with $\sigma \to 0^-$.

As an example we calculate the contributions of the first-order diagrams shown in Fig.10.5. We have

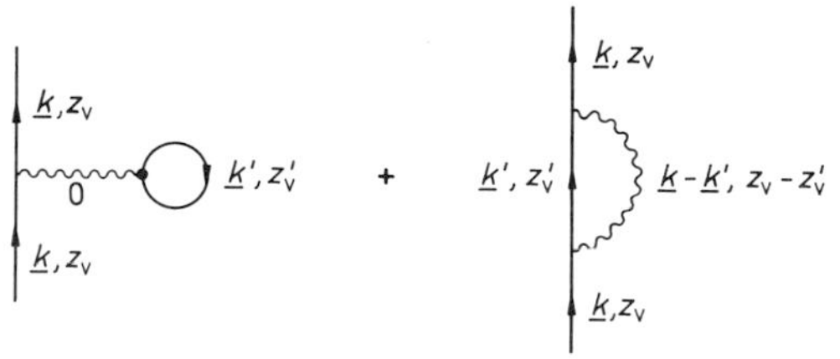

Fig.10.5. First-order Feynman diagrams for $G(\underline{k}, z_\nu)$

$$G_0^2(\underline{k}, z_\nu)\left[\mp v(0) \int \frac{d^3k'}{(2\pi)^3} \frac{1}{\beta} \sum_{\nu'} e^{-z'_\nu\sigma} G_0(\underline{k}', z'_\nu) - \int \frac{d^3k'}{(2\pi)^3} v(\underline{k} - \underline{k}') \frac{1}{\beta} \sum_{\nu'} e^{-z'_\nu\sigma} G_0(\underline{k}', z'_\nu)\right] \quad ,$$

which with the help of (9.53) and $A_0(\underline{k}, \omega) = 2\pi\delta(\omega - \varepsilon_k^0)$, becomes

$$G_0^2(\underline{k}, z_\nu)\left[v(0) \int \frac{d^3k'}{(2\pi)^3} f_\mp(\varepsilon_{k'}^0) \pm \int \frac{d^3k'}{(2\pi)^3} v(\underline{k} - \underline{k}') f_\mp(\varepsilon_{k'}^0)\right] \quad .$$

(upper sign for bosons; lower sign for fermions).

It should be noted that the above rules for finite temperatures are appropriate for an interaction of the form (10.40). For other interactions, such as the electron-phonon interactions, the rules must be modified [8.3, 9.1].

Before we conclude this section we remind the reader that the calculational schemes we have presented do not cover the case of bosons at low temperatures. The reason is that the unperturbed system for $T<T_c$ undergoes the phenomenon of Bose condensation where a finite fraction of the particles occupy a single quantum state, the $\underline{k} = 0$ state. This effect requires a special treatment which is presented in [8.3] and [9.1]. Special treat-

ment is also required for fermion systems which are not normal (superfluid or superconducting systems). These questions will not be covered in the present work.

10.4 Partial Summations. Dyson's Equation

It is very rare that a small number of lowest-order diagrams would be a good approximation for g. For this reason in practical calculations we either try to obtain certain general results without employing approximations, or, whenever specific results are sought, we try to find a class of diagrams such that the contribution of the whole class is both calculable and dominant. Of course, it is not so often that such a happy situation occurs. Anyway, our calculational task is facilitated by reorganizing our expansion through certain partial summations. These partial summations can be performed in a graphical way. An example of such a summation is shown in Fig.10.6. The square 1 denotes a part of the diagram connected to the rest by two directed lines; the square 2 denotes a different part connected to the rest by two directed lines. The sum of the two diagrams can be found by calculating first the contributions of part 1 and part 2 according to the rules, summing these two contributions, and using this result for the part 1 + 2. This idea can be developed further so that starting from a few simple diagrams (called skeletons) and performing all possible summations in the lines and vertices (putting flesh to the skeletons [9.5]) we obtain g. Such a procedure requires care to ensure that all diagrams have been included and no diagram was counted more than once. In order to examine this question, we need to introduce some definitions and present some relations.

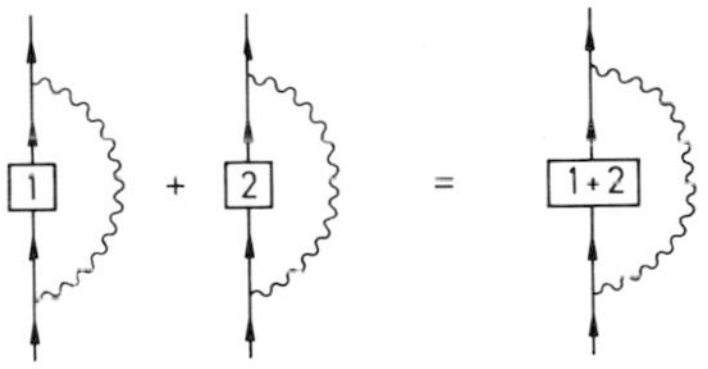

Fig.10.6. An example of graphical summation

We call self-energy Σ the sum of the contributions from all parts which are connected to the rest by two directed lines (one in and one out). From the structure of the diagrams for g it follows that

$$G(\underline{k},z_\nu) = G_0(\underline{k},z_\nu) + G_0(\underline{k},z_\nu)\,\Sigma(\underline{k},z_\nu)\,G_0(\underline{k},z_\nu) \quad . \qquad (10.55)$$

Equation (10.55) can be written in a graphical way (Fig.10.7) which shows that in x-space it would acquire an integral form. Note the similarity of (10.55) and (4.17). We define next the proper self-energy, Σ^*, which involves only those self-energy parts that cannot be separated into two pieces by cutting a single particle line. It is not difficult to see that

$$\begin{aligned}\Sigma &= \Sigma^* + \Sigma^* G_0 \Sigma^* + \Sigma^* G_0 \Sigma^* G_0 \Sigma^* + \dots \\ &= \Sigma^* + \Sigma^* G_0 \Sigma \\ &= \Sigma^* + \Sigma G_0 \Sigma^* \quad . \end{aligned} \tag{10.56}$$

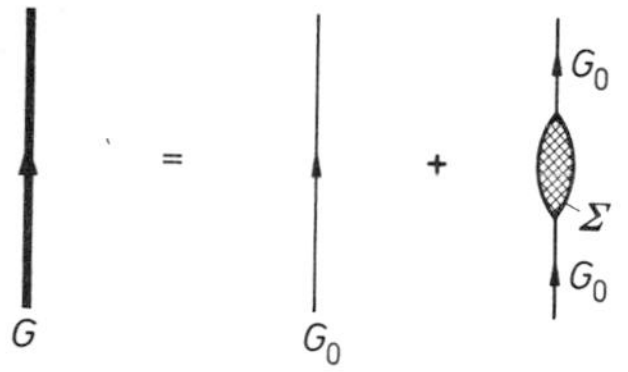

Fig.10.7. Relation between the Green's function G (thick line), the unperturbed Green's function G_0 (thin line) and the self energy Σ

Note the analogy of (10.56) and (4.13,15,16). Combining (10.55) and (10.56), we obtain

$$\begin{aligned} G &= G_0 + G_0 \Sigma^* G \\ &= G_0 + G \Sigma^* G_0 \end{aligned} \tag{10.57}$$

which is the analog of (4.6,7). In $\underline{k},\omega$-space (10.57) can be rewritten as

$$G(\underline{k},\omega) = \frac{G_0(\underline{k},\omega)}{1 - G_0(\underline{k},\omega)\Sigma^*(\underline{k},\omega)} = \frac{1}{\omega - \varepsilon_k^0 - \Sigma^*(\underline{k},\omega)} \quad , \tag{10.58}$$

where we have taken into account that $G_0(\underline{k},\omega) = (\omega - \varepsilon_k^0)^{-1}$. In Fig.10.8 we show (10.57) in a diagrammatic way.

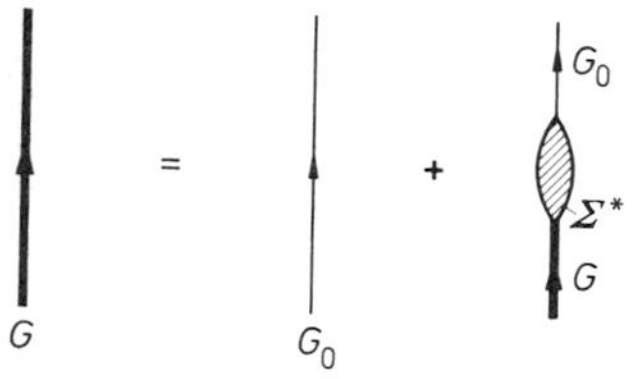

Fig.10.8. Relation between G (thick line), G_0 (thin line) and the proper self-energy Σ^*

It is clear that (10.58) has not solved the problem since Σ^* involves the summation of infinite diagrams. One could try to connect Σ^* with G by employing the idea of partial summations in the diagrams for Σ^*; these partial summations would replace the G_0-lines by G-lines. It turns out that one cannot express Σ^* with a few diagrams involving G-lines and interaction lines. This is to be expected since, as we have seen in the equation of motion method, one cannot obtain a system with a finite number of equations for a finite number of unknown functions which are related to the many body g. However, one can express Σ^* in terms of G with the help of a new quantity Γ, which is called the vertex part, and is defined as the sum of the contributions from all parts which are connected with four (two in and two out) directed lines (G_0-lines) to the rest and which cannot be decomposed into disconnected parts. Γ is related to the two particle Green's function g_2, as shown in Fig.10.9, i.e.,

$$g_2(x_1,x_2;x_1',x_2') = g(x_1,x_1')g(x_2,x_2') \pm g(x_1,x_2')g(x_2,x_1') + \int d\tilde{x}_1 d\tilde{x}_2 d\tilde{x}_1' d\tilde{x}_2' \cdot \Gamma(\tilde{x}_1,\tilde{x}_2;\tilde{x}_1',\tilde{x}_2')g(x_1,\tilde{x}_1)g(x_2,\tilde{x}_2)g(\tilde{x}_1',x_1')g(\tilde{x}_2',x_2') \quad (10.59)$$

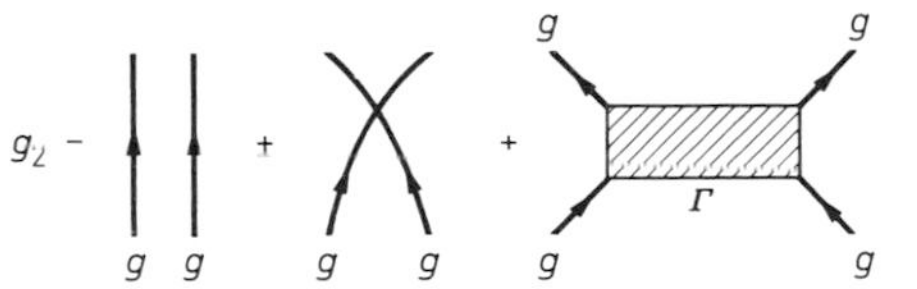

Fig.10.9. Relation between g_2, g and the vertex part Γ

One can show [9.1] that Σ^* is given in terms of G and Γ as shown in Fig. 10.10a. One can easily prove that the relation shown in Fig.10.10a is equivalent to the equation of motion (10.1). The proof makes use of (10.59) and (10.57). Fig.10.10 shows that Σ^* can be calculated by "putting flesh" to the four skeleton diagrams shown in Fig.10.10b, i.e., by replacing the G_0-lines by G-lines and one of the two interaction lines in the last two diagrams by the vertex part. (If both interaction lines in the same diagram are replaced by vertex parts, one would double count diagrams.) As expected, one cannot express Γ in a closed form involving G and Γ, i.e., the skeleton diagrams for Γ (to be fleshed by the substitution $G_0 \to G$ and $v \to \Gamma$) are infinite in number. Usually we stop this infinite hierarchy of relations by approximately expressing Γ in a closed form in terms of G and Γ. The simplest such approximation is to put $\Gamma = 0$

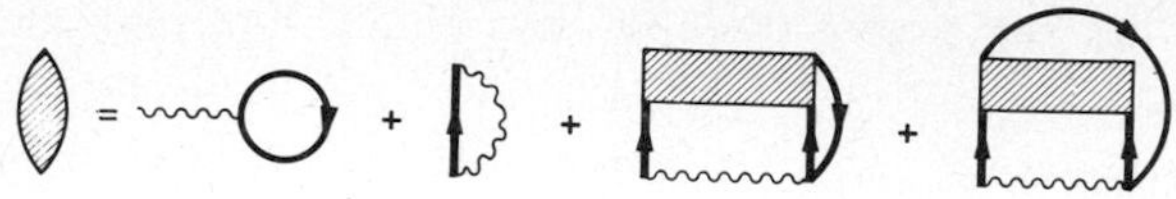

a

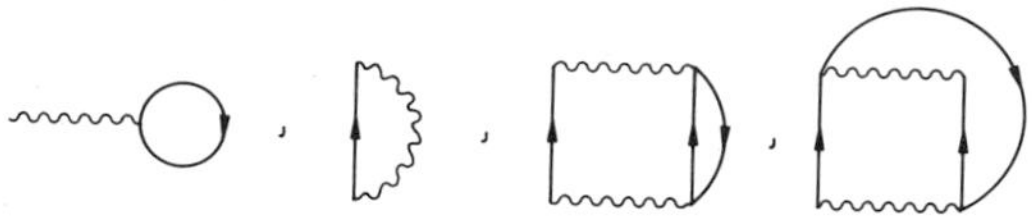

b

Fig.10.10. The proper self-energy, Σ^* can be calculated from the four diagrams (a) resulting by "fleshing" the skeleton diagrams (b)

in the relation shown in Fig.10.10a. This is the Hartree-Fock approximation. We obtain then for $\Sigma^*(\underline{k},z_\nu)$ the expression

$$\begin{aligned}\Sigma^*(\underline{k},z_\nu) &= \mp v(0) \int \frac{d^3k'}{(2\pi)^3} \frac{1}{\beta} \sum_{\nu'} e^{-z'_\nu\sigma} G(\underline{k}',z'_\nu) - \int \frac{d^3k'}{(2\pi)^3} v(\underline{k}-\underline{k}') \frac{1}{\beta} \\ &\qquad \sum_{\nu'} e^{-z'_\nu\sigma} G(\underline{k}',z'_\nu) \\ &= v(0)\int \frac{d^4q'}{(2\pi)^4} A(\underline{k}',\omega')f_{\mp}(\omega') \pm \int \frac{d^4q'}{(2\pi)^4} v(\underline{k}-\underline{k}')A(\underline{k}',\omega')f_{\mp}(\omega') \\ &= \Sigma^*(\underline{k}) \quad . \end{aligned} \tag{10.60}$$

Combining (10.60) with (10.58) we obtain (10.19,20). The Hartree-Fock approximation is shown diagrammatically in Fig.10.11.

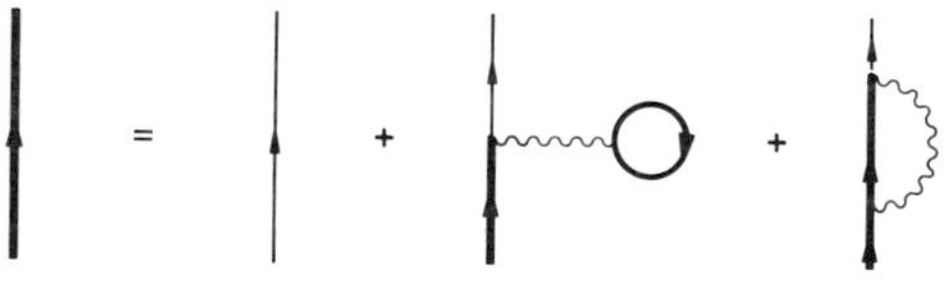

Fig.10.11. The Hartree-Fock approximation

The basic equation (10.58) combined with the equation connecting Σ^* with G and Γ (Fig.10.10a) is called Dyson's equation.

We introduce also the concept of an effective interparticle interaction v_e as the sum of v plus the contributions (apart from a factor i or -1 for the T = 0 or T ≠ 0 case, respectively) from all parts which have two external interaction lines. The effective interaction v_e can be expressed as shown in Fig.10.12a in terms of the polarization Π; the latter is defined as the sum of the contributions (apart from the above factors) of all parts which are connected to the rest by two interaction lines. Following an analysis similar to the one given for the self-energy, we can easily express the polarization Π in terms of the proper polarization Π^*, as shown in Fig.10.12b. The proper polarization Π^* is the sum of the contributions (apart from a factor -i or -1 for the T = 0 or T ≠ 0 case, respectively) of all parts which are connected to the rest by two interaction lines and which cannot be separated into two pieces by cutting a single interaction line.

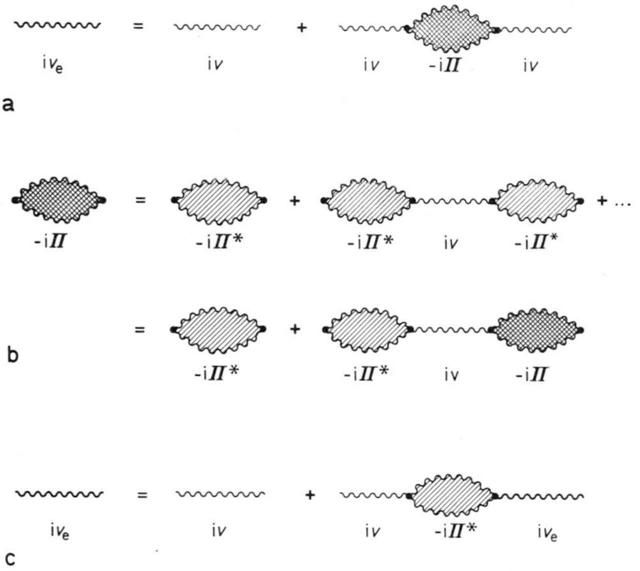

Fig.10.12. Relation between: (a) v_e, v and Π; (b) v, Π and Π^*; (c) v_e, v and Π^* for fermions at T = 0. For the T ≠ 0 imaginary time case in $\underline{k}, z_\nu$-space iv_e, iv, $-i\Pi$ and $-i\Pi^*$ must be replaced by $-v_e$, $-v$, $-\Pi$ and $-\Pi^*$, respectively

For a translationally invariant system and in the q-representation the relations of Fig.10.12 become

$$v_e(q) = v(q) + v^2(q)\Pi(q) \tag{10.61}$$

$$\Pi(q) = \Pi^*(q) + \Pi^*(q)v(q)\Pi^*(q) + \ldots = \Pi^*(q) + \Pi^*(q)v(q)\Pi(q) \quad , \tag{10.62}$$

which leads to

$$\Pi(q) = \frac{\Pi^*(q)}{1-v(q)\Pi^*(q)} \tag{10.62'}$$

and

$$v_e(q) = v(q) + v(q)\Pi^*(q)v_e(q) \quad , \tag{10.63}$$

which can be solved for $v_e(q)$ to yield

$$v_e(q) = \frac{v(q)}{1-v(q)\Pi^*(q)} \quad . \tag{10.63'}$$

If we define a dielectric function $\varepsilon^c(\underline{k},\omega)$ from the relation

$$v_e(q) = v(q)/\varepsilon^c(q) \quad , \tag{10.64}$$

we obtain by taking into account (10.61) and (10.63')

$$\frac{1}{\varepsilon^c(q)} = 1 + v(q)\Pi(q) = \frac{1}{1-v(q)\Pi^*(q)} \quad . \tag{10.65}$$

In order to find the relation between the usual (retarded) dielectric function defined in Chap.9 and $\varepsilon^c(\underline{k},\omega)$, we observe first that the inverse Fourier transform of the polarization Π (q) equals to D(x,x') where

$$D(x,x') \equiv -i\langle T[\tilde{n}_H(x)\tilde{n}_H(x')]\rangle \tag{10.66}$$

and

$$\tilde{n}_H(x) = n_H(x) - \langle n_H(x)\rangle = \psi_H^\dagger(x)\psi_H(x) - \langle\psi^\dagger(x)\psi(x)\rangle \quad . \tag{10.67}$$

The proof of the relation $D = \Pi$ follows from a diagrammatic expansion of (10.66). The causal Green's function D(x,x') is connected with the retarded Green's function $D^R(x,x')$, which was introduced in Chap.9 in the usual way. Thus for fermions at T = 0

$$\mathrm{Re}\{D^R(q)\} = \mathrm{Re}\{D(q)\} = \mathrm{Re}\{\Pi(q)\}$$

$$\mathrm{Im}\{D^R(q)\} = \bar{\varepsilon}(\omega)\mathrm{Im}\{D(q)\} = \bar{\varepsilon}(\omega)\mathrm{Im}\{\Pi(q)\} \tag{10.68}$$

where

$$\bar{\varepsilon}(\omega) = 1 \quad \text{for} \quad \omega > 0$$

$$= -1 \quad \text{for} \quad \omega < 0 \quad .$$

Taking into account (10.65), (9.45), (10.68) and the fact that $v(q) = 4\pi e^2/q^2$ (for Coulomb interaction), we obtain that

$$\mathrm{Re}\{\varepsilon(\underline{k},\omega)\} = \mathrm{Re}\{\varepsilon^C(\underline{k},\omega)\} \tag{10.69}$$

$$\mathrm{Im}\{\varepsilon(\underline{k},\omega)\} = \bar{\varepsilon}(\omega)\mathrm{Im}\{\varepsilon^C(\underline{k},\omega)\} \quad . \tag{10.70}$$

In the $T \neq 0$ imaginary time formalism, the sum of all polarization diagrams give $-\Pi(\underline{k},z_\nu)$ from which, by analytic continuation, one obtains $\Pi(\underline{k},z)$. The dielectric function $\varepsilon(\underline{k},\omega)$ is then given by

$$\frac{1}{\varepsilon(\underline{k},\omega)} = 1 + v(\underline{k}) \lim \Pi(\underline{k},z) \quad , \tag{10.71}$$

as z approaches ω from above the real ω-axis. The conclusion is that from the polarization diagrams one can obtain the dielectric function either for $T = 0$ or for $T \neq 0$.

Of course, the polarization Π or the proper polarization Π^* involve an infinite number of diagrams, which cannot be summed exactly. Thus the calculation of Π or Π^* must be done approximately. The approximation is usually implemented either by keeping a few diagrams for Π^* or by expressing Π in terms of g and Γ and then approximating Γ. An example will be examined in the next chapter.

10.5 Summary

In all practical cases the total Hamiltonian H can be decomposed into an unperturbed part H_0, such that the corresponding Green's functions can be easily calculated, and a perturbation H_1.

In the equation of motion method one writes a differential equation for one of the g's; this equation also contains a term involving H_1 and higher-order Green's function. The calculation is done by approximately expressing the higher-order Green's function in terms of g. This method is applicable to g^R, g^A (for T = 0 or T ≠ 0) and to g (for T = 0 only). It can be applied to the T ≠ 0 causal g if complex times are considered.

In the perturbative diagrammatic method g is expressed as a complicated series each term of which involves products of g_0 and H_1. Keeping track of the terms in the series is facilitated by associating diagrams with each term. The diagrams help in intermediate algebraic manipulations and permit also a physical interpretation of each term (this last feature can be used in order to develop meaningful approximations). The diagrammatic method is applicable to the causal Green's function g at T = 0; it is applicable to the imaginary time causal g for T ≠ 0.

The calculational effort is facilitated by defining the summation of certain type of parts of diagrams. Thus, the self-energy Σ and the proper self-energy Σ^* are defined; their relation to the causal Green's function are shown in Figs.10.7 and 10.8. We defined also the vertex part Γ which is related to g_2 as shown in Fig.10.9. The proper self-energy Σ^* can be expressed in terms of Γ and G. The system of equations can be closed by approximating Γ in terms of Γ and G. Finally, the polarization part Π and the proper polarization part Π^* are defined and related with the effective interaction, as shown in Fig.10.12. The quantity Π (or Π^*) are directly related with the dielectric function.

11. Applications

Some applications of the many-body Green's functions are briefly presented. They include the study of a high-density electronic system moving in a positive background; this model describes approximately electrons in metals. A low-density Fermi system with short-range interactions is also examined. The employment of the Green's functions formalism to justify the widely-used independent particle approximation is emphasized.

11.1 Normal Fermi Systems. Landau Theory

In Chap.9 we defined a normal Fermi system as one for which the a.c. $\{g(\underline{k},\omega)\}$ has a single pole with a trajectory (as k varies) which crosses the real ω-axis at $\omega = \mu$ (for $k = k_F$) with zero slope. Taking into account that

$$g(\underline{k},\omega) = [\omega - \varepsilon_k^0 - \Sigma^*(\underline{k},\omega)]^{-1} \quad , \tag{11.1}$$

it follows that for a normal Fermi system $\Sigma_I^* = \mathrm{Im}\{\Sigma^*(\underline{k},\omega)\}$ must behave as

$$\Sigma_I^*(\underline{k},\omega) = \bar{\varepsilon}(\mu - \omega)\, C_k|\omega - \mu|^n \quad ; \qquad C_k \geqq 0 \quad , \quad \text{real} \tag{11.2}$$

for $\omega \approx \mu$ with $n > 1$. LUTTINGER [11.1] has proved that n = 2 is a consistent solution of the equations for g and Σ^*. The outline of the proof is the following. We substitute (11.2) into (11.1), and the resulting expression is used in the equation shown in Fig.10.10a. The first two terms give $\Sigma_I^* = 0$. In the next two terms Γ is replaced by an infinite series of diagrams involving the bare interaction v and the dressed propagator g. One can then show [11.1] that each of the resulting diagrams gives an imaginary part which behaves as

$$(\omega - \mu)^{2m} \quad ; \quad m = 1,2,3, \ldots$$

in the limit $\omega \to \mu^{\pm}$. Hence,

$$\lim_{\omega \to \mu} \Sigma_I^*(\underline{k},\omega) \propto (\omega - \mu)^2 \tag{11.3}$$

as it was assumed at the beginning. Thus the expression (11.1) subject to the condition (11.3) is a solution of the equations for g, which means that the normal state of a strongly interacting Fermi system is a possible one. The pole $z_{\underline{k}} = \varepsilon_{\underline{k}} + i\gamma_{\underline{k}}$ is a solution of

$$z_{\underline{k}} - \varepsilon_{\underline{k}}^0 - \Sigma^*(\underline{k},z_{\underline{k}}) = 0 \quad , \tag{11.4}$$

which for $k \approx k_F$ gives

$$\varepsilon_{\underline{k}} = \mu + \left|\frac{\partial \varepsilon_{\underline{k}}^0}{\partial \underline{k}} + \frac{\partial \Sigma_R^*}{\partial \underline{k}}\right| w_{\underline{k}} \cdot (k - k_F) \tag{11.5}$$

$$w_{\underline{k}} = \left[1 - \partial \Sigma_R^* / \partial \omega\right]^{-1} \tag{11.6}$$

$$\gamma_{\underline{k}} = \bar{\varepsilon}(\mu - \varepsilon_{\underline{k}}) w_{\underline{k}} C_{\underline{k}} (\varepsilon_{\underline{k}} - \mu)^2 \tag{11.7}$$

where the derivatives are calculated for $k = k_F$ and $\omega = \mu$.

The Green's function g can be written as

$$g(\underline{k},\omega) = \frac{w_{\underline{k}}}{\omega - \varepsilon_{\underline{k}} - i\gamma_{\underline{k}}} + g_b(\underline{k},\omega) \quad , \tag{11.8}$$

where the smooth background contribution $g_b(\underline{k},\omega)$ can be omitted for $\omega \approx \varepsilon_{\underline{k}}$. We see that the normal solution (11.8) is not only a possible solution of the interacting Fermi system, but it also goes continuously to the unperturbed $g_o(\underline{k},\omega)$, since, as the coupling constant goes to zero, $C_{\underline{k}} \to 0$, $w_{\underline{k}} \to 1$, $\varepsilon_{\underline{k}} \to \varepsilon_{\underline{k}}^0$ and $g_b(\underline{k},\omega) \to 0$. Thus, if we assume that the actual state of the interacting system develops in a unique and *continuous* way from the unperturbed state as the interactions are turned on, we can conclude that the normal solution (11.8,5,6,7) is the actual one and that our system is a normal one. Obviously, if the perturbation expansion converges, the above assumptions are correct and the system is a normal one. However, it should be stressed that the continuity assumption may be valid while the expansion in powers of the coupling constant may diverge, which means that a system may be normal even if the perturbation expansion diverges. On the other hand, there are physical systems (a superconducting material is an example) where the perturbed state does not develop from the unperturbed in a continuous way as the interactions are turned on; such systems are not normal, and the actual Green's function does not have the form (11.8,5,7).

Equation (11.5) shows that the dispersion relation ε_k for $k \approx k_F$ is characterized by a single quantity $v_F = |\partial\varepsilon_k^0/\partial\underline{k} + \partial\Sigma_R^*/\partial\underline{k}^-|w_k$; the expansion coefficient v_F is the velocity of the quasi-particles at the Fermi level, and it is usually expressed as

$$v_F = k_F/m^* \quad . \tag{11.9}$$

Equation (11.9) can be considered as the definition of the effective mass m^*. For a noninteracting system $m^* = m$.

For a normal system characterized by (11.5-8) with short-range forces one can prove (see, e.g., [9.1]) by analyzing the diagrams for the vertex part $\Gamma(p_1,p_2;k)$ that

$$\Gamma(p_1,p_2;k) \simeq \Gamma^\omega(p_1,p_2) + \frac{w_{k_F}^2 k_F^2}{(2\pi)^3 v_F} \int \Gamma^\omega(p_1,q)\Gamma(q,p_2;k)\,\frac{\underline{v}\cdot\underline{k}}{\omega-\underline{v}\cdot\underline{k}}\,dO \quad . \tag{11.10}$$

In (11.10) $p_1 = \underline{p}_1, \omega_1$ and $p_2 = \underline{p}_2, \omega_2$ are the four momenta at the two "in" points of the vertex part Γ; $p_1 + k$, $p_2 - k$ are the four momenta at the two "out" points of the vertex part Γ; $k = \underline{k},\omega$ is the four momentum transfer; $\Gamma^\omega \equiv \lim_{\omega\to 0}\lim_{\underline{k}\to 0}\Gamma(k)$ (note the order of the limits); $\underline{v}$ is a vector directed along $\underline{q}$ with $|\underline{v}| = v_F$; the four momentum q equals to $\underline{q},\mu$ with $|\underline{q}| = k_F$; $\int dO$ denotes an integration over the direction of $\underline{q}$. Equation (11.10) is valid for small ω and $|\underline{k}|$.

Using (11.10) together with some relations connecting the derivatives $\partial g^{-1}/\partial\omega$, $\partial g^{-1}/\partial\underline{k}$, $\underline{k}\partial g^{-1}/\partial\underline{k}$, $\partial g^{-1}/\partial\mu$ with the vertex part Γ (see [9.1]), one can prove the following two important relations first derived by LANDAU [11.2].

The Fermi momentum k_F, which was defined by the equation $\varepsilon_{k_F} = \mu$, satisfies the relation

$$\frac{4\Pi}{3}\,\frac{k_F^3}{(2\pi)^3} = \frac{N}{\Omega} \quad , \tag{11.11}$$

where N is the total number of particles and Ω is the volume. We see that k_F for a normal interacting Fermi system satisfies the same basic equation as for the noninteracting system. Thus, for a normal system, turning on the interactions not only retains the concept of a Fermi surface; it leaves the Fermi surface unchanged.

The other relation connects the bare mass m with the effective mass m^* and the vertex part Γ^ω

$$\frac{1}{m^*} = \frac{1}{m} - \frac{k_F w_{k_F}^2}{2(2\pi)^3} \int \Gamma^\omega(p_1,p_2)\cos\theta d\mathcal{O} \quad , \tag{11.12}$$

where $p_1 \equiv \underline{p}_1,\mu$, $p_2 \equiv \underline{p}_2,\mu$, $|p_1| = |p_2| = k_F$,

θ is the angle between $\underline{p}_1$ and $\underline{p}_2$ and the integration is over all directions of $\underline{p}_1$ or $\underline{p}_2$ (the interaction is isotropic). For an alternative derivation of (11.11) and (11.12) the reader is referred to [9.3].

From (11.10) one can prove [9.1] that $\Gamma(p_1,p_2;k)$ develops a pole as a function of ω. Since $\Gamma(p_1,p_2;k)$ is directly related to the polarization part $\Pi(k)$, it follows that the pole of $\Gamma(p_1,p_2;k)$ appears as a pole in $\Pi(k) \equiv \Pi(\underline{k},\omega)$. As we have mentioned in Chap.9 the pole of $\Pi(\underline{k},\omega)[\equiv D(\underline{k},\omega)]$ corresponds to a collective Bose excitation which is an elementary density wave. Thus a normal system possesses collective Bose excitations. For short-range interactions at T = 0 these excitations are given as the poles of the solution of (11.10). It turns out [8.3,9.1] that the eigenfrequency $\omega_{\underline{k}}$ of this collective excitation is of the form

$$\omega_{\underline{k}} = c_0|\underline{k}| \quad , \tag{11.13}$$

i.e., it resembles sound waves. For this reason it is called zero sound. Zero sound is clearly distinguished from ordinary sound because the former, in contrast to the latter, does not correspond to a local equilibrium condition (i.e., the variation in the distribution, $\delta f(\underline{r},\underline{k})$, is not spherical in $\underline{k}$-space). This question was discussed by ABRIKOSOV et al. [9.1] and FETTER and WALECKA [8.3].

We close this section by mentioning that the leading term in the difference $\Phi(T) - \Phi(0)$ as $T \to 0^+$, where Φ is any thermodynamic quantity (such as specific heat, entropy or energy), is given by replacing the bare mass m in the corresponding free particle expression by the effective mass m^* [9.1]. In other words, for a normal Fermi system and for the purpose of calculating the temperature dependence of thermodynamic quantities in the limit $T \to 0$, one can replace the strongly interacting particles by noninteracting quasi-particles of an effective mass m^*.

In this section we have outlined how the many-body Green's function formalism can be used to justify and quantify the fundamental idea that an interacting system of particles can be replaced by a weakly interacting system of quasi-particles. For many purposes the interactions among quasi-particles can be omitted and consequently the so-called one-body approximation is justified. For normal systems there is one-to-one correspondence between particles and quasi-particles, and each quasi-particle reduces to a real particle as the interactions are turned off. Note though that the interactions produce a Bose type excitation (a density wave) which is absent in a noninteracting system.

We mention once again that there are systems (such as the superconducting materials or superfluid He^3) where the interactions play a fundamental role in reorganizing the low-lying states to a new configuration which cannot result from the unperturbed states in a continuous way. For such a system the Green's function formalism requires some modifications. The reader is referred to the book by MATTUCK [9.5] for an introduction. More detailed treatment is given in [8.3] and [9.1]. The modified Green's function formalism finds one of the most important applications in the study of these nonnormal fermion systems.

11.2 High-Density Electron Gas

In this section we consider a particular but important fermion system, the so-called jellium model: a high-density electronic system moving in the background of positive charge to ensure overall electrical neutrality. The jellium model is an idealization of a real metal; this idealization omits the discrete character of the positive background, its dynamics (phonons), and the interaction of the ion motion with the electrons. The high-density jellium model, although it is a normal Fermi system, has certain peculiarities associated with the long-range character of the electron-electron interaction. The $1/r^2$ nature of the Coulomb force gives that the average potential energy per particle is proportional to $n^{1/3}$; on the other hand, the average kinetic energy per particle, as a result of the Pauli principle, is proportional to $k_F^2/2m$, i.e., proportional to $n^{2/3}$, where $n = N/\Omega$ is the density. Thus, in contrast to ordinary gases, in the high-density limit the kinetic energy of the electron system dominates the potential energy, and therefore the system behaves as a gas; on the other hand, in the low-density limit the potential energy is the dominant one, and the system becomes a (Wigner) solid; for intermediate densities we have a liquid. We can express formally this feature by introducing the dimensionless interparticle spacing r_s, where

$$r_s = r_0/a_B \quad ; \tag{11.14}$$

r_0 is defined by $4\pi r_0^3/3 = \Omega/N$ and $a_B = \hbar^2/me^2$ is the Bohr radius. By introducing the dimensionless quantities Ω', $\underline{k}'$, $\underline{p}'$ and $\underline{q}$, where

$$\Omega' = \Omega/r_0^3, \quad \underline{k}' = r_0\underline{k} \quad , \quad \underline{p}' = r_0\underline{p} \quad , \quad \underline{q}' = r_0\underline{q} \quad , \tag{11.15}$$

we can rewrite the Hamiltonian of the jellium model (see Appendix C) as

$$H = \frac{e^2}{a_B r_s^2}\left[\frac{1}{2}\sum_{\underline{k}'} k'^2 a^\dagger_{\underline{k}'} a_{\underline{k}'} + \frac{r_s}{2\Omega'}\sum'_{\underline{k}'\underline{p}'\underline{q}'} \frac{4\pi}{q'^2} a^\dagger_{\underline{k}'+\underline{q}'} a^\dagger_{\underline{p}'-\underline{q}'} a_{\underline{p}'} a_{\underline{k}'}\right] \tag{11.16}$$

which shows that in the high-density limit ($r_s \to 0$) the potential energy becomes negligible. Note that the prime in the summation means that the $q' = 0$ term must be excluded.

By inspection of (11.16) one may suspect that terms of higher than second-order in the potential energy would give contributions which approach zero as $r_s \to 0$. Hence, it is reasonable to try to terminate the diagrammatic expansion in the second order. Because of the $q' = 0$ exclusion the diagrams involving a $q' = 0$ interaction line give no contribution and can be omitted. Thus the only diagrams which contribute to the proper self energy up to second order are those of Fig.11.1. The diagrams Σ_1^*, Σ_{2a}^* and Σ_{2b}^* can be calculated according to the rules and give no problems. Explicit results can be found in [8.3]. The diagram Σ_{2r}^* is troublesome and requires detailed consideration; we have

$$\Sigma_{2r}^*(\underline{k},\omega) = i\int \frac{d^3q}{(2\pi)^3}\frac{d\omega_1}{2\pi}\, g_0(\underline{k}-\underline{q},\, \omega-\omega_1) v^2(q)\, \Pi_0^*(\underline{q},\omega_1) \quad , \tag{11.17}$$

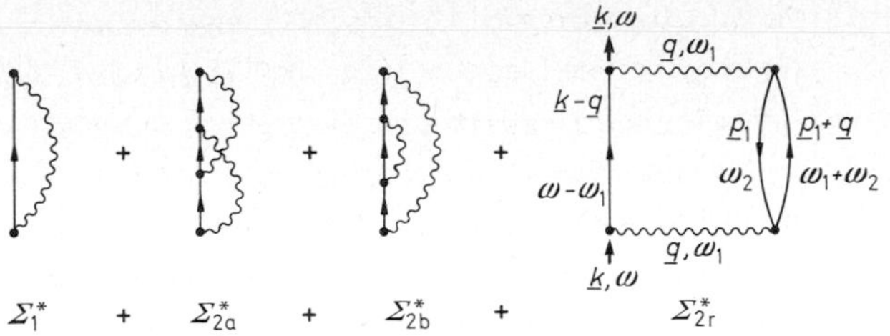

Fig.11.1. All the contributions of up to second order to the proper self energy of an electron gas

where $-i\Pi_0^*$ is the contribution of the bubble, i.e., the contribution of the lowest order to the proper polarization

$$\Pi_0^*(\underline{q},\omega_1) = -i\int \frac{d^3p_1}{(2\pi)^3}\frac{d\omega_2}{2\pi} g_0(\underline{p}_1,\omega_2)g_0(\underline{p}_1+\underline{q},\omega_1+\omega_2) \quad . \tag{11.18}$$

This quantity $\Pi_0^*(\underline{q},\omega_1)$ is calculated explicity in [8.3]. It turns out that

$$\lim_{\underline{q}\to 0} \Pi_0^*(\underline{q},|\underline{q}|x) = F(x) \quad , \tag{11.19}$$

where $F(x)$ is an integrable function of $x \equiv \omega_1/|\underline{q}|$. Substituting (11.19) in (11.17) and taking into account that $v(\underline{q}) = 4\pi e^2/\underline{q}^2$ we obtain that the integral over q behaves for small q as:

$$\Sigma_{2r}^*(\underline{k},\omega) \sim \int_0 \frac{d^3q}{q^4}\, q \sim \int_0 \frac{dq}{q} \quad , \tag{11.20}$$

i.e., it diverges logarithmically. This logarithmic divergence simply indicates that the perturbation expansion in powers of the Coulomb potential $4\pi e^2/\underline{q}^2$ does not converge. One way to avoid this difficulty is to replace the Coulomb potential by $e^2\exp(-\lambda r)/r$, which has a Fourier transform equal to

$$v_\lambda(\underline{q}) = \frac{4\pi e^2}{\underline{q}^2+\lambda^2} \quad , \tag{11.21}$$

with λ being a small quantity. One can then try to sum the perturbation series in powers of v_λ and take the limit $\lambda \to 0$ in the final result. In summing the power series expansion one must keep in mind that there are now two independent small quantities, namely r_s and λ, and hence one cannot terminate the series at the second order because some of the higher-order terms (which tend to zero as $r_s \to 0$) may approach infinity as $\lambda \to 0$. To handle this problem correctly one must find for each order of perturbation series (i.e., for each power of r_s) the leading term(s) in λ. In second order the leading term in λ is Σ^*_{2r} which behaves as $\ln|\lambda|$ in the limit $\lambda \to 0$. This behavior stems from the fact that the two interaction lines in Σ^*_{2r} have the same $\underline{q}$. It is easy to see that the leading term in λ of nth order will come from the diagram which has the same $\underline{q}$ in all interaction lines (see Fig.11.2). Such a diagram would give a contribution Σ^*_{nr} which is proportional to

$$\Sigma^*_{nr} \sim r_s^{n-2} \int_0 \frac{d^3q}{(\underline{q}^2+\lambda^2)^n}\, q \sim r_s^{n-2}\lambda^{2-n} \quad ; \quad n = 3,4, \dots \tag{11.22}$$

while all the other diagrams of nth order give a contribution of the order $r_s^{n-2}\lambda^{3-n}$ or less. Thus in the limit $\lambda \to 0$ the nth-order contribution to Σ^*, Σ^*_n, equals Σ^*_{nr}. We can conclude that in the limit of small r_s and small λ the proper self-energy is given by

$$\Sigma^* \simeq \Sigma^*_1 + \Sigma^*_{2a} + \Sigma^*_{2b} + \sum_{n=2}^{\infty} \Sigma^*_{nr} \quad . \tag{11.23}$$

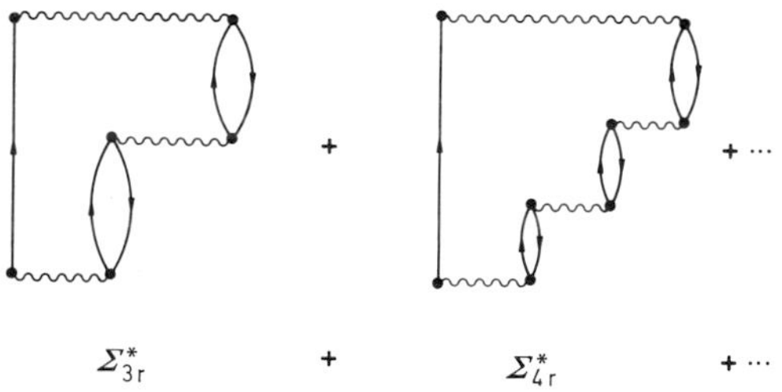

Fig.11.2. The leading diagrams of third, fourth, etc. order for the proper self-energy of an electron gas

The quantity $\Sigma_1^* + \Sigma_{2r}^* + \Sigma_{3r}^* + \ldots$ can be written as in Fig.11.3, where

$$v_{e,RPA} = v_\lambda + v_\lambda^2 \Pi_0^* + v_\lambda^3 \Pi^{*2} + \ldots = \frac{v}{1-v\Pi_0^*} \quad . \tag{11.24}$$

Fig.11.3. The above diagram is the sum $\Sigma_1^* + \Sigma_{2r}^* + \Sigma_{3r}^* + \ldots$; $v_{e,RPA}$ is the effective interaction corresponding to the approximation $\Pi^* = \Pi_0^*$

Thus we have for $\Sigma_r^* = \sum_{n=2}^{\infty} \Sigma_{nr}^*$

$$\Sigma_r^* = i \int \frac{d^3q}{(2\pi)^3} \frac{d\omega_1}{2\pi} g_0(\underline{k} - \underline{q}, \omega - \omega_1)[v_{e,RPA}(\underline{q},\omega) - v(\underline{q})]$$

$$= i \int \frac{d^3q}{(2\pi)^3} \frac{d\omega_1}{2\pi} g_0(\underline{k} - \underline{q}, \omega - \omega_1) \frac{v^2(\underline{q})\Pi_0^*(\underline{q},\omega_1)}{1-v(\underline{q})\Pi_0^*(\underline{q},\omega_1)} \quad . \tag{11.25}$$

Note that $v_{e,RPA}$ results from the general expression (10.63') by replacing the proper polarization Π^* by its lowest approximation Π_0^*. The approximation

$$\Pi^*(\underline{q},\omega_1) \simeq \Pi_0^*(\underline{q},\omega_1) \quad , \tag{11.26}$$

is known, for historical reasons, as the random phase approximation (RPA). It is also known as the summation of the ring diagrams. It must be stressed that in the right-hand side of (11.24) or (11.25) we can take the limit $\lambda \to 0$ without any divergence.

The approximation (11.26) coupled with the general equation (10.65) for the dielectric function gives

$$\varepsilon_{RPA}^c(\underline{q},\omega) = 1 - v(\underline{q})\Pi_0^*(\underline{q},\omega) \quad . \tag{11.27}$$

Having obtained Σ^* and ε^c we can calculate several quantities of physical interest:

a) *Quasi-particle's* ε_k *and* γ_k. The reader is referred to [9.5] and references therein.

b) *Ground-state energy*. Taking into account the general thermodynamic relation

$$\left\langle\frac{\partial H}{\partial\alpha}\right\rangle = \frac{\partial E}{\partial\alpha}\bigg|_{N,\Omega} \quad , \qquad (11.28)$$

(9.29), (9.7), Fig.9.1 and (10.58), we can express the ground-state energy E as follows

$$E = E_0 - \frac{i}{2}\int\frac{d\alpha}{\alpha}\sum_{\underline{k}}\int_C\frac{d\omega}{2\pi}\,\Sigma^*(\underline{k},\omega)g(\underline{k},\omega) \quad , \qquad (11.29)$$

where E_0 is the unperturbed ground-state energy, the contour C consists of the path shown in Fig.9.1a and a semicircle in the upper half plane and Σ^* and g correspond to an interaction potential equal to $\alpha v(\underline{q})$. Equation (11.29) is more convenient for calculations than (9.30). Substituting (11.23) in (11.29) and expanding g in powers of the small quantity Σ^*, we obtain that

$$g\Sigma^* \simeq (\Sigma_1^* + \Sigma_{2a}^* + \Sigma_{2b}^* + \Sigma_r^*)g_0 + (\Sigma_1^* g_0)^2 \quad , \qquad (11.30)$$

where the omitted terms in (11.30) would give a contribution to E which would approach zero as $r_s \to 0$. Combining (11.29) with (11.30) and performing the integrations, we obtain for the ground-state energy

$$E = \frac{e^2 N}{2a_B}\left[\frac{2.21}{r_s^2} - \frac{0.916}{r_s} + 0.0622\,\ln(r_s) - 0.094 + \ldots \quad .\right] \qquad (11.30')$$

where the first term in parenthesis is the unperturbed ground-state energy, the second corresponds to the first order $g_0\Sigma_1^*$ in (11.30), the third comes from $g_0\Sigma_r^*$ and the fourth from the sum $g_0(\Sigma_{2a}^* + \Sigma_{2b}^*) + (\Sigma_1^* g_0)^2$. The remaining terms approach zero as $r_s \to 0$. Equation (11.30') shows that we cannot obtain a series expansion of E in powers of r_s.

c) *Effective interaction* $v_{e,RPA}$. Using (11.27), we have

$$v_{e,RPA} = v(\underline{q})/\varepsilon^c_{RPA}(q) = v(\underline{q})/[1 - v(\underline{q})\Pi_0^*(q)] \quad , \qquad (11.31)$$

which for the static case, $\omega = 0$, becomes [8.3]

$$v_{e,RPA} = \frac{4\pi e^2}{\underline{q}^2 + \underline{q}_{TF}^2 f(q/k_F)} \qquad (11.32)$$

where the Thomas-Fermi screening length q_{TF}^{-1} is given by

$$q_{TF}^2 = \frac{4r_s}{\pi}\left(\frac{4}{9\pi}\right)^{1/3} k_F^2 \tag{11.33}$$

and the function f(x) equals to

$$f(x) = \frac{1}{2} - \frac{1}{2x}\left(1 - \frac{1}{4}x^2\right) \ln\left|\frac{2 - x}{2 + x}\right| \quad . \tag{11.34}$$

In the limit $q \to 0$, (11.32) becomes:

$$v_{e,RPA}(\underline{q},0) \simeq \frac{4\pi e^2}{\underline{q}^2 + q_{TF}^2} \quad \text{as} \quad q \to 0 \quad , \tag{11.35}$$

which shows clearly the effects of screening produced by the summation of the ring diagrams.

d) *Response to an external static point charge.* From $\varepsilon^c(q)$ we can immediately obtain with the use of (10.69,70) the retarded dielectric function $\varepsilon(q)$, which gives the following expression for the electronic charge density $\delta\rho(\underline{r})$ induced by a static point charge $Z|e|$ at the origin:

$$\delta\rho_r(\underline{r}) = -Z|e| \int \frac{d^3q}{(2\pi)^3} e^{i\underline{q}\cdot\underline{r}} \frac{q_{TF}^2 f(q/k_F)}{q^2 + q_{TF}^2 f(q/k_F)} \quad ; \tag{11.36}$$

the subscript r indicates that the ring approximation was employed. Some interesting consequences of (11.36) are discussed in [8.3]. Here we mention that

$$\delta\rho_r(\underline{r}) \xrightarrow[r \to \infty]{} \frac{Z|e|}{\pi} \frac{2\xi}{(4+\xi)^2} \frac{\cos(2k_F r)}{r^3} \quad ; \quad \xi \equiv q_{TF}^2/2k_F^2 \quad , \tag{11.37}$$

i.e., the induced charge does not decay exponentially, as one would conclude from the approximate expression (11.35), but in an oscillating power-law way. These Friedel oscillations [11.3] arise from the sharp Fermi surface, which produces the singularity at $q = 2k_F$ of the function $f(q/k_F)$.

e) *Collective (plasma) oscillations.* As it was mentioned in Chap.9, collective density oscillation modes appear as poles in the analytic continuation of $\Pi(\underline{k},\omega)$ or, in view of (10.71), as zeros in the analytic continuation of $\varepsilon^c(\underline{k},\omega)$ or $\varepsilon(\underline{k},\omega)$. Within the RPA $\varepsilon^c(\underline{k},\omega)$ is given by (11.27), so the collective mode of the electron gas has a complex eigenfrequency $z_q = \omega_q - i\bar{\varepsilon}(\omega_q)\gamma_q$ which satisfies the equation

$$1 = v(\underline{q})\text{a.c.}\{\Pi_0^*(\underline{q},z_q)\} \quad . \tag{11.38}$$

Using the properties of $\Pi_0^*(\underline{q},\omega)$ as $q \to 0$ and the fact that $v(\underline{q}) \sim \underline{q}^{-2}$ as $\underline{q} \to 0$, we can prove [8.3] that

$$z_q = \omega_q \simeq \omega_p\left[1 + \frac{9}{10}\left(\frac{q}{q_{TF}}\right)^2\right] \quad \text{as} \quad q \to 0 \quad , \tag{11.39}$$

where ω_p, the plasma frequency, is given by

$$\omega_p^2 = \frac{4\pi e^2 n}{m} \quad . \tag{11.40}$$

Equation (11.39) shows that the plasma has a dispersion relation which starts from a finite value ω_p at $q = 0$ and grows quadratically with q. Within the RPA the lifetime $1/\gamma_q$ is exactly infinite for small q. If the RPA is relaxed, γ_q becomes different from zero for all $q \neq 0$. It is worthwhile to note that if v(q) was approaching a constant v_0 as $q \to 0$, i.e., if $v(\underline{r})$ was of short range, then the solution of (11.38) would have a dispersion relation of the form

$$\omega_q = c_0 q \quad \text{as} \quad q \to 0 \tag{11.41}$$

where the velocity c_0 would satisfy the equation [8.3]

$$1 = \frac{v_0 m k_F}{\pi^2}\left[\frac{1}{2}\frac{c_0}{v_F^0}\,\ln\left|\frac{c_0/v_F^0 + 1}{c_0/v_F^0 - 1}\right| - 1 - \frac{i\pi}{2}\frac{c_0}{v_F^0}\,\theta\left(1 - \left|\frac{c_0}{v_F^0}\right|\right)\right] \quad , \tag{11.42}$$

where $v_F^0 = k_F/m$. Equation (11.42) shows that $c_0/v_F^0 > 1$; otherwise there is no real solution. In the weak coupling limit $v_0 m k_F \ll 1$, the solution of (11.42) is

$$c_0 \simeq v_F^0\left[1 + 2\exp\left(-\frac{2\pi^2}{mk_F v_0} - 2\right)\right] \quad , \tag{11.43}$$

while in the strong coupling limit $v_0 mk_F >> 1$, we have

$$c_0 \simeq \sqrt{nv_0/m} \quad . \tag{11.44}$$

As we have mentioned in Sect.11.1, the collective mode with the linear dispersion relation is called zero sound. For the Coulomb interaction case the collective mode is called plasmon and its dispersion is given by (11.39), i.e., it approaches a nonzero value ω_p as $q \to 0$. This nonzero ω_p implies a nonzero restoring force as $q \to 0$, which is a result of the long-range character of the Coulomb interaction. Note that the zero sound or the plasmon dispersion are given in general as solution of the same equation, $1 = v(\underline{q})\text{a.c.}\{\Pi^*(\underline{q},z)\}$ which in the RPA reduces to (11.38).

11.3 Dilute Fermi Gas

In this section we examine the case of a low-density Fermi system with repulsive short-range interactions. This is a particular example of the general theory outlined in Sect.11.1. Its interest lies in the fact that explicit results can be obtained which are of relevance to the studies of nuclear matter and He^3.

The small parameter in this case is $k_F a$, where a is the scattering length defined by the relation

$$f(\underline{k}_f,\underline{k}) \to -a \quad \text{as} \quad k_f = k \to 0 \quad , \tag{11.45}$$

where $f(\underline{k}_f,\underline{k})$ is the scattering amplitude (see Chap.4) corresponding to the potential $v(\underline{r})$. Thus the theory outlined here is appropriate for either a dilute system ($k_F \to 0$) or short-range interactions $a \to 0$. For nuclear matter $k_F a \simeq 1/3$.

Because $k_F a \to 0$, one could guess that the most important diagrams are those describing the interactions of pairs of particles to all orders. Quantitatively, this means that the vertex part Γ could be approximated by the so-called ladder diagrams shown in Fig.11.4a. One can prove [8.3,9.5] that the diagrams for Γ omitted in Fig.11.4a give contributions to Σ^* of higher order in $k_F a$. This statement can be demonstrated by comparing the two diagrams for Σ^* in Fig.11.5. Diagram (a) is proportional to $k_F a$ while diagram (b) is proportional to $(k_F a)^2$; in general the power of $k_F a$ is determined by the number of reverse running g_0-lines. It follows that the

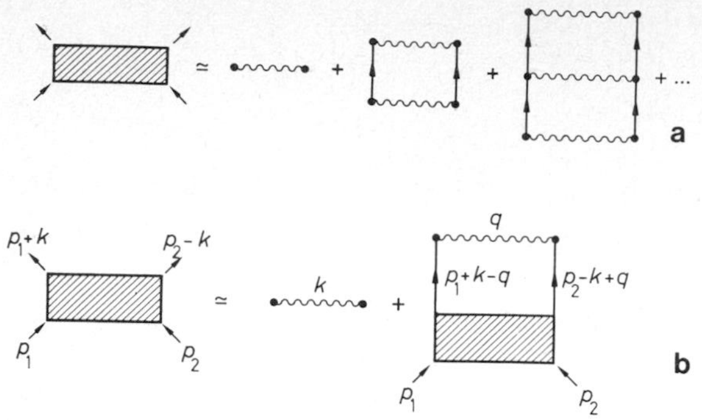

Fig.11.4. (a) The vertex part Γ in the ladder approximation; (b) The integral equation obeyed by Γ in the ladder approximation

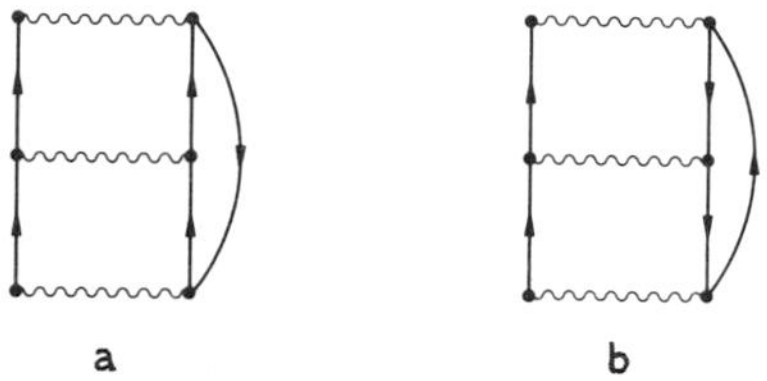

Fig.11.5. Diagram (a) was retained in the ladder approximation while diagram (b) was omitted

lowest order in k_Fa is obtained by keeping diagrams in Γ involving only parallel running g_0-lines, i.e., by keeping the ladder diagrams. The summation of the ladder diagrams is equivalent to an integral equation for Γ, as shown in Fig.11.4b. This integral equation is known as the ladder approximation to the Bethe-Salpeter equation; its kernel involves the product vg_0g_0. It is much more convenient to express the integral equation for Γ in terms of the scattering amplitude $f(\underline{k}',\underline{k})$ which is related to the potential $v(\underline{q})$ through (4.62). After some lengthy algebraic manipulations one obtains

$$\Gamma(\underline{p},\underline{p}';P) = \Gamma_0(\underline{p},\underline{p}';P) + m\int \frac{d^3k}{(2\pi)^3}\,\Gamma_0(\underline{p},\underline{k};P)\left[\frac{N(\underline{P},\underline{k})}{mE-\underline{k}^2+isN(\underline{P},\underline{k})} - \frac{1}{mE-\underline{k}^2+is}\right]\Gamma(\underline{k},\underline{p}',P) \quad , \tag{11.46}$$

with

$$\Gamma_0(\underline{p},\underline{p}';P) = -\frac{4\pi}{m} f(\underline{p},\underline{p}') + \frac{(4\pi)^2}{m} \int \frac{d^3k}{(2\pi)^3} f(\underline{p},\underline{k})\left[\frac{1}{mE-\underline{k}^2+is} + \frac{1}{\underline{k}^2-\underline{p}'^2-is}\right] f^*(\underline{p}',\underline{k}) \quad ; \quad (11.47)$$

in the above equations $s \to 0^+$; the four momentum P stands for $p_1 + p_2$; the four momentum p stands for $(p_1 - p_2)/2 + k$ and $p' = (p_1 - p_2)/2$; $E = P_0 - \underline{P}^2/4m$ is the total energy of the pair in the center of mass frame; $N(\underline{P},\underline{k}) = 1(-1)$ if both $\underline{P}/2 \pm \underline{k}$ are outside (inside) the Fermi sea; $N(\underline{P},\underline{k}) = 0$ otherwise. The quantity Γ_0 is the vertex part for two particles in the vacuum: $\Gamma \to \Gamma_0$ as $k_F \to 0$. The usefulness of (11.46,47) is that the quantity Γ is related to the observable scattering amplitude which remains finite even for infinite potentials such as the hard-sphere potential. Equations (11.46,47) were obtained first by GALITSKII [11.4] and BELIAEV [11.5].

The integration over $\underline{k}$ in (11.46) is essentially restricted within the Fermi sea, because the integrand vanishes when both $\underline{P}/2 \pm \underline{k}$ are outside the Fermi sea. Since we are interested in energies of the order ε_F, the last term in (11.46) is of the order $m\Gamma_0 k_F \Gamma$, so that $(\Gamma - \Gamma_0)/\Gamma \sim m\Gamma_0 k_F \sim fk_F \sim k_F a$. Thus (11.46) can be solved by iteration in the present case where $k_F a \ll 1$. In the same limit only small values of $\underline{p},\underline{p}'$ and $\underline{k}$ are important in (11.47) for which the scattering amplitude $f(\underline{k},\underline{k}')$ behaves as

$$f(\underline{k},\underline{k}') = -a + ika^2 + O(k^2a^3) \quad ; \quad |\underline{k}| = |\underline{k}'| \to 0 \quad . \quad (11.48)$$

Substituting (11.48) in (11.47), replacing Γ by Γ_0 in the last term of (11.46), and keeping terms of order a^2 we obtain Γ to order a^2. Replacing the result for Γ in the equation shown in Fig.10.10a (with the g-lines approximated by g_0-lines) we have the self-energy to order $(k_F a)^2$. For details the reader is referred to [8.3]. The final results for physical quantities are the following [8.3]

$$\gamma_k = \frac{k_F^2}{2m}\frac{2}{\pi}(k_F a)^2\left(\frac{k_F - k}{k_F}\right)^2 \bar{\varepsilon}(k_F - k) + \dots \quad (11.49)$$

$$\mu \equiv \varepsilon_{k_F} = \frac{k_F^2}{2m}\left[1 + \frac{4}{3\pi}k_F a + \frac{4}{15\pi^2}(11 - 2\ln 2)(k_F a)^2 + \dots\right] \quad (11.50)$$

$$\frac{m^*}{m} = 1 + \frac{8}{15\pi^2}(7\ln 2 - 1)(k_F a)^2 + \dots \quad (11.51)$$

$$\langle H \rangle = \frac{Nk_F^2}{2m}\left[\frac{3}{5} + \frac{2}{3\pi}k_F a + \frac{4}{35\pi^2}(11 - 2\ln 2)(k_F a)^2 + \dots\right] \quad . \tag{11.52}$$

The theory outlined above can be improved if in the equation shown in Fig. 11.4b the g_0-lines are replaced by g-lines. The resulting integral equation together with the equation of Fig.10.10a (which in view of the integral equation for Γ can be simplified as in Fig.11.6) and the equation $g^{-1} = \omega - \varepsilon_k^0 - \Sigma^*$ form a complete set of equations. For a discussion of this improvement the reader is referred to [8.3].

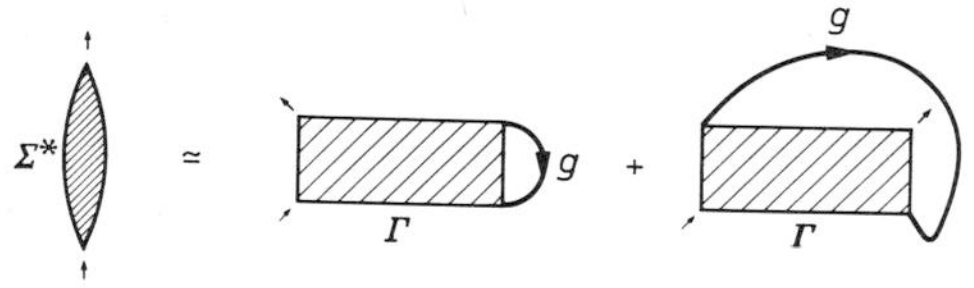

Fig.11.6. In the self-consistent (i.e., with the g_0-lines replaced by g-lines) ladder approximation Σ^* is related to Γ as shown

11.4 Summary

We first outlined the proof that the normal state of a Fermi system is a consistent solution of the equations, and it goes continuously to the free-particle state as the interactions are turned off. Thus, if the actual state of the interacting system develops in a continuous way as the interactions are turned on, the system is normal, i.e., it has a well-defined Fermi surface determined from the equation

$$\varepsilon_{k_F} = \mu$$

where ε_k is the quasi-particle energy. Moreover, the quasi-particle lifetime approaches infinity as $(k - k_F)^{-2}$ in the limit $k \to k_F$; and each quasi-particle is in one-to-one correspondence with the real particles.

We then indicated how the Green's function formalism can be used to justify Landau's theory. The most important results are: 1) The Fermi momentum k_F (as defined above) satisfies the simple equation

$$\frac{4\pi}{3}\frac{k_F^3}{(2\pi)^3} = \frac{N}{\Omega} \quad . \tag{11.11}$$

2) The effective mass m^* defined by the equation

$$m^* = k_F / |\partial\varepsilon_k/\partial k|_{k=k_F} \tag{11.9}$$

is related with the bare mass m as follows

$$\frac{1}{m^*} = \frac{1}{m} - \frac{k_F w_{k_F}^2}{2(2\pi)^3} \int \Gamma^{\omega}(p_1,p_2)\cos\theta d\mathcal{O} \tag{11.12}$$

where Γ^{ω} is a limiting value of the vertex part. 3) There are also collective excitations (such as zero sound) in an interacting Fermi system with an acoustic-like dispersion relation which is given by the poles of the vertex part Γ or the poles of the polarization part Π.

The above results are quite important because they provide (for normal Fermi systems) a justification for the one-body approximation which is widely used in solid-state and nuclear physics. Moreover, they give corrections and refinements; they also clarify the conditions under which the one-body approximation is valid.

We also examined two important specific examples for which explicit results were obtained: 1) The high-density electron gas moving in a uniform positive background. This is a normal Fermi system which approximates electrons in a metal. It exhibits some peculiarities associated with the long-range character of the Coulomb interaction; the most important consequence of this feature is that the collective excitation, which is called plasmon, has a dispersion relation of the form

$$\omega_q^2 = \omega_p^2 + Aq^2 \quad \text{as} \quad q \to 0 \quad . \tag{11.39}$$

2) A low-density Fermi system with short-range repulsive interactions. Explicit results for the quasi-particles, and the ground-state energy were obtained; these results are applicable to nuclear matter and with appropriate refinements to liquid He^3.

We must mention that the Green's function formalism is very useful in describing effects due to interactions among other elementary excitations in solids such as phonons and magnons.

Finally, we must stress the importance of the Green's function formalism in treating systems where the interactions trigger a profound change in the ground state (superconducting materials, superfluid He^3). These systems were not examined here. This omission by no means reflects upon the significance of the subject.

Appendix A: Analytic Behavior of G (z) Near a Band Edge

The Green's function G(z) can be expressed in terms of the discontinuity $\tilde{G}(E) \equiv G^+(E) - G^-(E)$ as follows

$$G(z) = \frac{i}{2\pi}\int_{-\infty}^{\infty} dE\,\frac{\tilde{G}(E)}{z-E} \quad . \tag{A.1}$$

The derivative of G(z) is obtained by differentiation under the integral. Integrating by parts we obtain

$$G'(z) = \frac{i}{2\pi}\int_{-\infty}^{\infty} dE\,\frac{\tilde{G}'(E)}{z-E} \quad , \tag{A.2}$$

where the prime denotes differentiation with respect to the argument. In obtaining (A.2) we have assumed that $\tilde{G}'(E)$ exists and is integrable and that $\tilde{G}(E)/E \to 0$ as $|E| \to \infty$. By taking the limits $\mathrm{Im}\{z\} \to 0^{\pm}$ and using (1.20) we obtain

$$G^{\pm}(E) = \frac{i}{2\pi}\,P\int_{-\infty}^{\infty} dE'\,\frac{\tilde{G}(E')}{E-E'} \pm \frac{1}{2}\,\tilde{G}(E) \tag{A.3}$$

$$\frac{d}{dE}\,G^{\pm}(E) = \frac{i}{2\pi}\,P\int_{-\infty}^{\infty} dE'\,\frac{\tilde{G}'(E')}{E-E'} \pm \frac{1}{2}\,\tilde{G}'(E) \tag{A.4}$$

where one can show under the above assumptions that $G'(z) \xrightarrow[\mathrm{Im}\{z\}\to 0^{\pm}]{} d\,G^{\pm}(E)/dE$.

We would like to connect the behavior of G(z) and G'(z) as $z \to E_0$ with the behavior of $\tilde{G}(E)$ around E_0. One can prove the following theorems [3.2] about F(z) given by

$$F(z) = \frac{i}{2\pi}\int_{-\infty}^{\infty} dx\,\frac{f(x)}{z-x} = \frac{1}{2\pi i}\int_{-\infty}^{\infty} \frac{f(x)dx}{x-z} \quad : \tag{A.5}$$

1) If f(x) satisfies the condition

$$|f(x_1) - f(x_2)| \leqq A|x_1 - x_2|^{\mu} \tag{A.6}$$

for all x_1 and x_2 in the neighborhood of x_0 where A and μ are positive constants, then

$$|F(z_1) - F(z_2)| \leqq A'|z_1 - z_2|^{\mu} \; ; \quad \text{when} \quad \mu < 1 \tag{A.7}$$

and

$$|F(z_1) - F(z_2)| \leqq A'|z_1 - z_2|^{1-\varepsilon} \; ; \quad \text{when} \quad \mu = 1 \tag{A.8}$$

for all z_1 and z_2 in the neighborhood of x_0 and $\mathrm{Im}\{z_1\}\mathrm{Im}\{z_2\} > 0$; A' is a positive constant and ε is an arbitrary positive number. Inequalities (A.7,8) are valid as one or both of z_1 and z_2 reach the real axis either from above or from below. Thus, e.g.,

$$|F^+(x_1) - F^+(x_2)| \leqq A'|x_1 - x_2|^{\mu} \; ; \quad \text{when} \quad \mu < 1 \tag{A.9}$$

with a similar relation for $F^-(x_1) - F^-(x_2)$.

2) If f(x) is discontinuous at $x = x_0$, i.e.,

$$f(x_0^+) - f(x_0^-) = D \neq 0 \tag{A.10}$$

but otherwise f(x) satisfies the condition (A.6) for both x_1 and x_2 being either to the left or to the right of x_0, then

$$F(z) = \frac{D}{2\pi i} \ln\left(\frac{1}{x_0 - z}\right) + F_0(z) \tag{A.11}$$

where $F_o(z)$ satisfies (A.7) or (A.8).

3) If f(x) behaves around x_0 as

$$f(x) = f_0(x) + f_1(x)(\pm x \mp x_0)^{-\gamma}\theta(\pm x \mp x_0) \tag{A.12}$$

where $f_0(x)$ and $f_1(x)$ satisfy the condition (A.6) and $0<\gamma<1$, then

$$F(z) = \pm \frac{e^{\pm\gamma\pi i}}{2i\sin\gamma\pi} \frac{f_1(x_0)}{(\pm z \mp x_0)^{\gamma}} + F_0(z) \quad , \tag{A.13}$$

where $(\pm z\mp x_0)^{-\gamma}$ is the branch which coincides with the branch $(\pm x\mp x_0)^{-\gamma}$ in (A.12) as z approaches x from the upper half plane and

$$|F_0(z)| < \frac{C}{|z-x_0|^{\gamma_0}} \quad , \tag{A.14}$$

where C and γ_0 are positive constants such that $\gamma_0 < \gamma$.

The above theorems allow us to obtain the analytic behavior of G(z) near a band edge. Thus,

1) When G(E) goes to zero as $(E - E_B)^{\mu}$ near the band edge E_B, then, according to theorem 1 above, G(z) and $G^{\pm}(E)$ are bounded in the neighborhood of E_B. The derivatives G'(z) and $G^{\pm}{}'(E)$ can be obtained using (A.2). Since $\tilde{G}'(E)$ behaves as $(E - E_B)^{\mu-1}$ inside the band and near the band edge, then, according to theorem 3 above the quantities G'(z) and $G^{\pm}{}'(E)$ behave as $(z - E_B)^{\mu-1}$ and $(E - E_B)^{\mu-1}$ as z and E approach E_B. For a free particle in 3-D, $\mu = 1/2$.

2) When $\tilde{G}(E)$ goes to zero in a discontinuous way at the band edge (as in the 2-D free particle case), one can apply theorem 2 above. Consequently G(z) and $G^{\pm}(E)$ exhibit a logarithmic singularity as shown in (A.11).

3) When $\tilde{G}(E)$ behaves as $|E - E_B|^{-\gamma}$ with $0<\gamma<1$ in the interior of the band and near the band edge E_B, then, according to theorem 3 above, G(z) and $G^{\pm}(E)$ behave as $(z - E_B)^{-\gamma}$ and $(E - E_B)^{-\gamma}$, respectively. For a free particle in 1-D, $\gamma = 1/2$.

Appendix B: The Renormalized Perturbation Expansion (RPE)

Consider the tight-binding Hamiltonian $H = H_0 + H_1$, where

$$H_0 = \sum_{\underline{\ell}} |\underline{\ell} > \varepsilon_{\underline{\ell}} < \underline{\ell}| \tag{B.1}$$

$$H_1 = V \sum_{\underline{\ell}\underline{m}}' |\underline{\ell}><\underline{m}| \tag{B.2}$$

and the summation in (B.2) extends over nearest-neighbor sites only. If we consider H_0 as the unperturbed Hamiltonian and H_1 as a perturbation, we can apply the formalism developed in Chap.4. Thus we have for $G(z) \equiv (z - H)^{-1}$

$$G = G_0 + G_0H_1G_0 + G_0H_1G_0H_1G_0 + \dots$$

or

$$G(\underline{\ell},\underline{m}) = G_0(\underline{\ell},\underline{m}) + \sum_{\underline{n}_1\underline{n}_2} G_0(\underline{\ell},\underline{n}_1)<\underline{n}_1|H_1|\underline{n}_2>G_0(\underline{n}_2\underline{m})$$

$$+ \sum_{\underline{n}_1\dots\underline{n}_4} G_0(\underline{\ell},\underline{n}_1)<\underline{n}_1|H_1|\underline{n}_2>G_0(\underline{n}_2,\underline{n}_3)<\underline{n}_3|H_1|\underline{n}_4>G_0(\underline{n}_4,\underline{m}) + \dots \tag{B.3}$$

It is clear from (B.1) that $G_0(\underline{n}_1,\underline{n}_2) = \delta_{\underline{n}_1\underline{n}_2} G_0(\underline{n}_1)$ where $G_0(\underline{n})$ is

$$G_0(\underline{n}) = \frac{1}{z - \varepsilon_{\underline{n}}} \quad . \tag{B.4}$$

Similarly, $<\underline{n}_1|H_1|\underline{n}_2>$ is different from zero only when $\underline{n}_1$ and $\underline{n}_2$ are nearest neighbors. Hence (B.3) can be simplified as follows:

$$G(\underline{\ell},\underline{m}) = \delta_{\underline{\ell}\underline{m}} G_0(\underline{\ell}) + G_0(\underline{\ell})VG_0(\underline{m})\delta_{\underline{\ell},\underline{m}+1} + \sum_{\underline{n}_1} G_0(\underline{\ell})VG_0(\underline{n}_1)VG_0(\underline{m}) + \dots \quad . \tag{B.5}$$

One way of keeping track of the various terms in the expansion (B.5) is to consider all possible paths in the lattice starting from site $\underline{\ell}$ and ending at site $\underline{m}$ with steps connecting one lattice site with a nearest neighbor site. There is a one-to-one correspondence between the terms in (B.5) and the set of all such paths. Each term in (B.5) can be obtained from the corresponding path by calculating a product according to the following rules: 1) For each lattice site $\underline{n}$ (including the initial $\underline{\ell}$ and the final $\underline{m}$) visited by the path include a factor $G_0(\underline{n})$; 2) For each step from one site to a nearest neighbor site include a factor V. Thus, the contribution to $G(\underline{\ell},\underline{m})$ corresponding to the path shown in Fig.B.1 is

$$G_0(\underline{\ell})VG_0(\underline{n}_1)VG_0(\underline{n}_2)VG_0(\underline{n}_1)VG_0(\underline{n}_2)VG_0(\underline{m}) \quad .$$

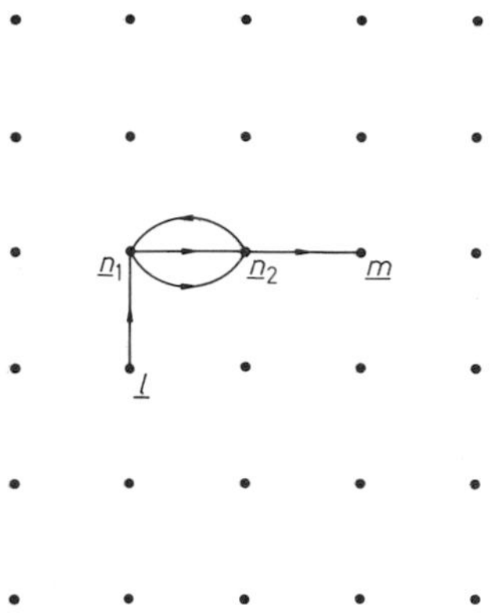

Fig.B.1. A path starting from site $\underline{\ell}$ and ending at site $\underline{m}$ involving five steps

The most general path starting from $\underline{\ell}$ and ending at $\underline{m}$ can be constructed by "decorating" a "skeleton" path. The latter is a self-avoiding (no site is visited more than once) path starting from $\underline{\ell}$ and ending at $\underline{m}$; the "decorations" consist of closed paths starting from and ending at sites visited by the self-avoiding path. The skeleton path in Fig.B1 is the self-avoiding path $\underline{\ell} \rightarrow \underline{n}_1 \rightarrow \underline{n}_2 \rightarrow \underline{m}$. There is one decoration $\underline{n}_1 \rightarrow \underline{n}_2 \rightarrow \underline{n}_1$ at the site $\underline{n}_1$; the same decoration can be considered as associated with site $\underline{n}_2$ (being $\underline{n}_2 \rightarrow \underline{n}_1 \rightarrow \underline{n}_2$). Because of this ambiguity one must be careful not to count the same decoration twice. The above remarks allow us to perform a partial summation. Consider the subset of all paths whose only difference is the decorations at site $\underline{\ell}$. The contribution of all these paths is

$$VG_0(\underline{n}_1) \; \ldots \; VG_0(\underline{m}) \sum_{\underline{\ell}}$$

where $\sum_{\underline{\ell}}$ is the sum of *all* the decorations of site $\underline{\ell}$, i.e., the sum of the contributions of all paths starting from and ending at $\underline{\ell}$, which equals to the Green's function $G(\underline{\ell},\underline{\ell})$. Thus, one can omit all decorations of site $\underline{\ell}$ if at the same time one replaces $G_0(\underline{\ell})$ by $G(\underline{\ell},\underline{\ell})$. The same is true for the next site $\underline{n}_1$ but with one difference: decorations of site $\underline{n}_1$ which visit site $\underline{\ell}$ must be omitted because these decorations have been included already as decorations associated with site $\underline{\ell}$. The decorations at site $\underline{n}_1$ can be omitted if $G_0(\underline{n}_1)$ is replaced by $G(\underline{n}_1,\underline{n}_1[\underline{\ell}])$ where the symbol $[\underline{\ell}]$ denotes that in evaluating $G(\underline{n}_1,\underline{n}_1[\underline{\ell}])$ the paths visiting site $\underline{\ell}$ must be excluded. This exclusion is obtained automatically if one assumes that $\varepsilon_{\underline{\ell}} = \infty$ [because then $G_0(\underline{\ell}) = 0$]. Hence $G(\underline{n}_1,\underline{n}_1[\underline{\ell}])$ is the $\underline{n}_1,\underline{n}_1$ matrix element of H with $\varepsilon_{\underline{\ell}} = \infty$. Similarly all the decorations at site $\underline{n}_2$ can be omitted if at the same time one replaces $G_0(\underline{n}_2)$ by $G(\underline{n}_2,\underline{n}_2[\underline{\ell},\underline{n}_1])$ where the symbol $[\underline{\ell},\underline{n}_1]$ denotes that both $\varepsilon_{\underline{\ell}}$ and $\varepsilon_{\underline{n}_1}$ are infinite. This way one avoids double counting decorations passing through $\underline{n}_2$ which have been already counted as decorations associated with site $\underline{\ell}$ or site $\underline{n}_1$. As a result of these partial summations one can write for $G(\underline{\ell},\underline{m})$

$$G(\underline{\ell},\underline{m}) = \sum G(\underline{\ell},\underline{\ell})VG(\underline{n}_1,\underline{n}_1[\ell])VG(\underline{n}_2,\underline{n}_2[\underline{\ell},\underline{n}_1])V$$

$$\ldots VG(\underline{m},\underline{m}[\ell,\underline{n}_1,\underline{n}_2,\ldots]) \tag{B.6}$$

where the summation is over all self-avoiding paths starting from $\underline{\ell}$ and ending at $\underline{m}$, $\underline{\ell} \to \underline{n}_1 \to \underline{n}_2 \to \ldots \to \underline{m}$. In particular, for the diagonal matrix elements $G(\underline{\ell},\underline{\ell})$ we have

$$G(\underline{\ell},\underline{\ell}) = G_0(\underline{\ell}) + \sum G(\underline{\ell},\underline{\ell})VG(\underline{n}_1,\underline{n}_1[\underline{\ell}])V \ldots G_0(\underline{\ell}) \tag{B.7}$$

where again the summation extends over all closed self-avoiding paths starting from and ending at $\underline{\ell}$. The last factor is $G_0(\underline{\ell})$ because all decorations of the final site $\underline{\ell}$ have been already counted as decoration of the initial site $\underline{\ell}$. Equation (B.7) can be rewritten as

$$G(\underline{\ell},\underline{\ell}) = G_0(\underline{\ell}) + G(\underline{\ell},\underline{\ell})\Delta(\underline{\ell})G_0(\underline{\ell}) \tag{B.8}$$

where $\Delta(\underline{\ell})$ is called the self-energy and is given by

$$\Delta(\underline{\ell}) = \sum VG(\underline{n}_1,\underline{n}_1[\underline{\ell}])V \ldots V \quad . \tag{B.9}$$

Equation (B.8) can be solved for $G(\underline{\ell},\underline{\ell})$ to give

$$G(\underline{\ell},\underline{\ell};z) = \frac{G_0(\underline{\ell})}{1-G_0(\underline{\ell})\Delta(\underline{\ell};z)} = \frac{1}{z-\varepsilon_{\underline{\ell}}-\Delta(\underline{\ell};z)} \quad . \tag{B.10}$$

The last step follows from (B.4). Equation (B.10) justifies the name self-energy for $\Delta(\underline{\ell})$. The expansion (B.6,7, and 9) for $G(\underline{\ell},\underline{m})$ $G(\underline{\ell},\underline{\ell})$ and $\Delta(\underline{\ell})$, respectively, are called the renormalized perturbation expansions (RPE) [B.1,2,3]. Their characteristic property is that they involve summation over terms which are in one-to-one correspondence with the *self-avoiding* paths in the lattice. The price for the simplification of having self-avoiding paths is that the factors associated with lattice sites are complicated Green's functions of the type $G(\underline{n},\underline{n}[\underline{\ell},\underline{n}_1, \ldots])$. These Green's functions can be evaluated by using again the RPE. One can iterate this procedure. Note that for a system involving only a *finite* number of sites the summations associated with the RPE involve only a *finite* number of terms. The iteration procedure *terminates* also, since with every new step in the iteration at least one additional site is excluded. Thus, the RPE iterated as indicated above gives a *closed* expression for the Green's functions of a system involving a finite number of sites. For a system having an infinite number of sites the terms in each summation and the steps in the iteration procedure are infinite; thus for an infinite system the question of the convergence of the RPE arises.

The RPE can be used very successfully to calculate Green's functions for Bethe lattices. For the double periodic case shown in Fig.5.8, we have for the self-energy $\Delta(\ell)$

$$\Delta(\ell) = (K+1)V^2G(\ell+1,\ell+1[\ell]) \tag{B.11}$$

since there are only K + 1 self-avoiding paths starting from and ending at the site ℓ (each one visiting a nearest neighbor). The quantity $G(\ell+1,\ell+1[\ell])$ can be written according to (B.10) as

$$G(\ell+1,\ell+1[\ell]) = \frac{1}{z-\varepsilon_{\ell+1}-\Delta(\ell+1[\ell])} \quad . \tag{B.12}$$

The self-energy $\Delta(\ell+1[\ell])$ can be written by using the RPE as

$$\Delta(\ell+1[\ell]) = KV^2G(\ell+2,\ell+2[\ell+1]) \quad . \tag{B.13}$$

We use K and not (K + 1) since the self-avoiding path involving site ℓ is excluded. By combining (B.12) and (B.13), we have

$$G(\ell + 1,\ell + 1[\ell]) = \frac{1}{z-\varepsilon_{\ell+1}-KV^2G(\ell+2,\ell+2[\ell+1])} \quad . \tag{B.14}$$

Similarly,

$$G(\ell + 2,\ell + 2[\ell + 1]) = \frac{1}{z-\varepsilon_{\ell+2}-KV^2G(\ell+3,\ell+3[\ell+2])} \quad . \tag{B.15}$$

Because of the periodicity, $G(\ell + 3,\ell + 3[\ell + 2]) = G(\ell + 1,\ell + 1[\ell])$. Thus (B.14,15) becomes a set of two equations for the two unknown quantities $G(\ell + 1,\ell + 1[\ell])$ and $G(\ell + 2,\ell + 2[\ell + 1]) = G(\ell,\ell[\ell + 1])$. By solving this system, we obtain

$$G(\ell+1,\ell+1[\ell]) = \frac{(z-\varepsilon_1)(z-\varepsilon_2)\pm\sqrt{(z-\varepsilon_1)(z-\varepsilon_2)[(z-\varepsilon_1)(z-\varepsilon_2)-4KV^2]}}{2K(z-\varepsilon_\ell)V^2} \tag{B.16}$$

$$G(\ell,\ell[\ell+1]) = \frac{(z-\varepsilon_1)(z-\varepsilon_2)\pm\sqrt{(z-\varepsilon_1(z-\varepsilon_2)[(z-\varepsilon_1)(z-\varepsilon_2)-4KV^2]}}{2K(z-\varepsilon_{\ell+1})V^2} \quad . \tag{B.17}$$

Substituting in (B.11) and (B.10) we obtain for $G(\ell,\ell)$ the expression given in (5.48).

For the calculation of the off-diagonal matrix element $G(\ell,m)$ one can apply (B.6). For a Bethe lattice there is just one self-avoiding diagram connecting site ℓ with site m; $G(\ell + 1,\ell + 1[\ell])$ has already been calculated and $G(\ell + 2,\ell + 2[\ell,\ell + 1])$ equals $G(\ell + 2,\ell + 2[\ell + 1])$ which has been evaluated. These two quantities alternate as we proceed along the sites of the unique path connecting ℓ with m. Thus one obtains the expression (5.50) for $G(\ell,m)$.

We mention without proof two theorems concerning products of the form

$$P = G(\underline{\ell},\underline{\ell};z)G(\underline{n}_1,\underline{n}_1[\underline{\ell}];z)G(\underline{n}_2,\underline{n}_2[\underline{\ell},\underline{n}];z)\dots G(\underline{m},\underline{m}[\underline{\ell},\underline{n},\underline{n}_2\dots];z) \quad .$$

The first states that

$$P = \frac{\prod_i (z - E_i^\alpha)}{\prod_j (z - E_j)} \quad , \tag{B.18}$$

where $\{E_i^{\alpha}\}$ are the eigenenergies of H with the sites $\underline{\ell}$, $\underline{n}_1$, $\underline{n}_2, \ldots$, $\underline{m}$ removed, i.e., $\varepsilon_{\underline{\ell}} = \varepsilon_{\underline{n}_1} = \varepsilon_{\underline{n}_2} = \ldots = \varepsilon_{\underline{m}} = \infty$ and $\{E_j\}$ are the eigenenergies of H. The second theorem states that

$$P = \det\{G\} \quad , \tag{B.19}$$

where G is a matrix with matrix elements $G(\underline{s},\underline{r};z)$ where both $\underline{s}$ and $\underline{r}$ belong to the set $\{\underline{\ell},\underline{n}_1,\underline{n}_2,\ldots,\underline{m}\}$.

Appendix C: Second Quantization

We consider here Schrödinger's equation describing the motion of a single particle in an external potential

$$\left(i\hbar\frac{\partial}{\partial t}+\frac{\hbar^2}{2m}\nabla^2-V\right)\psi(\underline{r},t)=0 \tag{C.1}$$

and the wave equation

$$\left(\nabla^2-\frac{1}{c^2}\frac{\partial^2}{\partial t^2}\right)u(\underline{r},t)=0\quad, \tag{C.1'}$$

where $u(\underline{r},t)$ is a real field.

Equations (C.1,1') can be derived respectively from the Lagrangian densities [4.2,5.2]

$$\ell = i\hbar\psi^*\dot{\psi}-\frac{\hbar^2}{2m}\nabla\psi^*\nabla\psi-V\psi^*\psi \tag{C.2}$$

$$\ell = \frac{1}{2c^2}\dot{u}^2-\frac{1}{2}(\nabla u)^2 \tag{C.2'}$$

by applying the principle of least action which leads to the basic equation

$$\frac{\partial\ell}{\partial\phi}-\frac{\partial}{\partial t}\frac{\partial\ell}{\partial\dot{\phi}}-\sum_{\nu=1}^{3}\frac{\partial}{\partial x_\nu}\frac{\partial\ell}{\partial(\partial\phi/\partial x_\nu)}=0\quad, \tag{C.3}$$

where the dot denotes differentiation with respect to t and ϕ stands for ψ,ψ^* or for u; x_1, x_2 and x_3 are the three cartesian coordinates of $\underline{r}$. Substituting (C.2,2') into (C.3) we obtain (C.1) (or its complex conjugate) and (C.1'), respectively.

The field momentum conjugate to the field variable $\phi(\phi=\psi,\psi^*$ or u) is obtained from the general relation

$$\pi = \partial\ell/\partial\dot{\phi} \quad . \tag{C.4}$$

We obtain for the field momenta conjugate to ψ and u, respectively,

$$\pi = i\hbar\psi^* \tag{C.5}$$

$$\pi = \dot{u}/c^2 \quad . \tag{C.5'}$$

The Hamiltonian densities are obtained from the general relation

$$h = \sum \pi\dot{\phi} - \ell \quad ; \tag{C.6}$$

the summation in (C.6) is over all independent fields $\phi(\psi,\psi^*$ or u). We obtain by substituting in (C.6) from (C.5,5') and (C.2,2'), respectively,

$$h = -\frac{i\hbar}{2m}(\nabla\pi)(\nabla\psi) - \frac{i}{\hbar}V\pi\psi \tag{C.7}$$

$$h = \frac{1}{2}c^2\pi^2 + \frac{1}{2}(\nabla u)^2 \quad . \tag{C.7'}$$

The most general solution of (C.1,1') can be written as a linear superposition of the corresponding eigensolutions. Thus, we have

$$\psi(\underline{r},t) = \sum_n a_n\psi_n(\underline{r})\, e^{-iE_nt/\hbar} \tag{C.8}$$

$$u(\underline{r},t) = \frac{1}{\sqrt{\Omega}}\sum_{\underline{k}}\left(P_{\underline{k}}\, e^{i(\omega_kt-\underline{k}\underline{r})} + N_{\underline{k}}\, e^{-i(\omega_kt-\underline{k}\underline{r})}\right) \quad , \tag{C.8'}$$

where $(-\hbar^2\nabla^2/2m + V)\psi_n = E_n\psi_n$ and $\omega_k = ck$. For V = 0, $\psi_n = \exp(i\underline{k}\underline{r})/\sqrt{\Omega}$ and $E_n = \hbar k^2/2m$. Note that in (C.8'), in contrast to (C.8), there are two terms for each $\underline{k}$, one associated with the positive frequency ω_k and the other with the negative frequency $-\omega_k$. This feature is a consequence of the second-order (in time) nature of the wave equation. Note also that as a consequence of u being real, we have the relation

$$P_{\underline{k}} = N_{\underline{k}}^* \quad . \tag{C.9}$$

Substituting (C.8,8') in (C.7,7'), taking into account (C.5,5'), and integrating over the whole volume Ω, we obtain the total energy of the system in terms of the coefficients a_n and $P_{\underline{k}}$, $N_{\underline{k}}$, respectively. We have explicitly

$$H = \sum_n a_n^* a_n E_n \qquad (C.10)$$

$$H = \sum_{\underline{k}} \frac{\omega_{\underline{k}}^2}{c^2} (N_{\underline{k}}^* N_{\underline{k}} + N_{\underline{k}} N_{\underline{k}}^*) \quad ; \qquad (C.10')$$

we can rewrite (C.10') by introducing the quantities

$$b_{\underline{k}} = \left(\frac{2\omega_{\underline{k}}}{\hbar c^2}\right)^{\frac{1}{2}} N_{\underline{k}} \qquad (C.11)$$

as follows

$$H = \frac{1}{2} \sum_{\underline{k}} \hbar\omega_{\underline{k}} (b_{\underline{k}}^* b_{\underline{k}} + b_{\underline{k}} b_{\underline{k}}^*) \quad . \qquad (C.12)$$

A general way to quantize a field ϕ is to replace $\phi(\underline{r},t)$ by an operator $\phi(\underline{r},t)$ which satisfies the commutation relation

$$\phi(\underline{r},t)\pi(\underline{r}',t) \mp \pi(\underline{r}'t)\phi(\underline{r},t) = i\hbar\delta(\underline{r} - \underline{r}') \quad , \qquad (C.13)$$

where the upper sign corresponds to the case where ϕ describes bosons (i.e., particles with integer spin) and the lower sign corresponds to the case where ϕ describes fermions (i.e., particles with integer plus 1/2 spin). The field u, being a classical field, corresponds to integer spin and as such is associated with the upper sign in (C.13). The field ψ may describe either fermions (e.g., electrons) or bosons (e.g., He^4 atoms). The operator $\phi(\underline{r},t)$ commutes (anticommutes) with $\phi(\underline{r}',t)$; similarly, $\pi(\underline{r},t)$ commutes (anticommutes) with $\pi(\underline{r}',t)$. Combining this last statement with (C.13) and expressing everything in terms of ψ or u, we have the equal time commutation (anticommutation) relation.

$$\psi(\underline{r},t)\psi^\dagger(\underline{r}',t) \mp \psi^\dagger(\underline{r}',t)\psi(\underline{r},t) = \delta(\underline{r} - \underline{r}')$$

$$\psi(\underline{r},t)\psi(\underline{r}',t) \mp \psi(\underline{r}',t)\psi(\underline{r},t) = 0$$

$$\psi^\dagger(\underline{r},t)\psi^\dagger(\underline{r}',t) \mp \psi^\dagger(\underline{r}',t)\psi^\dagger(\underline{r},t) = 0 \qquad (C.14)$$

$$u(\underline{r},t)\dot{u}(\underline{r}',t) - \dot{u}(\underline{r}',t)u(\underline{r},t) = i\hbar c^2\delta(\underline{r} - \underline{r}')$$

$$u(\underline{r},t)u(\underline{r}',t) - u(\underline{r}',t)u(\underline{r},t) = 0$$

$$\dot{u}(\underline{r},t)\dot{u}(\underline{r}'t) - \dot{u}(\underline{r}',t)\dot{u}(\underline{r},t) = 0 \quad , \qquad (C.14')$$

where the quantity ψ^*, which is the complex conjugate of ψ, has been replaced by the operator $\psi^\dagger$, which is the adjoint of the operator ψ. Since ψ and u became operators the coefficients a_n, $b_{\underline{k}}$ are operators obeying certain commutation (or anticommutation) relations. To find these relations we substitute (C.8,8') and (C.11) in (C.14,14'). After some straightforward algebra we obtain

$$a_n a_{n'}^\dagger \mp a_{n'}^\dagger a_n = \delta_{nn'}$$

$$a_n a_{n'} \mp a_{n'} a_n = 0$$

$$a_n^\dagger a_{n'}^\dagger \mp a_{n'}^\dagger a_n^\dagger = 0 \qquad (C.15)$$

$$b_{\underline{k}} b_{\underline{q}}^\dagger - b_{\underline{q}}^\dagger b_{\underline{k}} = \delta_{\underline{kq}}$$

$$b_{\underline{k}} b_{\underline{q}} - b_{\underline{q}} b_{\underline{k}} = 0$$

$$b_{\underline{k}}^\dagger b_{\underline{q}}^\dagger - b_{\underline{q}}^\dagger b_{\underline{k}}^\dagger = 0 \quad . \qquad (C.15')$$

From the above relations it follows that the operator $a_n^\dagger a_n$ has the eigenvalues 0, 1, 2, ... if the upper sign is taken and the eigenvalues 0, 1 if the lower sign is taken. The operator $b_{\underline{k}}^\dagger b_{\underline{k}}$ takes the eigenvalues 0, 1, 2, ... For this reason the operator $a_n^\dagger a_n$ or $b_{\underline{k}}^\dagger b_{\underline{k}}$ is called the number operator and is symbolized by n_n or $n_{\underline{k}}$; its eigenvalues show how many particles we have in the state $\psi_n(\underline{r})$ or how many quanta have been excited in the mode $\underline{k}$. One can easily show from (C.15,15') that the operator $a_n^\dagger$ or $b_{\underline{k}}^\dagger$ creates a particle or quantum in the state ψ_n or $|\underline{k}>$; similarly, we show that a_n or $b_{\underline{k}}$ annihilates a particle from this state. These operators can be used to create a complete set of states starting from the vacuum state $|0>$, i.e., from the state which has no particle or quanta. Obviously,

$$a_n|0> = 0 \tag{C.16}$$

$$b_{\underline{k}}|0> = 0 \quad . \tag{C.16'}$$

All the one-particle states can be obtained by applying the operator $a_n^\dagger$ (or $b_{\underline{k}}^\dagger$) to $|0>$, i.e., $a_n^\dagger|0>$ (or $b_{\underline{k}}^\dagger|0>$) where n (or $\underline{k}$) takes all possible values. By applying two creation operators, we can construct the two particle states, and so on. Note that the correct symmetry or antisymmetry of the state is automatically included in this formalism.

We can define different sets of operators $\{a_n\}$ and $\{a_n^\dagger\}$ depending on which complete set of one particle states $\{\psi_n\}$ we choose. Let us have another complete orthonormal set $\{\psi_m\}$ to which a different set of operators $\{a_m\}$ and $\{a_m^\dagger\}$ corresponds. Taking into account that

$$|m> = \sum_n <n|m> \; |n>$$

and that $|m> = a_m^\dagger|0>$ and $|n> = a_n^\dagger|0>$, we obtain

$$a_m^\dagger = \sum_n <n|m>a_n^\dagger \quad , \tag{C.17}$$

from which it follows that

$$a_m = \sum_n <m|n>a_n \quad . \tag{C.18}$$

Usually the set $\{\psi_n\}$ is: 1) the position eigenstates $|\underline{r}>$; 2) the momentum eigenstates $|\underline{k}>$; 3) the eigenstates of the Hamiltonian $-\hbar^2\nabla^2/2m + V$. Of course, when V = 0, the set (2) is identical to (3). Let us apply (C.18) when the set $|\{m>\}$ is the set $\{|\underline{r}>\}$ and the set $|\{n>\}$ is the set (3) above. We obtain

$$a_{r} = \sum_n <r|n> a_n = \sum_n \psi_n(\underline{r}) a_n \quad ;$$

comparing with (C.8), we see that $a_{\underline{r}} = \psi(\underline{r},0)$, i.e., $\psi(\underline{r},0)$ $[\psi^\dagger(r,0)]$ is an annihilation (creation) operator annihilating (creating) a particle at the point $\underline{r}$. Similarly $\psi^\dagger(\underline{r},0)\psi(\underline{r},0)$ is the number operator at the point $\underline{r}$, i.e., the density operator $\rho(\underline{r})$

$$\rho(\underline{r}) = \psi^\dagger(\underline{r},0)\psi(\underline{r},0) \quad . \tag{C.19}$$

The relation between the operators $\psi(\underline{r}),\psi^\dagger(\underline{r})$ and $a_{\underline{k}}$, $a_{\underline{k}}^\dagger$ is, according to (C.17,18),

$$\psi(\underline{r}) = \sum_{\underline{k}} <\underline{r}|\underline{k}> a_{\underline{k}} = \frac{1}{\sqrt{\Omega}} \sum_{\underline{k}} e^{i\underline{k}\underline{r}} \, a_{\underline{k}} \tag{C.20}$$

$$a_{\underline{k}} = \int d^3r <\underline{k}|\underline{r}> \psi(\underline{r}) = \frac{1}{\sqrt{\Omega}} \int d^3r \, e^{-i\underline{k}\underline{r}} \, \psi(\underline{r}) \quad . \tag{C.21}$$

All the operators can be expressed in terms of creation and annihilation operators. We have already seen that

$$\rho(\underline{r}) = \psi^\dagger(\underline{r})\psi(\underline{r}) \quad .$$

Hence, the total number operator is

$$N = \int d^3r \, \psi^\dagger(\underline{r})\psi(\underline{r}) = \sum_{\underline{k}} a_{\underline{k}}^\dagger a_{\underline{k}} \quad . \tag{C.22}$$

The last relation follows with the help of (C.20). The kinetic energy operator is obtained by putting V = 0 in (C.7) and integrating over $\underline{r}$:

$$T = \frac{\hbar^2}{2m} \int \nabla\psi^\dagger \nabla\psi d^3r = -\frac{\hbar^2}{2m} \int d^3r \, \psi^\dagger \nabla^2 \psi = \sum_{\underline{k}} \frac{\hbar^2 k^2}{2m} a_{\underline{k}}^\dagger a_{\underline{k}} \quad . \tag{C.23}$$

The second step follows by integrating by parts the first expression; the last expression is obtained with the help of (C.20).

The potential energy in the presence of an external potential $V(\underline{r})$ is obtained by integrating over $\underline{r}$ the second term on the right-hand side of (C.7). We have

$$V_e = \int V(\underline{r})\psi^\dagger(\underline{r})\psi(\underline{r})d^3r = \int V(\underline{r})\rho(\underline{r})d^3r \quad , \tag{C.24}$$

which is what was expected. The operators ρ, N, T and V_e involve the product of one creation and one annihilation operator. This is a general feature for operators which correspond to additive quantities. On the other hand, operators involving summation over pair of particles (such as the Coulomb interaction energy of a system of particles) contain a product of two creation and two annihilation operators. For example, we consider the interaction energy

$$V_i = \frac{1}{2}\sum_{ij} v(\underline{r}_i,\underline{r}_j) \quad . \tag{C.25}$$

This expression can be written in terms of the density ρ as follows

$$V_i = \frac{1}{2}\int \rho(\underline{r})v(\underline{r},\underline{r}')\rho(\underline{r}')d^3rd^3r' \quad ;$$

by using (C.19) we obtain

$$V_i = \frac{1}{2}\int d^3rd^3r'\psi^\dagger(\underline{r})\psi^\dagger(\underline{r}')v(\underline{r},\underline{r}')\psi(\underline{r}')\psi(\underline{r}) \quad . \tag{C.26}$$

Note that in our derivation of (C.26) there is an uncertainty in the ordering of the four creation and annihilation operators. One can verify that the ordering in (C.26) is the correct one by evaluating the matrix elements of V_i among states with a fixed total number of particles [8.1,3].

The total Hamiltonian of a system of particles interacting with each other via the pairwise potential $v(\underline{r},\underline{r}')$ and placed in an external potential $V(\underline{r})$ is given by

$$H = T + V_e + V_i \quad , \tag{C.27}$$

where the quantities T, V_e and V_i are given by (C.23,24,26), respectively. Note that the Hamiltonian (C.27) cannot be brought to the form (C.10) because of the presence of the interaction term V_i. Furthermore, the field operator $\psi(\underline{r},t)$ does not satisfy the simple equation (C.1). There are two ways to find the equation obeyed by ψ: 1) by applying the general equation (C.3) with ℓ containing an extra term $-V_i$; 2) one can also use the general equation

$$i\hbar \frac{\partial\psi}{\partial t} = [\psi, H]$$

which leads to

$$\left[i\hbar \frac{\partial}{\partial t} + \frac{\hbar^2\nabla^2}{2m} - V(\underline{r})\right]\psi(\underline{r},t) = \int d^3r' v(\underline{r},\underline{r}')\psi^\dagger(\underline{r}',t)\psi(\underline{r}'t)\psi(\underline{r},t) \qquad (C.28)$$

with a similar equation for $\psi^\dagger(\underline{r},t)$. For the particular but important case of a system of electrons moving in a positive background (to ensure overall electrical neutrality) and repelling each other through $v(\underline{r},\underline{r}') = e^2/|\underline{r}-\underline{r}'|$, the total Hamiltonian H can be expressed in terms of $a^\dagger_{\underline{k}}, a_{\underline{k}}$ as follows

$$H = \sum_{\underline{k}} \frac{\hbar^2k^2}{2m} a^\dagger_{\underline{k}} a_{\underline{k}} + \frac{e^2}{2\Omega} \sum_{\underline{kpq}\neq 0} \frac{4\pi}{q^2} a^\dagger_{\underline{k}+\underline{q}} a^\dagger_{\underline{p}-\underline{q}} a_{\underline{p}} a_{\underline{k}} \quad . \qquad (C.29)$$

We can also express various operators associated with the field $u(\underline{r},t)$ in terms of the field operator $u(\underline{r},t)$ or in terms of the operators $b_{\underline{k}}$ and $b^\dagger_{\underline{k}}$. To be more specific, we consider the particular case of the longitudinal vibrations of an isotropic continuous solid. In this case each operator $b^\dagger_{\underline{k}}$ creates a quantum of a plane wave longitudinal oscillation which carries a momentum equal to $\hbar\underline{k}$ and an energy equal to $\hbar\omega_k$. This quantum is called longitudinal acoustic (LA) phonon. The theory we have presented for the field $u(\underline{r},t)$ requires some modifications in order to be applicable to the longitudinal vibrations of a continuum; the reason is that the latter are described by a *vector* field, namely the displacement $\underline{d}(\underline{r},t)$ from the equilibrium position; since we are dealing with longitudinal vibrations, it follows that $\nabla\times\underline{d} = 0$. The various quantities of physical interest can be expressed in terms of LA phonon creation and annihilation operators as follows (for detailed derivations see [8.3]). The displacement operator $\underline{d}(\underline{r},t)$ is

$$\underline{d}(\underline{r},\underline{t}) = -i \sum_{\underline{k}} \left(\frac{\hbar}{2\omega_k\rho\Omega}\right)^{\frac{1}{2}} \frac{\underline{k}}{k}[b_{\underline{k}}\, e^{-i(\omega_k t - \underline{kr})} - b^\dagger_{\underline{k}}\, e^{i(\omega_k t - \underline{kr})}] \qquad (C.30)$$

where ρ is the constant equilibrium mass density. The conjugate momentum $\underline{\pi}(\underline{r},t)$ is

$$\underline{\pi}(\underline{r},t) = \rho \frac{\partial \underline{d}}{\partial t} = - \sum_{\underline{k}} \left(\frac{\hbar\omega_k\rho}{2\Omega}\right)^{\frac{1}{2}} \frac{\underline{k}}{k} (b_{\underline{k}}\, e^{-i(\omega_k t - \underline{k}\underline{r})} + b^{\dagger}_{\underline{k}}\, e^{i(\omega_k t - \underline{k}\underline{r})}) \ . \qquad (C.31)$$

The Lagrangian and Hamiltonian densities are

$$\ell = \frac{1}{2}\rho\dot{\underline{d}}^2 - \frac{1}{2}B(\nabla\cdot\underline{d})^2 \qquad (C.32)$$

$$h = \frac{1}{2\rho}\underline{\pi}^2 + \frac{1}{2}B(\nabla\cdot\underline{d})^2 \qquad (C.33)$$

where

$$B = \rho c^2 \qquad (C.34)$$

is the adiabatic bulk modulus, $B = -\Omega(\partial P/\partial\Omega)_S$. The Hamiltonian expressed in terms of $b^{\dagger}_{\underline{k}}$, $b_{\underline{k}}$ has exactly the form (C.12) which can be rewritten as

$$H = \sum_{\underline{k}} \hbar\omega_k (b^{\dagger}_{\underline{k}} b_{\underline{k}} + \frac{1}{2}) \qquad (C.35)$$

by taking into account (C.15'). Equation (C.35) means that the phonons are noninteracting. If the harmonic approximation is relaxed, the Lagrangian (C.32) will contain extra terms involving powers of $\nabla\cdot\underline{d}$ higher than the second. Since $\underline{d}$ is a linear combination of $b^{\dagger}_{\underline{k}}$ and $b_{\underline{k}}$, these third- and higher-order terms will add to the Hamiltonian (C.35) an H_i which will involve terms of the form $b_{\underline{k}}b_{\underline{q}}b_{\underline{p}}$, $b^{\dagger}_{\underline{k}}b_{\underline{q}}b_{\underline{p}}$, $b^{\dagger}_{\underline{k}}b^{\dagger}_{\underline{q}}b_{\underline{p}}$, $b^{\dagger}_{\underline{k}}b^{\dagger}_{\underline{q}}b^{\dagger}_{\underline{p}}$ and terms involving four, five, etc., creation and annihilation operators. Thus the term H_i represents complicated interactions among the phonons, and as a result the equation of motion (C.1') is modified; the new equation can be found from (C.3) with ℓ containing whatever anharmonic terms are present.

In a solid there are interactions between the electrons [which are described by the field $\psi(\underline{r},t)$] and the longitudinal phonons [which are described by the field $\underline{d}(\underline{r},t)$]. It is customary and convenient to describe the longitudinal phonons through the scalar field $\phi(\underline{r},t)$ which is defined as

$$\phi(\underline{r},t) = c\sqrt{\rho}\,\nabla\cdot\underline{d} \quad . \qquad (C.36)$$

Substituting in (C.36) from (C.30), we obtain for the phonon field $\phi(\underline{r},0)$ the following expression

$$\phi(\underline{r},0) = \sum_{\underline{k}} \left(\frac{\hbar\omega_k}{2\Omega}\right)^{\frac{1}{2}} (b_{\underline{k}}^{\dagger}\, e^{-i\underline{k}\underline{r}} + b_{\underline{k}}\, e^{i\underline{k}\underline{r}}) \quad ; \tag{C.37}$$

the summation over $\underline{k}$ is restricted by the condition $k \leq k_D$ where k_D is the Debye wave number. This restriction ensures that the degrees of freedom in our isotropic continuum are equal to the vibrational degrees of freedom of a real solid. In terms of the fields $\psi(\underline{r})$ and $\phi(\underline{r})$ the electron-phonon interaction Hamiltonian is [8.3]

$$H_{e-p} = \gamma \int d^3r\psi^{\dagger}(\underline{r})\psi(\underline{r})\phi(\underline{r}) \tag{C.38}$$

where

$$\gamma = \frac{\pi^2\hbar^2 z\rho}{mk_F M\sqrt{B}} \quad ; \tag{C.39}$$

z is the valence and M is the mass of each ion, m is the mass of the electron, and k_F is the Fermi momentum of the electron gas. If the various interactions discussed up to now are included, the total Hamiltonian H_t of an electron-phonon system can be written as follows

$$H_t = H_e + H_p + H_{e-e} + H_{p-p} + H_{e-p} \quad , \tag{C.40}$$

where H_e, H_p, H_{e-e} and H_{e-p} are given by (C.10,35,26,38), respectively; H_{p-p} contains terms involving the product of at least three phonon creation and annihilation operators as was discussed above. The time dependence of the operators $\psi(\underline{r},t)$ and $\phi(\underline{r},t)$ can be obtained from the general relation

$$i\hbar \frac{\partial\psi(\underline{r},t)}{\partial t} = [\psi(\underline{r},t), H_t] \tag{C.41}$$

$$i\hbar \frac{\partial\phi(\underline{r},t)}{\partial t} = [\phi(\underline{r},t), H_t] \quad . \tag{C.42}$$

If the interaction terms were absent, (C.41) would give (C.8) and (C.42) would give

$$\phi(\underline{r},t) = \sum_{\underline{k}} \left(\frac{\hbar\omega_k}{2\Omega}\right)^{\frac{1}{2}} (b_{\underline{k}}^{\dagger}\, e^{i(\omega_k t-\underline{k}\underline{r})} + b_{\underline{k}}\, e^{-i(\omega_k t-\underline{k}\underline{r})}) \quad . \tag{C.43}$$

References

Part I: Green's Functions in Mathematical Physics

Chapter 1 Time-Independent Green's Functions

1.1 J. Mathews, R.L. Walker: *Mathematical Method of Physics*, 2nd ed. (Benjamin, New York 1970)
1.2 F.N. Byron, Jr., R.W. Fuller: *Mathematics of Classical and Quantum Physics*, Vol. II (Addison-Wesley, Reading, Mass. 1969)
1.3 P.M. Morse, H. Feshbach: *Methods of Theoretical Physics*, Vols. I and II (McGraw-Hill, New York 1953)
1.4 W.R. Smythe: *Static and Dynamic Electricity* (McGraw-Hill, New York 1968)
1.5 J.D. Jackson: *Classical Electrodynamics* (Wiley and Sons, New York 1967)

Chapter 2 Time-Dependent Green's Functions

2.1 E. Merzbacher: *Quantum Mechanics* (Wiley and Sons, New York 1961)
2.2 N.N. Bogoliubov, D.V. Shirkov: *Introduction to the Theory of Quantized Fields* (Interscience, New York 1959)

Part II: Green's Functions in One-Body Quantum Problems

Chapter 3 Physical Significance of *G*. Application to the Free-Particle Case

3.1 M. Abramowitz, I.A. Stegun (eds.): *Handbook of Mathematical Functions* (Dover, London 1965)
3.2 N.I. Muskhelishvili: *Singular Integral Equations* (Noordhoff, Groningen-Holland 1958)

Chapter 4 Green's Functions and Perturbation Theory

4.1 A.L. Fetter, J.D. Walecka: *Quantum Theory of Many-Particle Systems* (McGraw-Hill, New York 1971)
4.2 L.I. Schiff: *Quantum Mechanics*, 2nd ed. (McGraw-Hill, New York 1955)
4.3 L.D. Landau, E.M. Lifshitz: *Quantum Mechanics* (Addison-Wesley, Reading, Mass. 1958)

Chapter 5 Green's Functions for Tight-Binding Hamiltonians

5.1 C. Kittel: *Introduction to Solid State Physics*, 5th ed. (Wiley and Sons, New York 1976)
5.2 C. Kittel: *Quantum Theory of Solids* (Wiley and Sons, New York 1973)
5.3 J.M. Ziman: *Principles of the Theory of Solids* (Cambridge University Press, London 1964)

5.4 J.C. Slater, G.F. Koster: Phys. Rev. *94*, 1498 (1954)
5.5 T. Morita: J. Math. Phys. *12*, 1744 (1971)
5.6 L. Van Hove: Phys. Rev. *89*, 1189 (1953); see also H.P. Rosenstock: Phys. Rev. *97*, 290 (1955)
5.7 J. Callaway: J. Math. Phys. *5*, 783 (1964)
5.8 S. Katsura, S. Inawashiro: J. Math. Phys. *12*, 1622 (1971)
5.9 T. Horiguchi: J. Math. Phys. *13*, 1411 (1972)
5.10 T. Horiguchi, C.C. Chen: J. Math. Phys. *15*, 659 (1974)
5.11 T. Morita, T. Horiguchi: J. Math. Phys. *12*, 981 (1971)
5.12 T. Horiguchi: J. Phys. Soc. Jpn. *30*, 1261 (1971)
5.13 G.S. Joyce: Philos. Trans. Roy. Soc. A*273*, 583 (1973)
5.14 T. Morita: J. Phys. A*8*, 478 (1975)
5.15 T. Horiguchi, T. Morita: J. Phys. C*8*, L232 (1975)
5.16 D.J. Austen, P.D. Loly: J. Comput. Phys. *11*, 315 (1973)
5.17 F.T. Hioe: J. Math. Phys. *19*, 1065 (1978)
5.18 M.A. Rashid: J. Math. Phys. *21*, 2549 (1980)
5.19 S. Inawashiro, S. Katsura, Y. Abe: J. Math. Phys. *14*, 560 (1973)
5.20 S. Katsura, S. Inawashiro, Y. Abe: J. Math. Phys. *12*, 895 (1971)
5.21 G.S. Joyce: J. Phys. A*5*, L65 (1972)
5.22 S. Katsura, T. Morita, S. Inawashiro, T. Horiguchi, Y. Abe: J. Math. Phys. *12*, 892 (1971)
5.23 T. Morita, T. Horiguchi: J. Math. Phys. *12*, 986 (1971)
5.24 K. Mano: J. Math. Phys. *15*, 2175 (1974); *16*, 1726 (1975)
5.25 G.S. Joyce: J. Math. Phys. *12*, 1390 (1971)
5.26 M. Inoue: J. Math. Phys. *15*, 704 (1974)
5.27 W.A Harrison: *Pseudopotentials in the Theory of Metals* (Benjamin, New York 1966)
5.28 V. Heine, D. Weaire: In *Solid State Physics*, Vol.24, ed. by H. Ehrenreich, F. Seitz, D. Turnbull (Academic, New York 1970) p.249
5.29 W.A. Harrison: *Electronic Structure and the Properties of Solids* (Freeman, San Francisco 1980)
5.30 J.C. Phillips, L. Kleinman: Phys. Rev. *116*, 287 (1959)
5.31 M.H. Cohen, V. Heine: Phys. Rev. *122*, 1821 (1961)
5.32 M.L. Cohen, V. Heine: In *Solid State Physics*, Vol.24, ed. by H. Ehrenreich, F. Seitz, D. Turnbull (Academic, New York 1970) p.37
5.33 J. Hafner, H. Nowotny: Phys. Status Solidi B*51*, 107 (1972)
5.34 J.R. Chelikowsky, M.L. Cohen: Phys. Rev. Lett. *36*, 229 (1976); Phys. Rev. B*14*, 556 (1976)
5.35 D.J. Chadi, M.L. Cohen: Phys. Status Solidi B*68*, 405 (1975)
5.36 J.C. Phillips: Rev. Mod. Phys. *42*, 317 (1970)
5.37 S.T. Pantelides, W.A. Harrison: Phys. Rev. B*11*, 3006 (1975)
5.38 W.A. Harrison: Phys. Rev. B*8*, 4487 (1973)
5.39 S. Froyen, W.A. Harrison: Phys. Rev. B*20*, 2420 (1979); W.A. Harrison: Phys. Rev. B*24*, 5835 (1981)
5.40 E.O. Kane, A.B. Kane: Phys. Rev. B*17*, 2691 (1978)
5.41 J.D. Joannopoulos, M.A. Schlüter, M.L. Cohen: In *Proc. 12th Intern. Conf. Physics of Semiconductors (Stuttgart)*, ed. by M.H. Pilkuhn (Teubner, Stuttgart 1974)
5.42 R.M. Martin, G. Lucovsky, K. Helliwell: Phys. Rev. B*13*, 1383 (1976)
5.43 D.J. Chadi: Phys. Rev. B*16*, 790 and 3572 (1977)
5.44 K.C. Pandey, J.C. Phillips: Phys. Rev. B*13*, 750 (1976)
5.45 D.A. Papaconstantopoulos, E.N. Economou: Phys. Rev. B*22*, 2903 (1980)
5.46 S.T. Pantelides: Phys. Rev. B*11*, 5082 (1975)
5.47 J.R. Smith, J.G. Gay: Phys. Rev. B*12*, 4238 (1975)
5.48 L. Hodges, H. Ehrenreich: Phys. Lett. *16*, 203 (1965)
5.49 F.M. Mueller: Phys. Rev. *153*, 659 (1967)
5.50 J. Rath, J. Callaway: Phys. Rev. B*8*, 5398 (1973)

5.51 L.F. Mattheiss: Phys. Rev. B*5*, 290 and 306 (1972); B*6*, 4718 (1972) B*13*, 2433 (1976)
5.52 L.F. Mattheiss: Phys. Rev. B*12*, 2161 (1975)
5.53 B.M. Klein, L.L. Boyer, D.A. Papaconstantopoulos, L.F. Mattheiss: Phys. Rev. B*18*, 6411 (1978)
5.54 J.D. Joannopoulos, F. Yndurain: Phys. Rev. B*10*, 5164 (1974); F. Yndurain, J.D. Joannopoulos: Phys. Rev. B*11*, 2957 (1975)
5.55 D.C. Allan, J.D. Joannopoulos: Phys. Rev. Lett. *44*, 43 (1980); J.D. Joannopoulos: J. Non Cryst. Solids *35-36*, 781 (1980)
5.56 M.F. Thorpe: In *Excitations in Disordered Systems* , ed. by M.F. Thorpe (Plenum, New York 1981) pp.85-107

Chapter 6 Single Impurity Scattering

6.1 G.F. Koster, J.C. Slater: Phys. Rev. *95*, 1167 (1954); Phys. Rev. *96*, 1208 (1954)
6.2 J. Callaway: J. Math. Phys. *5*, 783 (1964); Phys. Rev. *154*, 515 (1967)
6.3 J. Callaway, A. Hughes: Phys. Rev. *156*, 860 (1967); *164*, 1043 (1967)
6.4 J. Bernholc, S.T. Pantelides: Phys. Rev. B*18*, 1780 (1978); J. Bernholc, N.O. Lipari, S.T. Pantelides: Phys. Rev. B*21*, 3545 (1980)
6.5 M. Lannoo, J. Bourgoin: *Point Defects in Semiconductors I*, Springer Ser. Solid-State Sci., Vol.22 (Springer, Berlin, Heidelberg, New York 1981)
6.6 G.A. Baraff, M. Schlüter: Phys. Rev. B*19*, 4965 (1979)
6.7 G.A. Baraff, E.O. Kane, M. Schlüter: Phys. Rev. Lett. *43*, 956 (1979)
6.8 N.O. Lipari, J. Bernholc, S.T. Pantelides: Phys. Rev. Lett. *43*, 1354 (1979)
6.9 L.N. Cooper: Phys. Rev. *104*, 1189 (1956)
6.10 L.D. Landau, E.M. Lifshitz: *Statistical Physics* (Pergamon, London or Addison-Wesley, Reading, Mass., 1st ed. 1958) pp.202-206
6.11 J. Bardeen, L.N. Cooper, J.R. Schrieffer: Phys. Rev. *106*, 162 (1957); *108*, 1175 (1957)
6.12 J.R. Schrieffer: *Theory of Superconductivity* (Benjamin, Reading, Mass. 1964)
6.13 G. Rickayzen: *Theory of Superconductivity* (Wiley, New York 1964)
6.14 R.D. Parks (ed.): *Superconductivity* (Dekker, New York 1969)
6.15 E.N. Economou: In *Metal Hydrides*, ed. by G. Bambakidis (Plenum, New York 1981) pp.1-19
6.16 G. Grimvall: Phys. Scripta *14*, 63 (1976)
6.17 W.L. McMillan: Phys. Rev. *167*, 331 (1968)
6.18 P.B. Allen, R.C. Dynes: Phys. Rev. B*12*, 905 (1975)
6.19 J. Kondo: Progr. Theor. Phys. *32*, 37 (1964)
6.20 A.A. Abrikosov: Physics *2*, 5 (1965)
6.21 H. Suhl: Phys. Rev. *138*, A515 (1965)
6.22 Y. Nagaoka: Phys. Rev. *138*, A1112 (1965); Progr. Theor. Phys. *37*, 13 (1967)
6.23 J. Kondo: In *Solid State Physics*, Vol.23, ed. by H. Ehrenreich, F. Seitz, D. Turnbull (Academic, New York 1969) p.183
6.24 P.W. Anderson, G. Yuval: Phys. Rev. B*1*, 1522 (1970); J. Phys. C*4*, 607 (1971); P.W. Anderson, G. Yuval, D.R. Hamann: Phys. Rev. B*1*, 4464 (1970); Solid State Commun. *8*, 1033 (1970); P.W. Anderson: J. Phys. C*3*, 2436 (1970)
6.25 M. Fowler, A. Zawadowski: Solid State Commun. *9*, 471 (1971)
6.26 G.T. Rado, H. Suhl (eds.): *Magnetism*, Vol.5 (Academic, New York 1973)
6.27 K.G. Wilson: Rev. Mod. Phys. *67*, 773 (1975)
6.28 P. Nozières: In *Proc. 14th Intern. Conf. Low Temperature Physics*, ed. by M. Krusius, M. Vuorio (North-Holland, Amsterdam 1975) p.339

6.29 N. Andrei: Phys. Rev. Lett. *45*, 379 (1980);
N. Andrei, J.H. Lowenstein: Phys. Rev. Lett. *46*, 356 (1981)
6.30 P.B. Wiegmann: Pis'ma Zh. Eksp. Teor. Fiz. *31*, 392 (1980) [JETP Lett. *31*, 364 (1980)]

Chapter 7 Two or More Impurities; Disordered Systems

7.1 P. Lloyd: J. Phys. C*2*, 1717 (1969)
7.2 J.M. Ziman: *Models of Disorder* (Cambridge Unversity Press, London 1979)
7.3 R.J. Elliott, J.A. Krumhansl, P.L. Leath: Rev. Mod. Phys. *46*, 465 (1974)
7.4 H. Ehrenreich, L.M. Schwartz: In *Solid State Physics*, Vol.31, ed. by H. Ehrenreich, F. Seitz, D. Turnbull (Academic, New York 1976) p.150
7.5 L.M. Schwartz: In [Ref.5.56, pp.177-224]
7.6 A. Bansil: In [Ref.5.56, pp.225-240]
7.7 A. Bansil, L. Schwartz, H. Ehrenreich: Phys. Rev. B*12*, 2893 (1975)
7.8 P.E. Mijnarends, A. Bansil: Phys. Rev. B*19*, 2912 (1979); B*13*, 2381 (1976)
7.9 A. Bansil, R.S. Rao, P.E. Mijnarends, L. Schwartz: Phys. Rev. B*23*, 3608 (1981);
A. Bansil: Phys. Rev. B*20*, 4025 and 4035 (1979)
7.10 P.N. Sen: In [Ref.5.56, pp.647-659]
7.11 J.C. Maxwell: *A Treatise on Electricity and Magnetism* (1873) (Dover, New York 1945)
7.12 J. Hubbard: Proc. Roy. Soc. A*281*, 401 (1964)
7.13 D.W. Taylor: Phys. Rev. *156*, 1017 (1967)
7.14 P. Soven: Phys. Rev. *156*, 809 (1967)
7.15 P.L. Leath: Phys. Rev. *171*, 725 (1968); Phys. Rev. B*5*, 1643 (1972)
7.16 R.N. Aiyer, R.J. Elliott, J.A. Krumhansl, P.L. Leath: Phys. Rev. *181*, 1006 (1969)
7.17 B.G. Nickel, J.A. Krumhansl: Phys. Rev. B*4*, 4354 (1971)
7.18 S.F. Edwards: Philos. Mag. *3*, 1020 (1958)
7.19 J.S. Langer: Phys. Rev. *120*, 714 (1960)
7.20 R. Klauder: Ann. Phys. *14*, 43 (1961)
7.21 T. Matsubara, Y. Toyozawa: Progr. Theor. Phys. *26*, 739 (1961)
7.22 P.L. Leath, B. Goodman: Phys. Rev. *148*, 968 (1966); Phys. Rev. *175*, 963 (1968)
7.23 F. Yonezawa, T. Matsubara: Progr. Theor. Phys. *35*, 357 and 759 (1966); *37*, 1346 (1967)
7.24 P.L. Leath: Phys. Rev. B*2*, 3078 (1970)
7.25 T. Matsubara, T. Kaneyoshi: Progr. Theor. Phys. *36*, 695 (1966)
7.26 B. Velicky, S. Kirkpatrick, H. Ehrenreich: Phys. Rev. *175*, 747 (1968); B*1*, 3250 (1970)
7.27 Y. Onodera, Y. Toyozawa: J. Phys. Soc. Jpn. *24*, 341 (1968)
7.28 F. Yonezawa: Progr. Theor. Phys. *39*, 1076 (1968); *40*, 734 (1968)
7.29 W.H. Butler: Phys. Lett. A*39*, 203 (1972)
7.30 F. Brouers, M. Cyrot, F. Cyrot-Lackmann: Phys. Rev. B*7*, 4370 (1973)
7.31 F. Yonezawa, K. Morigaki: Progr. Theor. Phys. Suppl. *53*, 1 (1973)
7.32 F. Yonezawa, T. Odagaki: Solid State Commun. *27*, 1199 and 1203 (1978); J. Phys. Soc. Jpn. *47*, 388 (1979)
7.33 J.S. Faulkner: Phys. Rev. B*13*, 2391 (1976)
7.34 J.S. Faulkner, G.M. Stocks: Phys. Rev. B*23*, 5628 (1981); B*21*, 3222 (1980)
7.35 D.A. Papaconstantopoulos, B.M. Klein, J.S. Faulkner, L.L. Boyer: Phys. Rev. B*18*, 2784 (1978)
7.36 D.A. Papaconstantopoulos, E.N. Economou: Phys. Rev. B*24*, 7233 (1981)
7.37 B.L. Gyorffy, G.M. Stocks: In *Electrons in Disordered Metals and at Metallic Surfaces*, ed. by P. Phariseau, B.L. Gyorffy, L. Scheire (Plenum, New York 1979)
J.S. Faulkner: Progr. Mat. Sci. *27*, 1 (1982)

7.38 G.M. Stocks, W.M. Temmerman, B.L. Gyorffy: Phys. Rev. Lett. *41*, 339 (1978)
7.39 J. Korringa, R.L. Mills: Phys. Rev. B*5*, 1654 (1972)
7.40 L. Roth: Phys. Rev. B*9*, 2476 (1974)
7.41 E. Balanovski: Phys. Rev. B*20*, 5094 (1979)
7.42 L Schwartz, H. Krakauer, H. Fukuyama: Phys. Rev. Lett. *30*, 746 (1973)
7.43 T. Kaplan, M. Mostoller: Phys. Rev. B*9*, 1983 (1974)
7.44 A.B. Harris, P.L. Leath, B.G. Nickel, R.J. Elliott: J. Phys. C*7*, 1693 (1974)
7.45 R.A. Tahir-Kheli: Phys. Rev. B*6*, 2808 and 2826 (1972); R.A. Tahir-Kheli, T. Fujuwara, R.J. Elliott: J. Phys. C*11*, 497 (1978)
7.46 R.J. Elliott: In [Ref.5.56, pp.3-25]
7.47 H. Shiba: Progr. Theor. Phys. *46*, 77 (1971)
7.48 E-Ni Foo, A. Amar, M. Austloos: Phys. Rev. B*4*, 3350 (1971)
7.49 J.A. Blackman, D.M. Esterling, N.F. Berk: Phys. Rev. B*4*, 2412 (1971)
7.50 D.J. Whitelaw: J. Phys. C*14*, 2871 (1981)
7.51 P.L. Leath: In [Ref.5.56, pp.109-127]
7.52 E. Müller-Hartmann: Solid State Commun. *12*, 1269 (1973)
7.53 F. Ducastelle: J. Phys. C*4*, L75 (1971); C*7*, 1795 (1974)
7.54 W.H. Butler, B.G. Nickel: Phys. Rev. Lett. *30*, 373 (1973)
7.55 V. Ĉapek: Phys. Status Solidi B*43*, 61 (1971)
7.56 P.L. Leath: J. Phys. C*6*, 1559 (1973)
7.57 A. Gonis, J.W. Garland: Phys. Rev. B*16*, 1495 and 2424 (1977)
7.58 L.M. Falicov, F. Yndurain: Phys. Rev. B*12*, 5664 (1975)
7.59 R. Haydock: In [Ref.5.56, pp.29-57]
7.60 F. Cyrot-Lackmann, S.N. Khanna: In [Ref.5.56, pp.59-83]
7.61 K.F. Freed, M.H. Cohen: Phys. Rev. B*3*, 3400 (1971)
7.62 C.T. White, E.N. Economou: Phys. Rev. B*15*, 3742 (1977)
7.63 K. Aoi: Solid State Commun. *14*, 929 (1974)
7.64 T. Miwa: Progr. Theor. Phys. *52*, 1 (1974)
7.65 F. Brouers, F. Ducastelle: J. Phys. F*5*, 45 (1975)
7.66 A.R. Bishop, A. Mookerjee: J. Phys. C*7*, 2165 (1974)
7.67 P.R. Best, P. Lloyd: J. Phys. C*8*, 2219 (1975)
7.68 I. Takahashi, M. Shimizu: Progr. Theor. Phys. *51*, 1678 (1973)
7.69 S. Wu, M. Chao: Phys. Status Solidi B*68*, 349 (1975)
7.70 A.R. McGurn, R.A. Tahir-Kheli: J. Phys. C*10*, 4385 (1977)
7.71 H. Schmidt: Phys. Rev. *105*, 425 (1957)
7.72 R. Mills, P. Ratanavararaksa: Phys. Rev. B*18*, 5291 (1978)
7.73 A. Mookerjee: J. Phys. C*6*, L205 (1973); C*6*, 1340 (1973)
7.74 T. Kaplan, P.L. Leath, L.J. Gray, H.W. Diehl: Phys. Rev. B*21*, 4230 (1980)
7.75 L.J. Gray, T. Kaplan: Phys. Rev. B*24*, 1872 (1981)
7.76 T. Kaplan, L.J. Gray: In [Ref.5.56, pp.129-143]
7.77 J.M. Ziman: *Electrons and Phonons* (Oxford University Press, London 1960); B.R. Nag: *Electron Transport in Compound Semiconductors*, Springer Ser. Solid-State Sci., Vol.11 (Springer, Berlin, Heidelberg, New York 1980)
7.78 N.F. Mott, E.A. Davis: *Electronic Processes in Non-Crystalline Materials*, 2nd ed., (Clarendon, Oxford 1979)
7.79 J. Callaway: *Quantum Theory of the Solid State* (Academic, New York 1976)
7.80 D.J. Thouless: In *Ill-Condensed Matter*, ed. by R. Balian, R. Maynard, G. Toulouse (North-Holland, Amsterdam 1979) pp.1-62
7.81 R. Kubo: Can. J. Phys. *34*, 1274 (1956); J. Phys. Soc. Jpn. *12*, 570 (1957)
7.82 A.D. Greenwood: Proc. Phys. Soc. *71*, 585 (1958)
7.83 A.A. Abrikosov, L.P. Gor'kov, I.E. Dzyaloshinski: *Methods of Quantum Field Theory in Statistical Physics* (Prentice Hall, Englewood Cliffs, NJ 1963)
7.84 S. Doniach, E.H. Sondheimer: *Green's Functions for Solid State Physicists* (Benjamin, Reading, Mass. 1974)

7.85 E.N. Economou, C.M. Soukoulis: Phys. Rev. Lett. *46*, 618 (1981)
7.86 B.S. Andereck, E. Abrahams: J. Phys. C*13*, L383 (1980)
7.87 P.W. Anderson, D.J. Thouless, E. Abrahams, D.S. Fisher: Phys. Rev. B*22*, 3519 (1980)
7.88 A.A. Abrikosov: Solid State Commun. *37*, 997 (1981)
7.89 S.F. Edwards: Philos. Mag. *4*, 1171 (1959)
7.90 B. Velicky: Phys. Rev. *184*, 614 (1969)
7.91 D.C. Licciardello, E.N. Economou: Phys. Rev. B*11*, 3697 (1975)
7.92 M. Kaveh: Can. J. Phys. *60*, 746 (1982)
7.93 R. Landauer, J.C. Helland: J. Chem. Phys. *22*, 1655 (1954)
7.94 N.F. Mott, W.D. Twose: Adv. Phys. *10*, 107 (1961)
7.95 R.E. Borland: Proc. Phys. Soc. *77*, 705 (1961); *78*, 926 (1961); Proc. Roy. Soc. A*274*, 529 (1963); Proc. Phys. Soc. *83*, 1027 (1964)
7.96 F.J. Wegner: Z. Phys. B*22*, 273 (1975)
7.97 B.I. Halperin: Adv. Chem. Phys. *13*, 123 (1967)
7.98 J. Hori: *Spectral Properties of Disordered Chains and Lattices* (Pergamon, London 1968)
7.99 E.H. Lieb, D.C. Mattis: *Mathematical Physics in One Dimension* (Academic, New York 1966)
7.100 K. Ishii: Progr. Theor. Phys. Suppl. *53*, 77 (1973)
7.101 H. Furstenberg: Trans. Am. Math. Soc. *108*, 377 (1963)
7.102 R. Landauer: Philos. Mag. *21*, 863 (1970)
7.103 E.N. Economou, C.M. Soukoulis: Phys. Rev. Lett. *47*, 973 (1981)
7.104 R. Landauer: Phys. Lett. A*85*, 91 (1981)
7.105 D.J. Thouless: Phys. Rev. Lett. *47*, 972 (1981)
7.106 C.M. Soukoulis, E.N. Economou: Solid State Commun. *37*, 409 (1981)
7.107 D.J. Thouless, S. Kirkpatrick: J. Phys. C*14*, 235 (1981)
7.108 V.I. Mel'nikov: Fiz. Tverd. Tela *22*, 2404 (1980) [Sov. Phys. Sol. St. *22*, 1398 (1980)]
7.109 M.Ya. Azbel: Phys.Rev. B*22*, 4045 (1980)
7.110 D. Lenstra, W. van Haeringen: J. Phys. C*14*, L819 (1981)
7.111 J. Sak, B. Kramer: Phys. Rev. B*24*, 1761 (1981)
7.112 C.M. Soukoulis, E.N. Economou: Phys. Rev. B*24*, 5698 (1981)
7.113 A. Douglas Stone, J.D. Joannopoulos: Phys. Rev. B*24*, 3592 (1981); B*25*, 1431 (1982)
7.114 P.W. Anderson: Phys. Rev. B*23*, 4828 (1981)
7.115 P.W. Anderson, D.J. Thouless, E. Abrahams, D.S. Fisher: Phys. Rev. B*22*, 3519 (1980)
7.116 V.L. Berezinksii: Zh. Eksp. Teor. Fiz. *65*, 1251 (1973) [Sov. Phys. JETP *38*, 620 (1974)]
7.117 A.A. Gogolin: Sov. Phys. JETP *50*, 827 (1979)
7.118 A.A. Abrikosov, I.A. Ryzhkin: Adv. Phys. *27*, 147 (1978)
7.119 P. Erdös, R.C. Herndon: Adv. Phys. *31*, 65 (1982)
7.120 G. Theodorou, M.H. Cohen: Phys. Rev. B*13*, 4597 (1976)
7.121 L. Fleishman, D. Licciardello: J. Phys. C*10*, L125 (1977)
7.122 B.Y. Tong: Phys. Rev. A*1*, 52 (1970)
7.123 D.J. Thouless: J. Phys. C*5*, 77 (1972)
7.124 M. Kappus, F. Wegner: Z. Phys. B*45*, 15 (1981)
7.125 S. Sarker: Phys. Rev. B*25*, 4304 (1982)
7.126 E.N. Economou, C.T. Papatriantafillou: Phys. Rev. Lett. *32*, 1130 (1974); C. Papatriantafillou, E.N. Economou, T.P. Eggarter: Phys. Rev. B*13*, 910 (1976);
C. Papatriantafillou, E.N. Economou: Phys. Rev. B*13*, 920 (1976)
7.127 E. Abrahams, P.W. Anderson, D.C. Licciardello, T.V. Ramakrishnan: Phys. Rev. Lett. *42*, 673 (1979)
7.128 D.C. Licciardello, D.J. Thourless: J. Phys. C*11*, 925 (1978)

7.129 S. Sarker, E. Domany: Phys. Rev. B*23*, 6018 (1981)
7.130 N.F. Mott: Philos. Mag. B*44(2)*, 265 (1981)
7.131 A. Kawabata: Solid State Commun. *38*, 823 (1981); J. Phys. Soc. Jpn. *49*, 628 (1980)
7.132 F.J. Wegner: Z. Phys. B*25*, 327 (1976); Z. Phys. B*35*, 207 (1979)
7.133 L. Schäfer, F.J. Wegner: Z. Phys. B*38*, 113 (1980)
7.134 A. Houghton, A Jevicki, R.D. Kenway, A.M.M. Pruisken: Phys. Rev. Lett. *45*, 394 (1980)
7.135 K.B. Efetov, A.I. Larkin, D.E. Kheml'nitskii: Zh. Eksp. Teor. Fiz. *79*, 1120 (1980) [Sov. Phys. JETP *52(3)*, 568 (1981)]
7.136 G.A. Thomas, Y. Ootuka, S. Katsumoto, S. Kobayashi, W. Sasaki: Phys. Rev. B*25*, 4288 (1982);
G.A. Thomas, Y. Ootuka, S. Kobayashi, W. Sasaki: Phys. Rev. B*24*, 4886 (1981)
7.137 Y. Imry: Phys. Rev. Lett. *44*, 469 (1980)
7.138 W.L. McMillan: Phys. Rev. B*24*, 2739 (1981)
7.139 D. Vollhardt, P. Wölfle: Phys. Rev. Lett. *48*, 699 (1982); Phys. Rev. B*22*, 4666 (1980)
7.140 S. Hikami: Phys. Rev. B*24*, 2671 (1981)
7.141 T.F. Rosenbaum, K. Andres, G.A. Thomas, R.N. Bhatt: Phys. Rev. Lett. *45*, 1723 (1980);
T.F. Rosenbaum, K. Andres, G.A. Thomas, P. Lee: Phys. Rev. Lett. *46*, 568 (1981)
7.142 M.J. Vren, R.A. Davis, M. Kaveh, M. Pepper: J. Phys. C*14*, 5737 (1981)
7.143 G.J. Dolan, D.D. Osheroff: Phys. Rev. Lett. *43*, 721 (1979)
7.144 L. Van den Dries, C. Van Haesendonck, Y. Bruynseraede, G. Deutscher: Phys. Rev. Lett. *46*, 565 (1981); Physica *107*, B+C, 7 (1981)
7.145 D.J. Bishop, D.C. Tsui, R.C. Dynes: Phys. Rev. Lett. *44*, 1153 (1980); *46*, 360 (1981)
7.146 Y. Kawaguchi, S. Kawaji: J. Phys. Soc. Jpn. *48*, 699 (1980)
7.147 G. Bergmann: Phys. Rev. Lett. *49*, 162 (1982); Phys. Rev. B*25*, 2937 (1982); Solid State Commun. *42*, 815 (1982)
7.148 N. Giordano, W. Gilson, D.E. Prober: Phys. Rev. Lett. *43*, 725 (1979);
N. Giordano: Phys. Rev. B*22*, 5635 (1980);
J.T. Masden, N. Giordano: Physica *107*, B+C, 3 (1981)
7.149 P. Chaudhari, H.V. Habermeier: Solid State Commun. *34*, 687 (1980); Phys. Rev. Lett. *45*, 930 (1980);
P. Chaudhari, A.N. Broers, C.C. Chi, R. Laibowitz, E. Spiller, J. Viggiano: Phys. Rev. Lett. *45*, 930 (1980)
7.150 A.E. White, M. Tinkham, W.J. Skocpol, D.C. Flanders: Phys. Rev. Lett. *48*, 1752 (1982)
7.151 T.F. Rosenbaum, R.F. Milligan, G.A. Thomas, P.A. Lee, T.V. Ramakrishnan, R.N. Bhatt, K. DeConde, H. Hess, T. Perry: Phys. Rev. Lett. *47*, 1758 (1981)
7.152 S. Hikami, A.I. Larkin, Y. Nagaoka: Progr. Theor. Phys. *63*, 707 (1980)
7.153 S. Maekawa, H. Fukuyama: J. Phys. Soc. Jpn. *50*, 2516 (1981)
7.154 S. Yoshino, M. Okazaki: J. Phys. Soc. Jpn. *43*, 415 (1977)
7.155 F. Yonezawa: J. Noncrystall. Solids *35+36*, 29 (1980)
7.156 J. Stein, U. Krey: Solid State Commun. *27*, 797 and 1405 (1978)
7.157 D. Weaire, V. Srivastava: J. Phys. C*10*, 4309 (1977);
D. Weaire: J. Noncrystall. Solids *35+36*, 9 (1980) and in [Ref.5.56, pp.535-551]
7.158 P. Prelovsek: Phys. Rev. B*18*, 3657 (1978); Solid State Commun. *31*, 179 (1979)
7.159 P. Lee: Phys. Rev. Lett. *42*, 1492 (1978); J. Noncrystall. Solids *35+36*, 21 (1980)
7.160 J.L. Pichard, G. Sarma: J. Phys. C*14*, L127 and L617 (1981)

7.161 A. MacKinnon, B. Kramer: Phys. Rev. Lett. *47*, 1546 (1981)
7.162 C.M. Soukoulis, I. Webman, G.S. Crest, E.N. Economou: Phys. Rev. B*26*, 1838 (1982)
7.163 P.B. Allen: J. Phys. C*13*, L667 (1980)
7.164 M. Kaveh, N.F. Mott: J. Phys. C*14*, L177 (1981); N.F. Mott, M. Kaveh: J. Phys. C*14*, L659 (1981)
7.165 H.G. Schuster: Z. Phys. B*31*, 99 (1978)
7.166 R. Haydock: In [Ref.5.56, pp.553-563]; J. Phys. C*14*, 229 (1981); Philos. Mag. B*43*, 203 (1981)
7.167 W. Götze: Solid State Commun. *27*, 1392 (1978); J. Phys. C*12*, 1279 (1979)
7.168 P.W. Anderson: Phys. Rev. *109*, 1492 (1958)
7.169 J.M. Ziman: J. Phys. C*2*, 1230 (1969)
7.170 M. Kikuchi: J. Phys. Soc. Jpn. *29*, 296 (1970)
7.171 D.C. Herbert, R. Jones: J. Phys. C*4*, 1145 (1971)
7.172 D.J. Thouless: J. Phys. C*3*, 1559 (1970)
7.173 E.N. Economou, M.H. Cohen: Phys. Rev. Lett. *25*, 1445 (1970); Phys. Rev. B*5*, 2931 (1972)
7.174 R. Abou-Chacra, P.W. Anderson, D.J. Thouless: J. Phys. C*6*, 1734 (1973)
7.175 C.M. Soukoulis, E.N. Economou: Phys. Rev. Lett. *45*, 1590 (1980)
7.176 B.L. Altshuler, A.G. Aronov, P.A. Lee: Phys. Rev. Lett. *44*, 1288 (1980); B.L. Altshuler, D. Khmel'nitzkii, A.I. Larkin, P.A. Lee: Phys. Rev. B*22*, 5142 (1980)
7.177 B.L. Altshuler, A.G. Aronov: Solid State Commun. *30*, 115 (1979); Zh. Eksp. Teor. Fiz. *77*, 2028 (1979) [Sov. Phys. JETP *50*, 968 (1979)]
7.178 A. Schmid: Z. Phys. *271*, 251 (1974)
7.179 J.S. Langer, T. Neal: Phys. Rev. Lett. *16*, 984 (1966)
7.180 L.P. Gor'kov, D. Khmel'nitzkii, A.I. Larkin: Pis'ma Zh. Eksp. Teor. Fiz. *30*, 248 (1979) [Sov. Phys. JETP Lett. *30*, 228 (1979)]
7.181 H. Fukuyama: J. Phys. Soc. Jpn. *48*, 2169 (1980); *49*, 644 (1980); Progr. Theor. Phys. *69*, 220 (1980)
7.182 Y. Nagaoka, H. Fukuyama (eds.): *Anderson Localization*, Springer Ser. Solid-State Sci., Vol.39 (Springer, Berlin, Heidelberg, New York 1982)
7.183 J. Bernasconi, T. Schneider (eds.): *Physics in One Dimension*, Springer Ser. Solid-State Sci., Vol.23 (Springer, Berlin, Heidelberg, New York 1981)
7.184 F. Yonezawa (ed.): *Fundamental Physics of Amorphous Semiconductors*, Springer Ser. Solid-State Sci., Vol.25 (Springer, Berlin, Heidelberg, New York 1980)
7.185 W.A. Phillips (ed.): *Amorphous Solids*, Topics in Current Physics, Vol.24 (Springer, Berlin, Heidelberg, New York 1980)

Part III: Green's Functions in Many-Body Systems

Chapter 8 Definitions

8.1 G. Baym: *Lectures on Quantum Mechanics* (Benjamin, New York 1969)
8.2 S.S. Schweber: *An Introduction to Relativistic Quantum Field Theory* (Harper and Row, New York 1961)
8.3 A.L. Fetter, J.D. Walecka: *Quantum Theory of Many-Particle Systems* (McGraw-Hill, New York 1971)
8.4 P.L. Taylor: *Quantum Approach to the Solid State* (Prentice Hall, Englewood Cliffs, NJ 1970)
8.5 G. Rickayzen: *Green's Functions and Condensed Matter* (Academic, London 1980)

8.6 D.N. Zubarev: Double Time Green's Functions in Statistical Physics, Uspekhi Fiz. Nauk *71*, 71 (1960); English translation, Sov. Phys. Uspekhi *3*, 320 (1960)
8.7 N.N. Bogoliubov, S.V. Tyablikov: Retarded and Advanced Green's Functions in Statistical Physics, Dokl. Akad. Nauk SSSR *126*, 53 (1959)

Chapter 9 Properties and Use of the Green's Functions

9.1 A.A. Abrikosov, L.P. Gorkov, I.E. Dzyaloshinski: *Methods of Quantum Field Theory in Statistical Physics* (Prentice Hall, Englewood Cliffs, NJ 1963)
9.2 L.P. Kadanoff, G. Baym: *Quantum Statistical Mechanics* (Benjamin, New York 1962)
9.3 P. Nozieres: *Theory of Interacting Fermi Systems* (Benjamin, New York 1964)
9.4 S. Doniach, E.H. Sondheimer: *Green's Functions for Solid State Physicists* (Benjamin, Reading, Mass. 1974)
9.5 R.D. Mattuck: *A Guide to Feynman Diagrams in the Many-Body Problem* (McGraw-Hill, New York 1967)
9.6 L.D. Landau, E.M. Lifshitz: *Statistical Physics* (Pergamon Press, London 1959) pp.39 and 70
9.7 See, e.g., R. Kubo: J. Phys. Soc. Jap. *12*, 570 (1957); K.M. Case: Transp. Th. Stat. Phys. *2*, 129 (1972)

Chapter 11 Applications

11.1 J.M. Luttinger: Phys. Rev. *121*, 942 (1961)
11.2 L.D. Landau: Sov. Phys.-JETP *3*, 920 (1956); *5*, 101 (1957)
11.3 J. Friedel: Phil. Mag. *43*, 153 (1952); Nuovo Cimento *7*, 287 (1958)
11.4 V.M. Galitskii: Sov. Phys.-JETP *7*, 104 (1958)
11.5 S.T. Beliaev: Sov. Phys.-JETP *7*, 299 (1958)

Appendix B The Renormalized Perturbation Expansion (RPE)

B.1 E. Feenberg: Phys. Rev. *74*, 206 (1948)
B.2 K.M. Watson: Phys. Rev. *105*, 1388 (1957)
B.3 P.W. Anderson: Phys. Rev. *109*, 1492 (1958)

Subject Index

O. Madelung

Introduction to Solid-State Theory

Translated from the German by B.C. Taylor

1978. 144 figures. XI, 486 pages. (Springer Series in Solid-State Sciences, Volume 2).
ISBN 3-540-08516-5

H. Haken

Synergetics

An Introduction

Nonequilibrium Phase Transitions and Self-Organization in Physics, Chemistry and Biology

3rd revised and enlarged edition. 1983. 161 figures. Approx. 400 pages. (Springer Series in Synergetics, Volume 1). ISBN 3-540-12356-3

Contents: Goal. – Probability. – Information. – Chance. – Necessity. – Chance and Necessity. – Self-Organisation. – Physical Systems. – Chemical and Biochemical Systems. – Applications to Biology. – Sociology and Economics. – Chaos. – Some Historical Remarks and Outlook. – References, Further Readingand Comments. – Subject Index.

Y.L. Klimontovich

The Kinetic Theory of Electromagnetic Processes

Translated from the Russian by A. Dobroslavsky

1983. XI, 364 pages. (Springer Series in Synergetics, Volume 10).
ISBN 3-540-11458-0

Contents: Introduction. – Classical Theory: Free Charged Particles and a Field. Atoms and Field. The Kinetic Equations for a System of Free Charged Particles and a Field. Brownian Motion. Kinetic Equations for an Atom-Field System. – Quantum Theory: Microscopic Equations. The Kinetic Equations for Partially Ionized Plasma; The Coulomb Approximation. – Kinetic Equations for Partially Ionized Plasma; The Processes Conditioned by a Transverse Electromagnetic Field. Spectral Emission Line Broadening of Atoms in Partially Ionized Plasma. – Fluctuations and Kinetic Processes in Systems Composed of Strongly Interacting Particles. Fluctuations in Quantum Self-Oscillatory Systems. – Phase Transitions in a System Composed of Atoms and a Field. Conclusion. – References. – Subject Index.

H. Haken

Advanced Synergetics

Instability Hierarchies of Self-Organizing Systems and Devices
1983. 104 figures. Approx. 450 pages. (Springer Series in Synergetics, Volume 20)
ISBN 3-540-12162-5

Contents: Introduction. – Linear Ordinary Differential Equations. – Linear Ordinary Differential Equations With Quasiperiodic Coefficients. – Stochastic Nonlinear Differential Equations. – The World of Coupled Nonlinear Oscillators. – Nonlinear Coupling of Oscillators: The Case of Persistence of Quasiperiodic Motion. – Nonlinear Equations. The Slaving Principle. – Nonlinear Equations. Qualitative Macroscopic Changes. – Spatial Patterns. – The Inclusion of Noise. – Discrete Noisy Maps. – Example of an Unsolvable Problem in Dynamics. – Appendix: Moser's Proof of His Theorem.

Springer-Verlag
Berlin
Heidelberg
New York
Tokyo